Caliciviruses

Molecular and Cellular Virology

Edited by

Grant Stuart Hansman

National Institute of Infectious Diseases
Department of Virology II
Musashimurayama
Tokyo
Japan

Xi Jason Jiang

Division of Infectious Diseases
Cincinnati Children's Hospital Medical Center
Cincinnati, OH
USA

and

Kim Y. Green

Norovirus Gastroenteritis Unit
Laboratory of Infectious Diseases
NIAID, NIH
Department of Health and Human Services
Bethesda, MD
USA

W Caister Academic Press

Copyright © 2010

Caister Academic Press
Norfolk, UK

www.caister.com

British Library Cataloguing-in-Publication Data
A catalogue record for this book is available from the British Library

ISBN: 978-1-904455-63-9

Cover image: The structure of feline calicivirus, solved at 9 Å resolution by cryo-electron
microscopy and three-dimensional image reconstruction. Dr David Bhella, Medical Research
Council and Glasgow University Centre for Virus Research, UK.

Printed and bound in Great Britain

Contents

Contributors

Dalan Bailey
Department of Virology
Faculty of Medicine
Imperial College London
London
UK

d.bailey@imperial.ac.uk

Rowena A. Bull
Molecular Microbiology Laboratory
School of Biotechnology and Biomolecular Sciences
Faculty of Science
The University of New South Wales
Sydney, NSW
Australia

r.bull@unsw.edu.au

Kyeong-Ok Chang
Department of Diagnostic Medicine and Pathobiology
College of Veterinary Medicine
Kansas State University
Manhattan, KS
USA

kchang@vet.k-state.edu

Brian D. Cooke
Invasive Animals CRC
University of Canberra
Canberra, ACT
Australia

brian.cooke@canberra.edu.au

Erwin Duizer
National Institute for Public Health and the
  Environment
Laboratory for Infectious Diseases and Screening
Bilthoven
The Netherlands

edwin.duizer@rivm.nl

Ian Goodfellow
Department of Virology
Faculty of Medicine
Imperial College London
London
UK

i.goodfellow@imperial.ac.uk

Gail E. Greening
Environmental and Food Virology
Institute of Environmental Science and Research
Porirua
New Zealand

gail.greening@esr.cri.nz

Xi Jiang
Division of Infectious Diseases
Cincinnati Children's Hospital Medical Center
College of Medicine University of Cincinnati
Cincinnati, OH
USA

jason.jiang@CCHMC.org

Stephanie M. Karst
Department of Microbiology and Immunology
Louisiana State University Health Sciences Center
Shreveport, LA
USA

skarst@lsuhsc.edu

Yunjeong Kim
Department of Diagnostic Medicine and Pathobiology
College of Veterinary Medicine
Kansas State University
Manhattan, KS
USA

ykim@vet.k-state.edu

Marion P.G. Koopmans
National Institute for Public Health and the
 Environment
Laboratory for Infectious Diseases and Screening
Bilthoven
The Netherlands

marion.koopmans@rivm.nl

Christine Luttermann
Institut für Immunologie
Friedrich-Loeffler-Institut
Tübingen
Germany

christine.luttermann@fli.bund.de

Gregor Meyers
Institut für Immunologie
Friedrich-Loeffler-Institut
Tübingen
Germany

gregor.meyers@fli.bund.de

Kenneth K.S. Ng
Department of Biological Sciences
University of Calgary
Calgary, AB
Canada

ngk@ucalgary.ca

Francisco Parra
Departamento de Bioquímica y Biología Molecular
Instituto Universitario de Biotecnología de Asturias
Universidad de Oviedo
Oviedo
Spain

fparra@uniovi.es

Akos Putics
Department of Virology
Faculty of Medicine
Imperial College London
London
UK

a.putics@imperial.ac.uk

J. Joukje Siebenga
National Institute for Public Health and the
 Environment
Laboratory for Infectious Diseases and Screening
Bilthoven
The Netherlands

joukje.siebenga@rivm.nl

Stanislav V. Sosnovtsev
Norovirus Gastroenteritis Unit
National Institute of Allergy and Infectious Diseases
National Institutes of Health
Bethesda, MD
USA

ss216m@nih.gov

Tanja Strive
Commonwealth Scientific and Industrial Research
 Organisation
Canberra, ACT
Australia

tanja.strive@csiro.au

Ming Tan
Division of Infectious Diseases
Cincinnati Children's Hospital Medical Center
College of Medicine University of Cincinnati
Cincinnati, OH
USA

Ming.Tan@chmcc.org

Surender Vashist
Department of Virology
Faculty of Medicine
Imperial College London
London
UK

s.vashist@imperial.ac.uk

Vernon K. Ward
Department of Microbiology & Immunology
School of Medical Sciences
University of Otago
Dunedin
New Zealand

vernon.ward@stonebow.otago.ac.nz

Sandro Wolf
Technische Universität Dresden
Institut für Mikrobiologie
01062 Dresden
Germany

sandro.wolf@tu-dresden.de

Peter A. White
Molecular Microbiology Laboratory
School of Biotechnology and Biomolecular Sciences
Faculty of Science
The University of New South Wales
Sydney, NSW
Australia

p.white@unsw.edu.au

Preface

In this book, a panel of experts distil the most important up-to-date research findings in their calicivirus field of study, producing timely and comprehensive reviews. Each chapter gives the reader a brief introduction to the topic followed by a descriptive discussion of the past and present research areas. At the time of writing, the virus family *Caliciviridae* consisted of four genera, *Norovirus, Sapovirus, Lagovirus* and *Vesivirus*. Since then, a fifth genus has been added, *Nebovirus,* and several other genera have been proposed.

Caliciviruses infect a broad range of animals, including human, pig, cow, rabbit and mice. Caliciviruses cause a variety of symptoms in the different animal hosts, including diarrhoea, vomiting, fever, nausea, liver necrosis, and urinary tract infection. In this book, research results from a number of well-studied prototype caliciviruses are explained and discussed, and comparisons among different caliciviruses are included whenever possible. Examples include human Norwalk virus, a norovirus, thought to be responsible for roughly 90% of epidemic, non-bacterial outbreaks of gastroenteritis around the world; feline calicivirus, a vesivirus, which infect cats and causes a variety of diseases including oral and upper respiratory tract infections; porcine enteric calicivirus, a sapovirus, which infect pigs of all ages and has been found in numerous countries; murine norovirus, a norovirus, which infect mice and has the potential to be an important replication model; and rabbit haemorrhagic disease virus, a lagovirus, which is a highly pathogenic virus in rabbits and has an interesting research history in Australia and New Zealand.

Topics include norovirus epidemiology; calicivirus contamination of the environment; genome organization and recombination, proteolytic cleavage and viral proteins; viral protein structures; virus–host interactions; calicivirus reverse genetics and replicon systems; feline calicivirus; swine calicivirus; murine norovirus pathogenesis and immunity; murine norovirus translation, replication and reverse genetics; and lagoviruses.

Grant Hansman

Other Books of Interest

Caister Academic Press   www.caister.com

Norovirus Epidemiology

J. Joukje Siebenga, Erwin Duizer and Marion P.G. Koopmans

Abstract

Noroviruses are the dominant cause of outbreaks as well as sporadic community cases of viral gastroenteritis in the world. Their very low infectious dose, combined with high levels of shedding and long persistence in the environment make noroviruses extremely infectious. Although norovirus-related illness is generally regarded as mild and self-limiting, more severe outcomes are increasingly described among elderly and immunocompromised patients. The combination of large and difficult to control outbreaks and severe illness in some patients leads to major problems in healthcare settings, such as hospitals and nursing homes. Additionally, some large and diffuse, multi-national and even multi-continent, foodborne outbreaks have been described for norovirus, affecting up to thousands of people. With structured outbreak surveillance running in a number of regions across the world for the past 10 years, it has become clear that the spread of noroviruses is global, although important information from developing countries is missing. At present, norovirus strains belonging to genogroup II genotype 4 (GII.4) are dominant worldwide. In the last ten years, at least three global pandemics involving GII.4 strains of different genetic variants occurred. Although a straightforward culturing method remains lacking for noroviruses, important progress has been made in immunological studies using virus-like particles. Thus, it has been shown that the subsequent genetic variants of GII.4 are antigenically distinct, and that the GII.4 noroviruses evolved and continue to do so by a process known as epochal evolution, in which periods of genetic stasis are interrupted by rapid accumulation of mutations and the subsequent emergence of novel genetic variants. In norovirus evolution, this process is directed by population or herd immunity.

Introduction

Norovirus is one of the four recognized genera of the family *Caliciviridae*, together with *Lagovirus* (known host species: rabbits, brown hares), *Sapovirus* (known host species: humans, swine, mink) and *Vesivirus* (known host species: sea lions, other marine animals, swine, cats, dogs, fish, seals, cattle and primates). Additionally, two other potential genera comprised of viruses detected in cattle and rhesus macaques have been described (Farkas *et al.*, 2008; Oliver *et al.*, 2006; Wei *et al.*, 2008). The animal caliciviruses have been associated with a range of clinical syndromes, including lesions, stomatitis, upper respiratory tract infections, and systemic diseases with severe haemorrhagic syndromes. Noroviruses infect humans, but have also been detected in swine, bovine, ovine, murine, feline and canine species (Hsu *et al.*, 2007; Karst *et al.*, 2003; Martella *et al.*, 2007, 2008; Oliver *et al.*, 2007; Sugieda and Nakajima, 2002; Wang *et al.*, 2005a). In humans, noroviruses cause acute gastroenteritis, mild gastroenteritis and asymptomatic infections, whereas in animals they cause asymptomatic infections, as well as mild gastroenteritis symptoms. In immunocompromised mice, lacking components of the innate system, murine norovirus causes encephalitis, vasculitis, pneumonia and hepatitis, indicating a broad tissue tropism (Karst *et al.*, 2003).

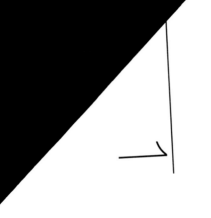

popularly
)ecause of
ιks follow
breaks in
common
Southern
ιce is not
has been
ιlia) (Tu
Zealand)
ɾ norovi-
stomach

.....ge on the evolution of these viruses show that this popular name – while unintended – may be more accurate than previously thought (Siebenga *et al.*, 2007). Outbreaks with high media impact among vacationers on cruise ships have also yielded the name 'cruise ship virus'.

The unavailability of a cell culture system has been a continuing handicap in advancing knowledge of the human noroviruses (Duizer *et al.*, 2004b). The breakthrough cloning and sequencing of the prototype strain, Norwalk virus (Jiang *et al.*, 1990) in 1990, led to the development of molecular tools to study the epidemiology of the noroviruses. Since then, numerous studies have shown the importance of noroviruses as a cause of illness in people of all age groups.

This chapter will focus on the state-of-the-art knowledge of norovirus epidemiology. Included will be discussions of norovirus properties, natural history, transmission, strain diversity, evolution, and fitness. Estimates of the norovirus disease burden in various populations and the economic impact will be reviewed. Active surveillance with sophisticated diagnostic tools and global networking will continue to provide critical information in the prevention and management of these important viruses.

History

In 1968, students and teachers of an elementary school in Norwalk, Ohio, fell ill with symptoms of acute gastroenteritis, including vomiting and diarrhoea. For this illness, which had previously been described as 'winter vomiting disease', no causative agent was found. Several attempts to identify an aetiology for this infectious form of acute gastroenteritis failed, until in 1972 Albert

Kapikian *et al.* described viral particles found in stool samples of volunteers experimentally infected with purified stool filtrate from a patient originally infected in the Norwalk outbreak (Kapikian *et al.*, 1972). Using immune electron microscopy, small 'picorna- or parvovirus-like' particles were detected, that elicited immune responses in volunteers as well as naturally infected individuals. In 1982, Kaplan *et al.* listed four criteria to identify a norovirus outbreak (Kaplan *et al.*, 1982). These criteria were (i) vomiting in more than half the affected persons; (ii) mean (or median) incubation period of 24–48 h; (iii) mean (or median) duration of illness lasting 12–60 h; and (iv) absence of a bacterial pathogen in stool cultures. Recently, these criteria were reassessed, and still found to be accurate, with 68% sensitivity and 99% specificity (Turcios *et al.*, 2006).

Factors that shape the success of noroviruses in the human population

Genome properties

The noroviruses contain a positive-sense RNA genome that is approximately 7.5 kb long (Jiang *et al.*, 1990). The genome encodes three open reading frames (ORFs) (Jiang *et al.*, 1993). ORF1 encodes a polyprotein comprising all non-structural proteins, which is autocatalytically cleaved to produce the non-structural proteins (see Chapters 3 and 4). ORF2 encodes the major capsid protein, referred to in the literature as VP1. ORF3 encodes a second, minor structural protein, VP2, about which little is known. Although in feline calicivirus VP2 was shown to be essential for productive replication (Sosnovtsev *et al.*, 2005), this was not the case for RHDV (Liu *et al.*, 2008). Alternatively, it was shown that VP2 mRNA regulates both the expression and stability of VP1 (Bertolotti-Ciarlet *et al.*, 2003). The norovirus RNA genome is replicated by the virus-encoded RNA-dependent RNA polymerase, which typically has no proofreading activity. The relatively high error-rate in genomic transcription makes norovirus, as other RNA viruses, highly flexible and increasingly diverse. Noroviruses can be divided into distinct genogroups, based on phylogenetic analyses of the capsid protein. To date, five norovirus genogroups have been recognized

(GI–GV). Viruses of GI, GII and rarely GIV are known to infect humans (Green *et al.*, 2001; Zheng *et al.*, 2006). GII viruses have additionally been detected in pigs (Han *et al.*, 2004; Mattison *et al.*, 2007; Sugieda *et al.*, 1998; Sugieda and Nakajima, 2002; Wang *et al.*, 2005b, 2006) and GIV viruses have been detected in a lion cub and a dog (Martella *et al.*, 2007, 2008). GIII viruses infect cattle and sheep (Wolf *et al.*, 2009) and GV viruses infect mice (Hsu *et al.*, 2007; Karst *et al.*, 2003). GI, GII and GIII are further subdivided into numerous genotypes (Green, 2007). It is likely that additional viruses belonging to novel genotypes will be detected in humans and animals. The genetic variability of the norovirus genome and the regular discovery of novel genotypes have caused some inconsistencies in the nomenclature used for norovirus genotypes. A striking example is the Alphatron strain, which is classified as a GIV.1 (genogroup IV genotype 1) virus in most typing schemes (Green, 2007; Zheng *et al.*, 2006), but as a GII.17 strain in another paper (Kageyama *et al.*, 2004). Similarly, GII.13 and GII.14 have gotten each others names in different publications, as did GII.15 and GII.16 (Kageyama *et al.*, 2004; Zheng *et al.*, 2006). Recombinant norovirus strains cause additional naming difficulties, since they cannot be adequately named with the currently used nomenclature system based solely on the capsid gene (ORF2). The development of a norovirus nomenclature scheme that includes genotyping designations for both ORF1 and ORF2 is being explored by a small group of laboratories involved in molecular surveillance. After completion of this discussion a typing tool applying the agreed-upon nomenclature will be available from http://www.rivm.nl/mpf/norovirus/typingtool.

Biochemical properties and disinfection

Noroviruses are particularly infectious and highly stable in the environment. These properties are determined by specific characteristics of the virion. The virus particles are approximately 28–32 nm in diameter and have no lipid envelope (Kapikian *et al.*, 1972). These particles have been shown to be highly stable in the environment (Duizer *et al.*, 2004a), although effective testing of the stability of human noroviruses has been hampered by the lack of an infectivity assay (Duizer

et al., 2004b). Inactivation studies have been performed using feline and canine caliciviruses as models for human norovirus. More recently the cultivable murine norovirus has been adopted as the surrogate of choice.

Studies using feline and canine caliciviruses as models for norovirus showed that heat inactivation of these two animal caliciviruses was highly comparable. Thermal inactivation is both temperature and time dependent with increasing survival at decreasing temperatures (Duizer *et al.*, 2004b). Based on the temperature dependent inactivation profiles it was suggested that a better inactivation of viruses may be expected from regular batch (63°C for 30 min) or classical pasteurization (70°C for 2 min) than from high temperature short time pasteurization (72°C for 15 s). Inactivation at 100°C is, however, complete within seconds (Cannon *et al.*, 2006; Duizer *et al.*, 2004a). The noroviruses are preserved well by refrigeration and freezing. Similarly, inactivation by UV radiation is dose and time dependent. The resistance of feline calicivirus in suspension to inactivation by UV 253.7 nm radiation is dose dependent and highly variable, most likely due to differences in composition (turbidity) of the irradiated suspensions (De Roda Husman *et al.*, 2004; Nuanualsuwan *et al.*, 2002; Thurston-Enriquez *et al.*, 2003). Feline calicivirus was not efficiently inactivated on environmental surfaces or in suspension by 1% anionic detergents, quaternary ammonium (1:10), hypochlorite solutions with < 300 ppm free chlorine, or less than 50% or more than 80% alcohol preparations (ethanol or 1- and 2-propanol) (Duizer *et al.*, 2004a; Gehrke *et al.*, 2004; Scott, 1980). Varying efficacies of 65–75% alcohol are reported, however, short contact times (< 1 min) rarely result in effective inactivation. Moreover, the presence of faecal or other organic material reduces the virucidal efficacy of many chemicals tested. The efficacy of seven commercial disinfectants for the inactivation of feline calicivirus on strawberries and lettuce was tested by Gulati and co-workers and they found that none of the disinfectants was effective when used at the FDA permitted concentration (Gulati *et al.*, 2001). These data were confirmed by Allwood and co-workers for sodium bicarbonate, chlorine bleach, peroxyacetic acid and hydrogen peroxide at FDA approved concentrations (Allwood *et al.*, 2004). More recently, Belliot *et al.* (2008) used murine norovirus as a surrogate in disinfectant tests. They reported that murine norovirus was sensitive to 60% alcohol, alcohol hand rubs, bleach and povidone iodine-based disinfectant in suspension tests using very low-level interfering substances. This would suggest a higher sensitivity of murine norovirus than of feline calicivirus for commercial disinfectants and most likely a higher sensitivity to hypochlorite for murine norovirus compared to human norovirus, since the RNA of human norovirus was much better protected during hypochlorite inactivation than the RNA of feline and canine caliciviruses (Duizer *et al.*, 2004a).

Shedding

Norovirus is shed in high quantities in the stool of infected persons; around 10^8 but up to 10^{10} RNA copies per gram of stool were reported for different GII viruses (Lee *et al.*, 2007b; Ludwig *et al.*, 2008; Tu *et al.*, 2008b). Similar virus concentrations were found after experimental infection with a GI virus (Atmar *et al.*, 2008). This leads to the theoretical possibility that one gram of stool of an infected person could be enough to infect up to five million individuals. Projectile vomiting, which is a very typical symptom for norovirus illness, was once calculated to have the potential to infect 300,000 to 3,000,000 people (Caul, 1995), and the distance of people at risk of infection to the place of vomiting was inversely related to the chance of them actually becoming infected (Marks *et al.*, 2000).

Shedding of virus continues after clinical recovery of the patient, and may last three or four weeks in otherwise healthy people, but it is especially long in young children (Atmar *et al.*, 2008; Rockx *et al.*, 2002). In a hospital study involving people of all ages, higher concentrations of virus in stool were found to be associated with older aged patients and also with prolonged diarrhoeal symptoms and increased severity of symptoms (Lee *et al.*, 2007a). In immunocompromised patients severely prolonged illness accompanied by prolonged shedding may last a very long time; up to several years (Carlsson *et al.*, 2009; Kirkwood and Streitberg, 2008; Ludwig *et al.*, 2008; Muller *et al.*, 2007; Siebenga *et al.*, 2008). It has however

not yet been shown if these prolonged shedders remain infectious for others.

Infectivity

The possibilities for infection experiments are limited by the unavailability of a cell culture model. Therefore, Teunis *et al.* (2008) used the results of volunteer studies with GI.1, Norwalk virus, to estimate infectivity. They showed norovirus to be extremely infectious, reportedly more than any other virus. The probability of infection of a single norovirus particle was estimated at 0.5, and the ID50 at 18 virus particles. One recent paper estimated the reproduction number (R0) of norovirus in an outbreak situation in the absence of hygiene measures at over 14 (Heijne *et al.*, 2009). This means that every primary case, on average, infects 14 secondary cases. Implementation of strict hygiene measures was shown by the same study to lower this R0 to around 2. For comparison, the R0 of the 1918 strain of pandemic influenza virus was estimated to be ~2–3 (Mills *et al.*, 2004). Poliovirus, designated 'highly contagious', and also an enteric virus transmitted through the faecal-oral route, has an R0 of ~5–7 (CDC, 2009).

Repeated infections

Even though there is no proof of the existence of different serotypes of norovirus by classical virus neutralization methods, the genetic diversity displayed by noroviruses probably translates into antigenic diversity, so that infection with a strain of one genotype may not confer immunity against strains of another genotype or even variants within a genotype (Chapter 6) (Hansman *et al.*, 2006; Lindesmith *et al.*, 2008; Siebenga *et al.*, 2007). Furthermore, volunteer studies have shown that protective immunity after infection may be absent or short-lived (Johnson *et al.*, 1990; Parrino *et al.*, 1977; Wyatt *et al.*, 1974). The combination of these two features, antigenic diversity and apparent lack of long term protective immunity, are the likely cause of the occurrence of norovirus infections in children, adults and the elderly. In effect one individual may suffer repeated infections, even with viruses belonging to the same genotype and therefore people of all ages are affected by norovirus illness, unlike with, for example, rotavirus. Antigenic differences between the subsequent GII.4 variants are the driving force behind the continued persistence of GII.4 as the dominant norovirus strain and the repeated occurrence of world-wide epidemics caused by emerging variants (see below) (Lindesmith et al., 2008; Siebenga et al., 2007).

Infections and exposure

Illness and pathogenesis

The illness caused by norovirus is usually described as mild and self-limiting. Incubation time is typically 12–72 h and symptoms may last 1–3 days, although longer times up to 5 days have been reported, particularly in young children and the elderly (Lopman et al., 2004b; Rockx et al., 2002). Diarrhoea is the most commonly reported symptom (87%), followed by vomiting (74%), abdominal pain (51%), cramps (44%) and nausea (49%). A fever was reported for 32–45% of cases and mucus in the stool was seen less commonly (19%) (Patel et al., 2008; Rockx et al., 2002). Illness in people with co-morbidity or in the elderly may be more severe and sometimes has very serious consequences, such as prolonged infections and excess deaths, as will be discussed below. Current information on the pathogenesis of norovirus infections has been derived from volunteers and analysing infections in gnotobiotic pigs. Only one of 48 pigs that developed illness had mild lesions in the proximal small intestine, and duodenal and jejunal enterocytes were shown by immunofluorescent microscopy to be infected (Cheetham et al., 2006). Binding of calicivirus-like particles to duodenal and buccal tissues was also shown in pigs (Cheetham et al., 2007). Volunteer studies had previously revealed villus atrophy in duodenal biopsies (Agus et al., 1973; Dolin and Baron, 1975). The study of duodenal biopsies of norovirus-infected people provided a basis for understanding the cause of diarrhoea, namely a combination of epithelial barrier dysfunction in the duodenum, a reduction of tight junctional proteins, increased apoptosis in duodenal epithelial cells and increased anion secretion (Troeger et al., 2008). Abdominal computed tomography (CT) scans of children with acute norovirus infections revealed wall thickening and enhancement in the different parts of the small intestine, namely the duodenum, jejunum and ileum, as well as fluid-filled bowel loops 2008).

Burden of disease estimates

The burden of hospital infections

Hospitals can be severely affected by norovirus outbreaks. The potential for person-to-person spread is great and the people at risk are often especially vulnerable to infections (Mattner et al., 2005; Tsang et al., 2008). Nosocomial norovirus infections, typically acquired after a patient was admitted to the hospital, are common. A study in a Dutch academic tertiary care hospital reported that over 51% of norovirus infections was hospital acquired, with a conservative definition of onset of symptoms at least 5 days after admission (Beersma et al., 2009). In high-risk hospitalized patients norovirus infections can have severe consequences such as prolonged illness and death. A patient who had undergone heart transplantation was reported to shed norovirus for several years, while remaining symptomatic with diarrhoea (Carlsson et al., 2009; Nilsson et al., 2003). Such prolonged infections have been reported more often (Colomba et al., 2006; Lee et al., 2008; Ludwig et al., 2008; Murata et al., 2007; Wood et al., 1988), and occur at alarmingly high prevalence. A retrospective 2-year survey of norovirus diagnostic records of a tertiary care hospital revealed eight patients with prolonged illness (8.4% of all identified norovirus-positive patients) for whom repeated diagnostics had been requested over a period of more than 3 weeks. Five of these eight patients were children below three years of age and all had a form of underlying illness, leaving them unable to clear the virus. Diarrhoeal symptoms remained for all but one patient, whereas vomiting was not mentioned often as a symptom. Two of the patients in this study died at least partially resulting from their norovirus infections, underlining the fact that norovirus illness is not always a 'mild disease' (Beersma et al., 2009; Siebenga et al., 2008). Similarly, an outbreak of necrotizing enterocolitis was reported in a neonatal unit with two of eight neonates dying (Turcios-Ruiz et al., 2008). No cause for necrotizing enterocolitis is known, but infections of the bowels are strongly suspected to play a role and the unit was struck by a norovirus outbreak at the time of the necrotizing enterocolitis outbreak.

ospital patients and staff
·ts, with rates among
76%) often at least as
tween 18% and 32%)
anerva et al., 2009;
al., 2007).

tions

...us outbreaks do not only
iorm a great health burden for patients and staff,
but are also associated with great financial costs.
Noroviruses were reported to have the highest
'closure rate' of all pathogens that may be responsible for nosocomial outbreaks; in 44.1% of norovirus outbreaks a ward had to be closed, against
38.5% for runner-up influenza and, for example,
25.9% for rotavirus and 11.8% for *Clostridium*
spp. (Hansen *et al.*, 2007).

One large outbreak in a 946-bed tertiary care
hospital in the US was calculated to have cost
over US$650,000 (Johnston *et al.*, 2007). These
costs included lost revenue associated with unit
closures, sick leave, replacement of supplies and
cleaning expenses. Several wards were affected
and 355 people fell ill with gastroenteritis. A
Swiss hospital of similar size (960 beds) analysed
the cost of a smaller outbreak, affecting only 45
people at two different wards and calculated that
this outbreak cost approximately US$40,000
(Zingg *et al.*, 2005).

The burden in long-term care facilities

Greig and Lee performed a review of enteric outbreak reports from long-term care facilities in several countries, and found that approximately 60%
of outbreaks were caused by norovirus (Greig and
Lee, 2009). More importantly, over 70% of the
symptomatic persons in all the studies reviewed
were in these norovirus outbreaks, with 16 of 60
reported deaths caused by noroviruses, exceeded
in numbers only by *Salmonella* outbreaks. That
norovirus has an extra great impact on nursing
homes is illustrated by the finding of the high rate
of illness among staff due to norovirus infections;
86% of all documented symptomatic staff were ill
resulting of a norovirus infection, whereas other
pathogens rarely caused illness among staff (Greig
and Lee, 2009). Not seldom as in hospitals, the
attack rate among staff is higher than among patients (Billgren *et al.*, 2002).

Greig and Lee reviewed recommendations
for containment measures in long-term care
facilities such as nursing homes. Options for
containment of an outbreak are limited by the
fact that the inhabitants of the facility usually live
there; therefore the possibilities for ward-closure,
a new-admissions stop or quarantining the affected patients are limited (Greig and Lee, 2009).
Limiting person-to-person spread was most
effective measure in controlling an outbreak and
included managing residents movements, isolation of ill residents when possible, and dedicating
specific staff to care for the ill. Also, especially
for norovirus outbreaks, vigorous cleaning was
important in containment. A study in Dutch nursing homes with different containment protocols
revealed that these different protocols were only
effective when implemented rapidly after onset
of the outbreak, and then positive effects often
only reduced the amount of sick among the staff
(Friesema *et al.*, 2009).

Economic cost of infections in long-term care facilities

The overall economic burden of gastroenteritis
outbreaks in long term care facilities is not known.
A study of the costs of illness among elderly
Americans found it difficult to calculate a reliable
estimate for gastrointestinal illness due to limited
availability of necessary data (Van Houtven *et
al.*, 2005). The annual costs for GI illness among
elderly (>65) in the US, excluding nursing home
costs and morbidity-related indirect costs were
nonetheless estimated to be around US$1 billion.

Host factors and attack rate

Differences in host-susceptibility for different
genotypes have been reported, and are based on
the presence or absence of specific virus receptors in the potential host. Although additional
research is needed to establish more detail, the
currently proposed receptors are encoded by
the human histo-blood group antigen (HBGA)
genes (see Chapter 6) (Cao *et al.*, 2007; Huang
et al., 2003, 2005a; Marionneau *et al.*, 2002; Tan
and Jiang, 2005, 2008). Because non-secretors
are rarely infected by any genotype of norovirus
– Lindesmith *et al.* reported a GII.2 infection in
one secretor negative individual (Lindesmith *et
al.*, 2005) – a 100% attack rate is equally rarely

reported in an outbreak, as they make up approximately 20% of the Caucasian population, and similar percentages in other people (Kelly et al., 1994). Some people may get infected by noroviruses and consequently shed virus, but show no symptoms. Such asymptomatic shedders probably contribute to the efficient spread of the virus.

Few systematic studies have reported on attack rates for outbreaks. An average attack rate of 34.7% was reported for 60 norovirus outbreaks caused by different genotypes in Catalonia, Spain, mostly ranging from 20% to 60% (Torner et al., 2008). An earlier study, with strains belonging to different GII genotypes, reported attack rates ranging between 17% and 100% (Vinje and Koopmans, 1996). Interestingly, a recent Japanese study reported a lower attack rate for GII.4 food-related outbreaks compared to other genotypes (Noda et al., 2008), even though GII.4 binds to the widest variety of HBGAs of all norovirus genotypes (Huang et al., 2005a). It should be kept in mind that it is often not known exactly who were exposed, and thus the given figures are estimates. Overall, reported attack rates in norovirus outbreaks vary greatly, and are probably influenced by the genotype causing the outbreak, the possible presence of immunity against this strain among the people at risk, their general health status, personal hygiene measures taken, the transmission mode and other factors.

Modes of transmission

Person-to-person transmission

The majority (88%) of reported norovirus outbreaks in a large European surveillance network (FBVE) study (Kroneman et al., 2008b) were associated with person-to-person transmission. In a review of enteric outbreaks in long-term care facilities by Greig et al., 71% of the described outbreaks in these settings were attributed to transmission by person-to-person contact (Greig and Lee, 2009). This route is therefore important to consider, especially in healthcare facilities, where care-giving personnel are in contact with many patients or residents. Thus, when they considered all containment measures available in these settings, both Greig et al. and Friesema et al. found that control measures aimed at reducing the person-to-person spread were most effective (Friesema et al., 2009; Greig and Lee, 2009).

Environmental transmission

Noroviruses are highly persistent in the environment. Viral particles may be transferred to new surfaces in a chain of relocation events, and during an outbreak, surfaces such as computer keyboards or mice, door handles, and cutting-boards in a kitchen may become contaminated (Clay et al., 2006). Contamination of surfaces in hospital wards was shown to decrease with the implementation of strict hygiene rules for staff and visitors (Gallimore et al., 2006, 2008). If a contaminated surface is not identified and cleaned, the outbreak may become protracted (Evans et al., 2002; Kuusi et al., 2002; Verhoef et al., 2008a). Projectile vomiting has been shown to play an important role in the contamination of unexpectedly large areas (Caul, 1995).

Food- and waterborne outbreaks

Noroviruses are a known cause of foodborne outbreaks. Studies vary in the proportion of reported outbreaks spread by foodborne transmission, probably reflecting differences in the surveillance focus between countries. Studies in Europe by the FBVE network of laboratories (Kroneman et al., 2008b) and certain studies in the US (Doyle et al., 2008) reported that approximately 10% of all reported outbreaks were associated with contaminated food. In Japan, up to 35% of norovirus outbreaks were linked to foodborne transmission (Hamano et al., 2005). In comparison, Lynch and co-workers reported that 55% of all foodborne disease with a known aetiological agent in the US between 1998 and 2002 were caused by several bacteria, whereas 33% of the outbreaks were caused by viruses. Noroviruses were the single most important known pathogen, causing the majority of diagnosed outbreaks (30%), with salmonella (27%) second in importance (Lynch et al., 2006). It has been estimated that up to 67% of all food-related illness is associated with noroviruses (Koopmans and Duizer, 2004; Mead et al., 1999). Foodborne norovirus outbreaks may be large, possibly affecting up to thousands of people, for example the contamination of cake icing by a baker in the US in 1982 (Kuritsky et al., 1984) and the importation of a contaminated batch of frozen raspberries in Denmark in 2005 (Falkenhorst et al., 2005). The source of contaminated outbreaks may transcend borders or even continents (de Wit *et al.*, 2007; Koopmans, 2005; Koopmans *et al.*, 2002, 2003; Ng *et al.*, 2005; Simmons *et al.*, 2007). The increasing global trade of food has increased the risk of international or even global foodborne outbreaks. Even though international surveillance systems have been implemented [e.g. FBVE (http://www.fbve.nl) or OzFoodNet (http://www.ozfoodnet.org.au/)], only a few large-scale outbreaks have been detected thus far. Recently, to facilitate the decision-making process for further in-depth investigation of a suspected foodborne outbreak, a selection tool was developed based on a 5-year history of outbreak reporting in Europe (Verhoef *et al.*, 2009). The background data on the outbreak under investigation would minimally include information on the outbreak setting, the number of cases involved at the time of the outbreak, the country, the intensity and focus of surveillance and the suspected norovirus genotype. This tool, that is provided online, provides an estimate of the probability that an outbreak was caused by foodborne transmission. By using this approach, the estimate of outbreaks attributed to foodborne contamination was increased to 40% (Verhoef *et al.*, 2009). Similarly, a reassessment of over 4000 food-related gastroenteritis outbreaks reported in the US between 1998 and 2002 resulted in an increase from 14% to 28% for noroviruses as the likely aetiological agent (Turcios *et al.*, 2006). Still, noroviruses are difficult to detect in food matrices; only specialized laboratories can perform this research and not many countries routinely screen for viruses in food-related outbreaks, as exemplified in a study in Japan (Hansman *et al.*, 2008). The food industry and governments pay little attention to possible viral contamination of products (Koopmans and Duizer, 2004).

Food is most often contaminated by the introduction of enteric pathogens present in faecal material that is derived from either an environmental source or an infected food handler. Filter feeding bivalve molluscs (oysters and mussels) are an important source of noroviruses in food-borne outbreaks. Molluscs filter water for nutrients and retain noroviruses in their gut by specific binding, often when their growth-beds have been contaminated with waste water effluents (Bon *et al.*, 2005; Gallimore *et al.*, 2005; Le Guyader *et al.*, 2006a,b; Ng *et al.*, 2005; Suffredini *et al.*, 2008; Tian *et al.*, 2006, 2007). Also, salads and soft fruits like berries can be contaminated by irrigation water or by washing with contaminated water (Makary *et al.*, 2009; Maunula *et al.*, 2004). Large-scale (drinking) water contaminations have also been reported (Hewitt *et al.*, 2007; Hoebe *et al.*, 2004; Krisztalovics *et al.*, 2006; Makary *et al.*, 2009; Maunula *et al.*, 2004; Nygard *et al.*, 2003; O'Reilly *et al.*, 2007). Food-handlers, who are ill or recently recovered at the time of their food-handling are also known sources of contamination. Numerous examples are known, such as the recently described norovirus illness among river rafters in the Grand Canyon (Malek *et al.*, 2009) and the employees of a government department who fell ill after attending a reception with a lunch buffet, where the bread rolls had been prepared by an ill baker (de Wit *et al.*, 2007).

Zoonotic transmission

Direct zoonotic transmission of noroviruses to humans is thought be rare. However, the detection of noroviruses closely related to human noroviruses in animals and especially the increasing numbers of detection reports in pigs indicates a potential risk (Mattison *et al.*, 2007; Wang *et al.*, 2005a,b, 2009). GIV strains relating to illness in humans are rarely reported, however, they have been detected in several different animals (Martella *et al.*, 2007, 2008). In Argentina in 2006, there was an unusually high number of GIV norovirus outbreaks in humans (Gomes *et al.*, 2007). Two genogroups (GIII and GV) comprised solely of animal noroviruses exist, of which the GIII viruses have been detected in cattle and sheep (Wolf *et al.*, 2009) and the GV viruses in mice. Cattle and swine have also been shown to be susceptible to human noroviruses (Cheetham *et al.*, 2006; Souza *et al.*, 2008) and a report of inter-genogroup recombination by Nayak and coworkers (Nayak *et al.*, 2008) opens a new possible source of norovirus diversity. On the other hand, the finding of strains related to the most highly prevalent genotype in humans (GII.4) in pig stools as well as in retail meat (Mattison *et al.*, 2007) raises the question whether this was the result of zoonoses, or, rather, anthroponoses.

Surveillance and detection

Surveillance history

In many countries, norovirus surveillance depends on outbreak reporting by peripheral institutions such as municipal health authorities to central laboratories, where diagnostic tests are performed to assign an aetiological agent in samples. Criteria that have typically been used to establish if the definition of a *viral* gastroenteritis outbreak is met are the previously mentioned Kaplan criteria (Kaplan *et al.*, 1982; Turcios *et al.*, 2006).

In the past, the lack of classical virological techniques for norovirus detection other than electron microscopy has greatly hampered norovirus surveillance. During the past decade, molecular detection and typing techniques for gastroenteritis viruses have become commonly used. As a result, awareness of the burden of norovirus outbreaks has grown. Sporadic outbreak investigations have clarified that outbreaks, which are generally reported in surveillance, only form the tip of the iceberg of all norovirus infections (de Wit *et al.*, 2001c; Kuusi *et al.*, 2004; Patel *et al.*, 2008; Wheeler *et al.*, 1999).

Surveillance systems typically cover healthcare settings, such as hospitals and nursing homes for the elderly, but not so much of the general community of otherwise healthy adults and children. Thus, gastrointestinal illness among the general population often goes unreported and norovirus illness has historically been mainly associated with the elderly and the frail. Nowadays the importance of norovirus as a cause of illness in young (hospitalized) children is increasingly recognized, in developed as well as developing countries (Dove *et al.*, 2005; Hansman *et al.*, 2004a,b; Hoebe *et al.*, 2004; Kirkwood and Streitberg, 2008; Martinez *et al.*, 2002; Murata *et al.*, 2007).

A need for internationalization of surveillance

Comparison of surveillance data has shown that circulating strains are generally highly uniform across the world, with the globally epidemic GII.4 strains predominant (Siebenga *et al.*, 2009). Less highly prevalent strains are also spread globally (Iritani *et al.*, 2008a). These observations indicate that norovirus illness and epidemiology should be regarded as an international global health problem. Yet, norovirus surveillance data have been difficult to compare among countries because case- and outbreak definitions and molecular detection techniques are not standardized (Kroneman *et al.*, 2008a). PCR-based or enzyme immunoassy (EIA) detection techniques vary in their abilities to detect different norovirus genotypes. Also, different population subgroups, for example children, elderly or hospital populations, are targeted as a result of local differences in reporting routes, e.g. directly to a central laboratories, through local laboratories or municipal health services, and differences in priorities, e.g. focusing on food-related outbreaks or health care related outbreaks or aiming to provide coverage of the complete population. This is a common problem, also known for other multinational disease reporting networks (http://www.ecdc.europa.eu/en/Publications/AER_report.aspx) (Amato-Gauci and Ammon, 2007), which will hopefully be overcome by increasing the effectiveness of international collaboration schemes. The initiation of several successful multi-institute and international joint surveillance efforts in the past few years has been an encouraging development towards this goal (www.fbve.nl, www.noronet.nl and ozfoodnet.org.au).

Detection methods
RT-PCR

Various reverse transcription-PCR (RT-PCR) assays have been designed to detect norovirus in clinical samples such as faecal samples or vomit, and also in environmental samples, such as surface swabs, or food and water. Several genomic regions may be targeted for detection. RT-PCR is relatively sensitive, offering the possibility of detecting a low quantity of virus, using degenerate primers targeting conserved genomic regions. However, due to the high degree of variation among noroviruses, some norovirus strains may not be detected. In clinical practice this may not be problematic given the dominance of a limited number of genotypes (see below) (Kroneman *et al.*, 2008b; Siebenga *et al.*, 2007; Tu *et al.*, 2007, 2008a). In other situations and in reference laboratories care should be taken to monitor test-performance against less common genotypes.

Real-time PCR is increasingly used, which is more sensitive and faster than RT-PCR. Using real-time PCR with virus-specific-primer and probe combinations in multiplex assays, the detection of multiple different viruses in one test has become feasible. Additionally, real-time assays are semi-quantitative, i.e. useful for measuring and comparing viral loads.

Many laboratories perform sequencing of norovirus-positive samples. Determining the genotype and possible signature mutations enables linking of patients or outbreaks, or finding a common source of infection. Again, several genomic regions can be analysed (Fig. 1.1) (Siebenga *et al.*, 2009; Vinje *et al.*, 2003). For surveillance purposes these partial genomic sequences are used to monitor trends, whereas the highly variable P2 domain of the capsid is sequenced for addressing questions regarding transmission routes, e.g. in assessing hospital epidemiology (Vinje *et al.*, 2004; Xerry *et al.*, 2008).

Enzyme immunoassay
EIAs have been developed for the detection of norovirus antigen in stool samples, and several of these tests are commercially available. Advantages of EIAs over PCR-based assays include simplicity (no specialized equipment or skilled personnel required) and speed (rapid bedside tests have been developed based on an EIA that promise results within 15 min; Khamrin *et al.*, 2009). The EIAs use either monoclonal or polyclonal antibodies specific for a limited number of antigenically distinct norovirus genotypes, which can be problematic in the detection of antigenic variants or emerging genotypes (Grey *et al.*, 2007). Knowledge of the local circulating norovirus genotypes is helpful in evaluating the efficiency

of the EIA in a particular setting (de Bruin *et al.*, 2006). Moreover, if outbreak samples are negative by the EIA, they should be further screened by RT-PCR. The low sensitivity of EIAs (between 44% and 59%) (Grey *et al.*, 2007) makes them unsuitable for diagnosing sporadic cases, unless negative results are, again, followed by RT-PCR analysis.

Outbreak algorithms
The availability of diagnostic tests for noroviruses has allowed the establishment of laboratory-based criteria for the designation of a norovirus outbreak. Duizer *et al.* proposed criteria that addresses the variation of sensitivity and specificity among diagnostic assays as well as the 'baseline prevalence' of noroviruses in the population (Duizer *et al.*, 2007). It has been estimated that, on average, 5.2% of the population is shedding noroviruses without symptoms at a given time, which is an important factor to consider when detecting norovirus-positive individuals in an outbreak investigation (de Wit *et al.*, 2001b). According to Duizer *et al.*, when RT-PCR is employed as the diagnostic assay, the number of norovirus-positive samples should be at least one when testing 2–4 samples, and at least two when testing 5–11 samples. When EIAs are utilized, the norovirus-positive samples should be at least one when testing 2–6 samples, or at least two when testing 7–11 samples.

Epidemiology

Outbreaks
Noroviruses' combined propensities of low infectious dose, high levels of shedding, high environmental persistence, long periods of asymptomatic

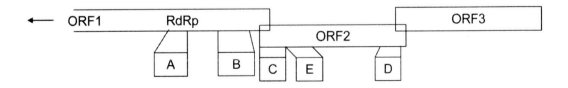

Figure 1.1 Schematic representation of the locations of the genomic regions of norovirus used for genotyping. Adapted from Vinjé *et al.* (2004) published in Siebenga *et al.* (2009). RdRp, RNA-dependent RNA polymerase.

shedding after clinical recovery, high genetic and antigenic variability and the lack of long-term immunity give the virus the ability to cause, possibly repeated, outbreaks in all possible settings of human contact. Such settings include hospitals and long-term care facilities, where outbreaks may have great impact on already weak persons. Other settings that are often associated with norovirus outbreaks are recreational facilities such as cruise ships (Koopmans et al., 2006; Verhoef et al., 2008a; Verhoef et al., 2008b; Widdowson et al., 2004), day care centres (Lyman et al., 2009), schools (Kapikian et al., 1972), restaurants and the army (Halperin et al., 2005). Also, hundreds of hurricane evacuees fell ill in evacuees housing in the aftermath of hurricane Katrina in the south of the US in 2005 (Yee et al., 2007). Hedlund et al. report that 50% and 26% of all norovirus outbreaks in Sweden between 1994 and 1998 took place in hospitals and nursing homes, respectively (Hedlund et al., 2000). Fifty-seven per cent of Dutch outbreaks were reported from nursing homes and 9% from hospitals (van Duynhoven et al., 2005). Forty-three per cent of outbreaks in the US were reported from nursing homes and hospitals (Fankhauser et al., 1998), although for a later period only 25% of outbreaks were reported for these settings (Fankhauser et al., 2002). However, the true prevalence of norovirus outbreaks is difficult to determine because not all outbreaks are reported whereas those in healthcare settings such as hospitals are almost always reported.

Sporadic cases

In the Netherlands, approximately 500,000 episodes of norovirus were estimated to occur annually, in a total population of 15.7 million (de Wit et al., 2001a). Even though norovirus is notorious for causing outbreaks, the major share of infections occurred in sporadic cases or in small family outbreaks that were not reported. Surveys of sporadic cases of gastroenteritis due to noroviruses showed that although prevalence in young children and the elderly are the highest, people of all ages experience norovirus episodes. A recent literature review provides a nice overview of current knowledge of the role of noroviruses in sporadic gastroenteritis (Patel et al., 2009). The studies included in this review reported a range of 5–36% of sporadic cases of diarrhoea to be caused by norovirus infection, with a pooled proportion of 12%.

A number of studies on sporadic norovirus cases have focused on illnesses in children. A Dutch population-based cohort study in 1999 indicated that 65% of all norovirus illness occurred in children less than 18 years of age (de Wit et al., 2001b). Patel et al. were able to estimate that approximately 200,000 yearly deaths result from norovirus infections among children under 5 years of age in developing countries. They also estimated that annually noroviruses cause approximately 900,000 episodes of gastroenteritis that require clinic visits among children under 5 years of age in high-income countries, and 64,000 children are hospitalized. Additionally, several recent studies comparing aetiological agents as causes of acute gastroenteritis in young children showed that noroviruses may be found as commonly as rotaviruses, a virus generally regarded as a major burden in childhood disease, and some of these even report similar clinical impact (Colomba et al., 2007; Iturriza Gomara et al., 2008; Reimerink et al., 2009; Sdiri-Loulizi et al., 2008).

The previously mentioned population study in the Netherlands found that 13% of norovirus illness was found among elderly over 65 years of age (de Wit et al., 2001b). Death among elderly resulting from norovirus infection has been reported for individual cases (Lopman et al., 2004b; Okada et al., 2006; Verhoef et al., 2008c). A syndromic surveillance study estimates excess deaths among elderly over 65 years of age in England at least 80 per year (Harris et al., 2008).

Molecular epidemiology

Many public health laboratories nowadays include norovirus genotyping in their standard diagnostic procedures, in combination with epidemiological background data. Strains belonging to GII were found in 75–100% of sporadic cases (for overview see Patel et al., 2008). From molecular surveillance data it has become clear that there is less diversity in outbreak-based studies than studies looking at sporadic cases, indicating differences in transmissibility between norovirus strains. GII.4 viruses are by far the most commonly detected strains in studies in recent years, whether describing outbreaks or sporadic cases, as illustrated by Fig. 1.2. Overall, GII.4 strains account for 60–70% of all reported outbreaks globally (Kroneman *et al.*, 2008a; Siebenga *et al.*, 2009). Among children and patients with prolonged shedding, GII.3 was the second most commonly reported strain, but these data were limited (Dove *et al.*, 2005; Siebenga *et al.*, 2008). Strains belonging to GII.4 were almost nine times more often associated with outbreaks in health-care settings than with outbreaks in other settings,

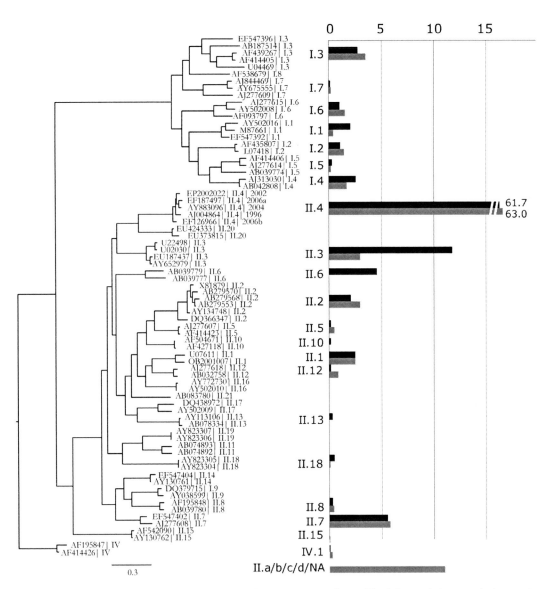

Figure 1.2 Human norovirus phylogenetic tree and genotype prevalence. The left panel shows a phylogenetic tree based on complete capsid nucleotide sequences, constructed using PHYML. The sequences selected to represent the genotypes have been derived from the sequence alignment used in the new norovirus typing tool. This alignment contains sequences aimed at covering optimal diversity within each genotype to enable reliable genotyping. Nomenclature refers to the current genotype nomenclature for which a consensus discussion is taking place (http://www.rivm.nl/mpf/norovirus/typingtool). To the right prevalence levels as percentage of all reported norovirus outbreaks are shown per genotype. These are based on the FBVE database, containing sequences of the outbreaks reported to the FBVE network, detected between 1990 and 2008 (described in Kroneman *et al.* 2008a,b). Typing results were obtained either by partial capsid sequencing (1472 sequences, represented by black bars) or partial polymerase sequencing (5089 sequences, grey bars). NA: not assigned.

and detected more often in outbreaks associated with person-to-person spread than for instance food-borne outbreaks (Gallimore *et al.*, 2004a; Kroneman *et al.*, 2008a). Genetic diversity of noroviruses is further increased through recombination, which occurs regularly in noroviruses (see Chapter 3). Recombination most often involves strains within a genogroup (Bull *et al.*, 2005; Bull *et al.*, 2007), but has also been reported between strains belonging to different genogroups (Nayak *et al.*, 2008). For a number of norovirus strains only ORF1 sequences have been detected with no known 'own' capsid genes. These 'promiscuous' ORF1 genes have been found associated with capsids of different genotypes. Since genotype assignment is based on capsid sequences, no genotype number but a letter has been assigned: GII.a, b and c. Especially strains with GII.b ORF1 have been found consistently in norovirus outbreak surveillance systems since their first detection, in France in 2000 (F. Bon, P. Pothier, C. Hemery, M. Cournot, C. Castor, H. de Valk, P. Gourier-Frery, P. Benhamida, R. Roques, L. Le Coustumier, F. Villeneuve, P. Megraud, P. Le Cann, E. Kohli, and A. Gallay, 21st Réunion Interdisciplinaire de Chimiothérapie Anti-infectieuse, Paris, France, 2001). It is unclear what the source of these non-structural genes has been, as their parental full-length viruses have never been detected through human disease surveillance. This may mean that the parental strains did not result in disease, or – theoretically – were from a reservoir other than humans. However, at present, there is no evidence for such interspecies transmission events.

Changing epidemiology; emerging GII.4 variants

For reasons not yet entirely understood, the number of reported outbreaks in many parts of the world has increased steeply since 2002 (Bruggink and Marshall, 2008; Hutson *et al.*, 2004; Siebenga *et al.*, 2007; Widdowson *et al.*, 2004). A syndrome surveillance study of norovirus illness showed that simultaneously with this rise both morbidity and mortality have increased (Van Asten, unpublished), observations that are supported by anecdotal reports from, for example, nursing staff. Such findings indicate that norovirus illness has become more severe during the past decade, possibly resulting from changes in the predominant norovirus GII.4 strain.

GII.4 variants displace each other in time

Characterization of a systematic sampling of outbreak strains showed that several global norovirus epidemics have occurred, all following the emergence of new genetic variants of GII.4 (Fig. 1.3). These variants displaced each other (Fig. 1.4). The first epidemic occurred in the winter of 1995–96 (Noel *et al.*, 1999; Vinje *et al.*, 1997), and after, what was in hindsight, a relatively long period of genetic stasis, a new predominant GII.4 strain emerged in the spring of 2002 (Lopman *et al.*, 2004a; Noel *et al.*, 1999; Vinje *et al.*, 1997; Widdowson *et al.*, 2004). This GII.4-2002 variant, also known as Farmington Hills, caused an unusual amount of off seasonal outbreaks during the spring and summer of 2002 in a number of European countries, and with a marked high prevalence on cruise ships, which was followed by a global epidemic in the winter of 2002–03. This scenario was repeated with the emergence of a new GII.4 variant in 2004–05 in Oceania (Bull *et al.*, 2006). In the spring of 2006 two genetically and epidemiologically distinct variants emerged. These strains co-circulated during the next winter, after which the 2006b variant became dominant over the other, 2006a variant. Currently, in 2009, new lineages of GII.4 strains have been identified, and detected at several locations around the world, but the 2006b variant remains dominant, interestingly already for three seasons in a row.

GII.4 variants, but not all, have a global distribution

Sequence comparison in a global prevalence study showed that four of these GII.4 variants had a true global distribution, namely the 1996, 2002, 2004 and 2006b variants. On the other hand, the 2006a variant was scarcely detected in Asia, and the 2003 Asian variant, conversely, was rarely detected outside Asia (Siebenga *et al.*, 2009). Besides the major variants that caused significant numbers of outbreaks, several less successful GII.4 strains circulated. These minor variants of GII.4 have been widely detected geographically, but at low prevalence. The cause for such differences in spread and prevalence currently remains open for speculation. Structural differences in

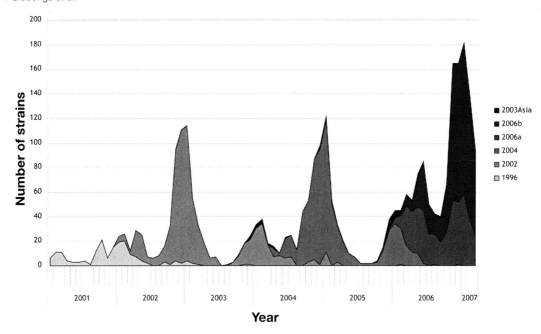

Figure 1.3 Norovirus GII.4 variants displaced each other through time. Data used for the first global Noronet GII.4 prevalence overview (Siebenga *et al.*, 2009). Strains (*n*=3089) detected between January 2001 and March 2007, contributions from Hungary, Hong Kong, Japan, Australia, Germany, Netherlands, the USA, Canada and New Zealand.

the capsid protein may limit the host range of the viruses, although based on the assessment of the currently identified ligand binding sites no differences could be identified that might cause this. Alternatively, the lack of a successful seeding event may have contributed to the lack of success of these variants in all geographic areas. However, this seems an unlikely explanation in light of the observed very rapid and efficient spread of other norovirus strains as well as other viruses across the globe (Iritani *et al.*, 2008a; Iritani *et al.*, 2008b; Siebenga *et al.*, 2009). Strain competition is also unlikely to have imposed a limitation to the introduction of an additional strain: GII.4-2006a and -2006b co-circulated in Europe, America and Oceania without restraints and even though the numbers of people infected with norovirus per season are impressive, it still concerns a minority of the population. Norovirus surveillance has a relatively limited global coverage, and especially from developing countries not much data is available. Thus, it remains largely unknown what strains circulate in Africa, South America and large parts of Asia. The sparse data from other continents do not provide a dense enough sampling to pinpoint a geographic origin

for emerging variants, although Oceania and Asia seem to have detected three of the global variants first, against one in Europe (Siebenga *et al.*, 2009).

GII.4 can remain dominant through rapid antigenic variant displacement
During inter-epidemic years, when GII.4 prevalence was lower, strains of other genotypes were reported in higher numbers, both in percentage as well as in absolute numbers (Siebenga *et al.*, 2007). For example, strains with GII.b ORF1 sequences were reported frequently in the winter of 2003–04 in Europe (Lindell *et al.*, 2005). It remains unknown what makes GII.4 so much more successful than other genotypes, although some plausible causes can be conceived. The GII.4 strains may utilize the widest range of HBGA molecules in host-cell binding (Huang *et al.*, 2005b), and alternatively factors like higher infectivity or more efficient transmission through higher stability, and more or longer shedding of viruses may play a role.

The leading explanation for the observed pattern of variant emergence, circulation and displacement is that the high prevalence of GII.4 in the population leads to population immunity

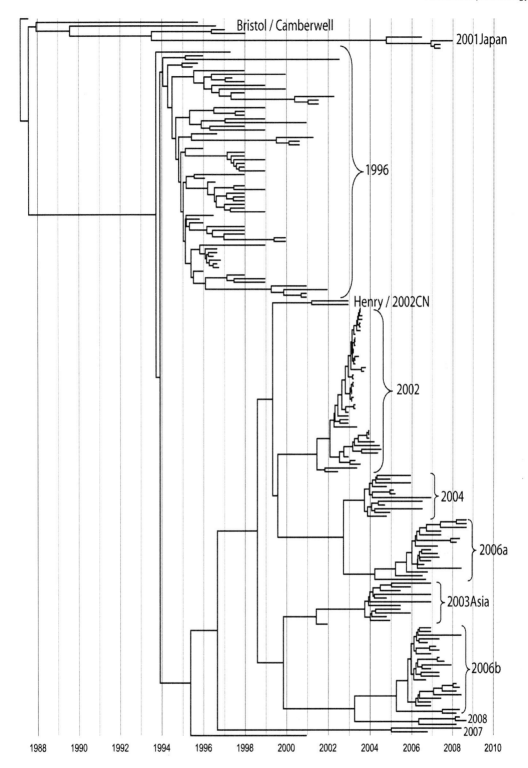

Figure 1.4 Maximum Clade Credibility tree of 194 norovirus GII.4 full capsid nucleotide sequences, detected worldwide. The tree is depicted on a time scale, with tips positioned on their detection dates and has been generated with BEAST. Variant clusters are indicated in the figure.

against the circulating dominant variant, based on short-term immunity that may occasionally be boostered by the same virus-variant while it still circulates, although no long-term immunity develops. This was confirmed by ligand-binding blocking experiments using virus-like particles of different GII.4 variants (Lindesmith *et al.*, 2008), showing that subsequent GII.4 variants are antigenic variants, in the absence of a method for neutralization assays. Additionally, changes in host-ligand binding properties were demonstrated between different GII.4 variants (Donaldson *et al.*, 2008), probably enabling the virus to switch between host populations. Although the capsid genes of epidemic variants did evolve during the time they circulated in the population, they were surprisingly stable. For influenza, the continuous circulation in densely populated cities in Asia is seen as a driving factor for strain evolution. Whether this is also the case for noroviruses as well remains to be seen.

The origin of emerging GII.4 variants remains unknown

The genetic diversity discerning between different antigenic variants of the GII.4 genotype may be large, up to 25 amino acid residues in the 541 amino acid capsid protein (~5%) (Siebenga *et al.*, 2007). They emerged in a stepwise manner, with periods of phenotypic stasis separated by the emergence of phenotypically, in this case genetically and antigenically distinct strains. This is strongly reminiscent of the evolutionary pattern displayed by Influenza A viruses, and is also known as epochal evolution (Koelle *et al.*, 2006). No intermediate strains, bridging between subsequent variants, have been detected, which may point to introductions from a non-human reservoir, but on the other hand the low intensity of surveillance worldwide does not make detection of low prevalence intermediate strains likely.

Another possible source of new variant strains may evolve from immunocompromised individuals. Prolonged infections and accompanying shedding of noroviruses have been reported in several studies. Immunocompromised patients are at especially high risk for such infections. In patients who showed prolonged shedding viruses belonging to GII.2, GII.3, GII.b-GII.3 and GII.4 have been found (Gallimore *et al.*, 2004b; Lee *et*

al., 2008; Ludwig *et al.*, 2008; Nilsson *et al.*, 2003; Siebenga *et al.*, 2008). Viruses detected in these patients accumulated mutations in their capsid proteins rapidly, up to 11 amino acid mutations during a 4-month period (Siebenga *et al.*, 2008). Nevertheless patients with most severe immune-dysfunction accumulated mutations at the lowest rate, and especially low general antibody levels in the blood seemed to be associated with low mutation accumulation rates (Siebenga *et al.*, 2008). In otherwise healthy children who showed no prolonged symptoms, shedding of norovirus was shown until 6 weeks after infection and long after clearance of symptoms (Murata *et al.*, 2007). Viruses shed in the stools of an otherwise healthy child were shown to accumulate mutations at the highest rate reported; 0.13 amino acid mutations were accumulated daily (~4 amino acids monthly) (Siebenga *et al.*, 2008). Among the general and immunocompetent population the number of infections that occur is magnitudes larger than the number of prolonged infections in immunocompromised patients, so if indeed the mutation accumulation rate is significantly higher in immunocompetent people than in immunocompromised patients, the general population is a more likely source for new variant strains.

Molecular epidemiology as a tool in linking outbreaks and tracing foodborne outbreaks

For prevention and containment purposes there is a need for tools to assess whether different outbreaks are connected to provide insight into transmission routes. Using genotyping data alone for such assessments presents the difficulty that GII.4 strains cause the great majority of all outbreaks, and within epidemic seasons genetic variation of outbreak strains may be minimal in the genetic regions assessed for diagnostic purposes. Lopman *et al.* proposed a method for assessing both molecular as well as statistical epidemiological data to verify if different outbreaks were actually linked (Lopman *et al.*, 2006). This method takes the probability of the occurrence of the observed patterns by pure chance into account, and provides statistical evidence that linked outbreaks share greater similarity in sequences detected, discernible even for GII.4 outbreaks.

If outbreaks are proven connected, preventative measures may be more effectively executed, for example if a common source of infection is involved. A further complication in detecting diffuse food-related norovirus outbreaks and their possible source lies in the high environmental stability of the virus; it implies that the contamination detected in the food may have occurred long before the outbreak (Duizer et al., 2004a). In addition, a practical problem is that the number of reported outbreaks may be too overwhelming for available infection control staff and the time required for proper outbreak investigation is insufficient. Again, based on statistical analyses assessing the probability of occurrence of norovirus in specific settings, Verhoef et al. proposed the previously described tool that helps in selecting the outbreaks for which foodborne infection is most likely (Verhoef et al., 2009). So, although some obstacles and difficulties remain in detecting common-source norovirus outbreaks, promising progress is made in understanding and distinguishing them.

Conclusion

Noroviruses are on the rise as a pathogen in humans, both in the number of infections they cause and at the level of recognition. While during the past decade different variants of one genotype, i.e. GII.4, have been globally dominant, it remains as yet unknown what factors determine the success of this genotype in comparison to other genotypes. Additional research is still needed to identify markers for antigenicity and virulence in norovirus, especially bearing in mind for example the wish to develop a vaccine against infection with this particular genotype, or the severe illnesses caused by other, related caliciviruses. Additionally, the role of noroviruses in developing countries and in sporadic infections needs to be elucidated in order to better understand the epidemiology.

References

Agus, S.G., Dolin, R., Wyatt, R.G., Tousimis, A.J., and Northrup, R.S. (1973). Acute infectious nonbacterial gastroenteritis: intestinal histopathology. Histologic and enzymatic alterations during illness produced by the Norwalk agent in man. Ann. Intern. Med. 79, 18–25.

Allwood, P.B., Malik, Y.S., Hedberg, C.W., and Goyal, S.M. (2004). Effect of temperature and sanitizers on the survival of feline calicivirus, Escherichia coli, and F-specific coliphage MS2 on leafy salad vegetables. J. Food Prot. 67, 1451–1456.Amato-Gauci, A., and Ammon, A. (2007). ECDC to launch first report on communicable diseases epidemiology in the European Union. Euro. Surveill. 12, E070607, 070602.

Atmar, R.L., Opekun, A.R., Gilger, M.A., Estes, M.K., Crawford, S.E., Neill, F.H., and Graham, D.Y. (2008). Norwalk virus shedding after experimental human infection. Emerg. Infect. Dis. 14, 1553–1557.

Beersma, M.F., Schutten, M., Vennema, H., Hartwig, N.G., Mes, T.H., Osterhaus, A.D., van Doornum, G.J., and Koopmans, M. (2009). Norovirus in a Dutch tertiary care hospital (2002–2007): frequent nosocomial transmission and dominance of GIIb strains in young children. J. Hosp. Infect. 71, 199–205.

Belliot, G., Lavaux, A., Souihel, D., Agnello, D., and Pothier, P. (2008). Use of murine norovirus as a surrogate to evaluate resistance of human norovirus to disinfectants. Appl. Environ. Microbiol. 74, 3315–3318.

Bertolotti-Ciarlet, A., Crawford, S.E., Hutson, A.M., and Estes, M.K. (2003). The 3′ end of Norwalk virus mRNA contains determinants that regulate the expression and stability of the viral capsid protein VP1: a novel function for the VP2 protein. J. Virol. 77, 11603–11615.

Billgren, M., Christenson, B., Hedlund, K.O., and Vinje, J. (2002). Epidemiology of Norwalk-like human caliciviruses in hospital outbreaks of acute gastroenteritis in the Stockholm area in 1996. J. Infect. 44, 26–32.

Bon, F., Ambert-Balay, K., Giraudon, H., Kaplon, J., Le Guyader, S., Pommepuy, M., Gallay, A., Vaillant, V., de Valk, H., Chikhi-Brachet, R., et al. (2005). Molecular epidemiology of caliciviruses detected in sporadic and outbreak cases of gastroenteritis in France from December 1998 to February 2004. J. Clin. Microbiol. 43, 4659–4664.

Bruggink, L., and Marshall, J. (2008). Molecular changes in the norovirus polymerase gene and their association with incidence of GII.4 norovirus-associated gastroenteritis outbreaks in Victoria, Australia, 2001–2005. Arch. Virol. 153, 729–732.

Bull, R.A., Hansman, G.S., Clancy, L.E., Tanaka, M.M., Rawlinson, W.D., and White, P.A. (2005). Norovirus recombination in ORF1/ORF2 overlap. Emerg. Infect. Dis. 11, 1079–1085.

Bull, R.A., Tanaka, M.M., and White, P.A. (2007). Norovirus recombination. J. Gen. Virol. 88, 3347–3359.

Bull, R.A., Tu, E.T., McIver, C.J., Rawlinson, W.D., and White, P.A. (2006). Emergence of a new norovirus genotype II.4 variant associated with global outbreaks of gastroenteritis. J. Clin. Microbiol. 44, 327–333.

Cannon, J.L., Papafragkou, E., Park, G.W., Osborne, J., Jaykus, L.A., and Vinje, J. (2006). Surrogates for the study of norovirus stability and inactivation in the environment: aA comparison of murine norovirus and feline calicivirus. J. Food Prot. 69, 2761–2765.

Cao, S., Lou, Z., Tan, M., Chen, Y., Liu, Y., Zhang, Z., Zhang, X.C., Jiang, X., Li, X., and Rao, Z. (2007).

Structural basis for the recognition of blood group trisaccharides by norovirus. J. Virol. *81*, 5949–5957.

Carlsson, B., Lindberg, A.M., Rodriguez-Diaz, J., Hedlund, K.O., Persson, B., and Svensson, L. (2009). Quasispecies dynamics and molecular evolution of human norovirus capsid P region during chronic infection. J. Gen. Virol. *90*, 432–441.

Caul, E.O. (1995). Hyperemesis hiemis – a sick hazard. J. Hosp. Infect. *30* Suppl., 498–502.

CDC (2009). History and Epidemiology of Global Smallpox Eradication. http://wwwbtcdcgov/agent/smallpox/training/overview/pdf/eradicationhistorypdf

Cheetham, S., Souza, M., McGregor, R., Meulia, T., Wang, Q., and Saif, L.J. (2007). Binding patterns of human norovirus-like particles to buccal and intestinal tissues of gnotobiotic pigs in relation to A/H histo-blood group antigen expression. J. Virol. *81*, 3535–3544.

Cheetham, S., Souza, M., Meulia, T., Grimes, S., Han, M.G., and Saif, L.J. (2006). Pathogenesis of a genogroup II human norovirus in gnotobiotic pigs. J. Virol. *80*, 10372–10381.

Clay, S., Maherchandani, S., Malik, Y.S., and Goyal, S.M. (2006). Survival on uncommon fomites of feline calicivirus, a surrogate of noroviruses. Am. J. Infect. Control *34*, 41–43.

Colomba, C., De Grazia, S., Giammanco, G.M., Saporito, L., Scarlata, F., Titone, L., and Arista, S. (2006). Viral gastroenteritis in children hospitalised in Sicily, Italy. Eur. J. Clin. Microbiol. Infect. Dis *25*, 570–575.

Colomba, C., Saporito, L., Giammanco, G.M., De Grazia, S., Ramirez, S., Arista, S., and Titone, L. (2007). Norovirus and gastroenteritis in hospitalized children, Italy. Emerg. Infect. Dis. *13*, 1389–1391.

de Bruin, E., Duizer, E., Vennema, H., and Koopmans, M.P. (2006). Diagnosis of Norovirus outbreaks by commercial ELISA or RT-PCR. J. Virol. Methods *137*, 259–264.

De Roda Husman, A.M., Bijkerk, P., Lodder, W., Van Den Berg, H., Pribil, W., Cabaj, A., Gehringer, P., Sommer, R., and Duizer, E. (2004). Calicivirus inactivation by nonionizing (253.7-nanometer-wavelength [UV]) and ionizing (gamma) radiation. Appl. Environ. Microbiol. *70*, 5089–5093.

de Wit, M.A., Koopmans, M.P., Kortbeek, L.M., van Leeuwen, N.J., Bartelds, A.I., and van Duynhoven, Y.T. (2001a). Gastroenteritis in sentinel general practices, The Netherlands. Emerg. Infect. Dis. *7*, 82–91.

de Wit, M.A., Koopmans, M.P., Kortbeek, L.M., van Leeuwen, N.J., Vinje, J., and van Duynhoven, Y.T. (2001b). Etiology of gastroenteritis in sentinel general practices in the netherlands. Clin. Infect. Dis. *33*, 280–288.

de Wit, M.A., Koopmans, M.P., Kortbeek, L.M., Wannet, W.J., Vinje, J., van Leusden, F., Bartelds, A.I., and van Duynhoven, Y.T. (2001c). Sensor, a population-based cohort study on gastroenteritis in the Netherlands: incidence and etiology. Am. J. Epidemiol. *154*, 666–674.

de Wit, M.A., Widdowson, M.A., Vennema, H., de Bruin, E., Fernandes, T., and Koopmans, M. (2007). Large outbreak of norovirus: The baker who should have known better. J. Infect. *55*, 188–193.

Dolin, R., and Baron, S. (1975). Absence of detectable interferon in jejunal biopsies, jejunal aspirates, and sera in experimentally induced viral gastroenteritis in man. Proc. Soc. Exp. Biol. Med. *150*, 337–339.

Donaldson, E.F., Lindesmith, L.C., Lobue, A.D., and Baric, R.S. (2008). Norovirus pathogenesis: mechanisms of persistence and immune evasion in human populations. Immunol. Rev. *225*, 190–211.

Dove, W., Cunliffe, N.A., Gondwe, J.S., Broadhead, R.L., Molyneux, M.E., Nakagomi, O., and Hart, C.A. (2005). Detection and characterization of human caliciviruses in hospitalized children with acute gastroenteritis in Blantyre, Malawi. J. Med. Virol. *77*, 522–527.

Doyle, T.J., Stark, L., Hammond, R., and Hopkins, R.S. (2008). Outbreaks of noroviral gastroenteritis in Florida, 2006–2007. Epidemiol. Infect. 1–9.

Duizer, E., Bijkerk, P., Rockx, B., De Groot, A., Twisk, F., and Koopmans, M. (2004a). Inactivation of caliciviruses. Appl. Environ. Microbiol. *70*, 4538–4543.

Duizer, E., Pielaat, A., Vennema, H., Kroneman, A., and Koopmans, M. (2007). Probabilities in norovirus outbreak diagnosis. J. Clin. Virol. *40*, 38–42.

Duizer, E., Schwab, K.J., Neill, F.H., Atmar, R.L., Koopmans, M.P., and Estes, M.K. (2004b). Laboratory efforts to cultivate noroviruses. J. Gen. Virol. *85*, 79–87.

Evans, M.R., Meldrum, R., Lane, W., Gardner, D., Ribeiro, C.D., Gallimore, C.I., and Westmoreland, D. (2002). An outbreak of viral gastroenteritis following environmental contamination at a concert hall. Epidemiol. Infect. *129*, 355–360.

Falkenhorst, G., Krusell, L., Lisby, M., Madsen, S.B., Bottiger, B., and Molbak, K. (2005). Imported frozen raspberries cause a series of norovirus outbreaks in Denmark, 2005. Euro. Surveill. *10*, E050922, 050922.

Fankhauser, R.L., Monroe, S.S., Noel, J.S., Humphrey, C.D., Bresee, J.S., Parashar, U.D., Ando, T., and Glass, R.I. (2002). Epidemiologic and molecular trends of 'Norwalk-like viruses' associated with outbreaks of gastroenteritis in the United States. J. Infect. Dis. *186*, 1–7.

Fankhauser, R.L., Noel, J.S., Monroe, S.S., Ando, T., and Glass, R.I. (1998). Molecular epidemiology of 'Norwalk-like viruses' in outbreaks of gastroenteritis in the United States. J. Infect. Dis. *178*, 1571–1578.

Farkas, T., Sestak, K., Wei, C., and Jiang, X. (2008). Characterization of a rhesus monkey calicivirus representing a new genus of Caliciviridae. J. Virol. *82*, 5408–5416.

Friesema, I.H., Vennema, H., Heijne, J.C., de Jager, C.M., Morroy, G., van den Kerkhof, J.H., de Coster, E.J., Wolters, B.A., Ter Waarbeek, H.L., Fanoy, E.B., *et al.* (2009). Norovirus outbreaks in nursing homes: the evaluation of infection control measures. Epidemiol. Infect. 1–12.

Gallimore, C.I., Cheesbrough, J.S., Lamden, K., Bingham, C., and Gray, J.J. (2005). Multiple norovirus genotypes characterised from an oyster-associated outbreak of gastroenteritis. Int. J. Food Microbiol. 103, 323–330.

Gallimore, C.I., Green, J., Richards, A.F., Cotterill, H., Curry, A., Brown, D.W., and Gray, J.J. (2004a).

Methods for the detection and characterisation of noroviruses associated with outbreaks of gastroenteritis: outbreaks occurring in the north-west of England during two norovirus seasons. J. Med. Virol. 73, 280–288.

Gallimore, C.I., Lewis, D., Taylor, C., Cant, A., Gennery, A., and Gray, J.J. (2004b). Chronic excretion of a norovirus in a child with cartilage hair hypoplasia (CHH). J. Clin. Virol. 30, 196–204.

Gallimore, C.I., Taylor, C., Gennery, A.R., Cant, A.J., Galloway, A., Iturriza-Gomara, M., and Gray, J.J. (2006). Environmental monitoring for gastroenteric viruses in a pediatric primary immunodeficiency unit. J. Clin. Microbiol. 44, 395–399.

Gallimore, C.I., Taylor, C., Gennery, A.R., Cant, A.J., Galloway, A., Xerry, J., Adigwe, J., and Gray, J.J. (2008). Contamination of the hospital environment with gastroenteric viruses: comparison of two pediatric wards over a winter season. J. Clin. Microbiol. 46, 3112–3115.

Gehrke, C., Steinmann, J., and Goroncy-Bermes, P. (2004). Inactivation of feline calicivirus, a surrogate of norovirus (formerly Norwalk-like viruses), by different types of alcohol in vitro and in vivo. J. Hosp. Infect. 56, 49–55.

Gomes, K.A., Stupka, J.A., Gomez, J., and Parra, G.I. (2007). Molecular characterization of calicivirus strains detected in outbreaks of gastroenteritis in Argentina. J. Med. Virol. 79, 1703–1709.

Gray, J.J., Kohli, E., Ruggeri, F.M., Vennema, H., Sanchez-Fauquier, A., Schreier, E., Gallimore, C.I., Iturriza-Gomara, M., Giraudon, H., Pothier, P., et al. (2007). European multicenter evaluation of commercial enzyme immunoassays for detecting norovirus antigen in fecal samples. Clin. Vaccine Immunol. 14, 1349–1355.

Green, K.Y. (2007). Caliciviridae: The Noroviruses. In Fields Virology, H.P. Knipe, and P.M. Howley, eds. (Philadelphia: Lippincott Williams & Wilkins), pp. 949–979.

Green, K.Y., Kapikian, A., and Chanock, R.M. (2001). Human caliciviruses. In Fields Virology, 4th edn, H.P. Knipe, and P.M. Howley, eds. (Philadelphia: Lippincott Williams & Wilkins), pp. 841–847.

Greig, J.D., and Lee, M.B. (2009). Enteric outbreaks in long-term care facilities and recommendations for prevention: a review. Epidemiol. Infect. 137, 145–155.

Gulati, B.R., Allwood, P.B., Hedberg, C.W., and Goyal, S.M. (2001). Efficacy of commonly used disinfectants for the inactivation of calicivirus on strawberry, lettuce, and a food-contact surface. J. Food Prot. 64, 1430–1434.

Halperin, T., Yavzori, M., Amitai, A., Klement, E., Kayouf, R., Grotto, I., Huerta, M., Hadley, L.A., Monroe, S.S., Cohen, D., et al. (2005). Molecular analysis of noroviruses involved in acute gastroenteritis outbreaks in military units in Israel, 1999–2004. Eur. J. Clin. Microbiol. Infect. Dis. 24, 697–700.

Hamano, M., Kuzuya, M., Fujii, R., Ogura, H., and Yamada, M. (2005). Epidemiology of acute gastroenteritis outbreaks caused by Noroviruses in Okayama, Japan. J. Med. Virol. 77, 282–289.

Han, M.G., Smiley, J.R., Thomas, C., and Saif, L.J. (2004). Genetic recombination between two genotypes of genogroup III bovine noroviruses (BoNVs) and capsid sequence diversity among BoNVs and Nebraska-like bovine enteric caliciviruses. J. Clin. Microbiol. 42, 5214–5224.

Hansen, S., Stamm-Balderjahn, S., Zuschneid, I., Behnke, M., Ruden, H., Vonberg, R.P., and Gastmeier, P. (2007). Closure of medical departments during nosocomial outbreaks: data from a systematic analysis of the literature. J. Hosp. Infect. 65, 348–353.

Hansman, G.S., Doan, L.T., Kguyen, T.A., Okitsu, S., Katayama, K., Ogawa, S., Natori, K., Takeda, N., Kato, Y., Nishio, O., et al. (2004a). Detection of norovirus and sapovirus infection among children with gastroenteritis in Ho Chi Minh City, Vietnam. Arch. Virol. 149, 1673–1688.

Hansman, G.S., Katayama, K., Maneekarn, N., Peerakome, S., Khamrin, P., Tonusin, S., Okitsu, S., Nishio, O., Takeda, N., and Ushijima, H. (2004b). Genetic diversity of norovirus and sapovirus in hospitalized infants with sporadic cases of acute gastroenteritis in Chiang Mai, Thailand. J. Clin. Microbiol. 42, 1305–1307.

Hansman, G.S., Natori, K., Shirato-Horikoshi, H., Ogawa, S., Oka, T., Katayama, K., Tanaka, T., Miyoshi, T., Sakae, K., Kobayashi, S., et al. (2006). Genetic and antigenic diversity among noroviruses. J. Gen. Virol. 87, 909–919.

Hansman, G.S., Oka, T., Li, T.C., Nishio, O., Noda, M., and Takeda, N. (2008). Detection of human enteric viruses in Japanese clams. J. Food Prot. 71, 1689–1695.

Harris, J.P., Edmunds, W.J., Pebody, R., Brown, D.W., and Lopman, B.A. (2008). Deaths from norovirus among the elderly, England and Wales. Emerg. Infect. Dis. 14, 1546–1552.

Hedlund, K.O., Rubilar-Abreu, E., and Svensson, L. (2000). Epidemiology of calicivirus infections in Sweden, 1994–1998. J. Infect. Dis. 181 Suppl. 2, S275–280.

Heijne, J.C., Teunis, P., Morroy, G., Wijkmans, C., Oostveen, S., Duizer, E., Kretzschmar, M., and Wallinga, J. (2009). Enhanced hygiene measures and norovirus transmission during an outbreak. Emerg. Infect. Dis. 15, 24–30.

Hewitt, J., Bell, D., Simmons, G.C., Rivera-Aban, M., Wolf, S., and Greening, G.E. (2007). Gastroenteritis outbreak caused by waterborne norovirus at a New Zealand ski resort. Appl. Environ. Microbiol. 73, 7853–7857.

Hoebe, C.J., Vennema, H., Husman, A.M., and van Duynhoven, Y.T. (2004). Norovirus outbreak among primary schoolchildren who had played in a recreational water fountain. J. Infect. Dis. 189, 699–705.

Hsu, C.C., Riley, L.K., and Livingston, R.S. (2007). Molecular characterization of three novel murine noroviruses. Virus Genes 34, 147–155.

Huang, P., Farkas, T., Marionneau, S., Zhong, W., Ruvoen-Clouet, N., Morrow, A.L., Altaye, M., Pickering, L.K., Newburg, D.S., LePendu, J., et al. (2003). Noroviruses bind to human ABO, Lewis, and secretor histo-blood group antigens: identification of 4 distinct strain-specific patterns. J. Infect. Dis. 188, 19–31.

Huang, P., Farkas, T., Zhong, W., Tan, M., Thornton, S., Morrow, A.L., and Jiang, X. (2005a). Norovirus and histo-blood group antigens: demonstration of a wide spectrum of strain specificities and classification of two major binding groups among multiple binding patterns. J. Virol. *79*, 6714–6722.

Huang, Z., Elkin, G., Maloney, B.J., Beuhner, N., Arntzen, C.J., Thanavala, Y., and Mason, H.S. (2005b). Virus-like particle expression and assembly in plants: hepatitis B and Norwalk viruses. Vaccine *23*, 1851–1858.

Hutson, A.M., Atmar, R.L., and Estes, M.K. (2004). Norovirus disease: changing epidemiology and host susceptibility factors. Trends Microbiol. *12*, 279–287.

Iritani, N., Kaida, A., Kubo, H., Abe, N., Murakami, T., Vennema, H., Koopmans, M., Takeda, N., Ogura, H., and Seto, Y. (2008a). Epidemic of genotype GII.2 noroviruses during spring 2004 in Osaka City, Japan. J. Clin. Microbiol. *46*, 2406–2409.

Iritani, N., Vennema, H., Siebenga, J.J., Siezen, R.J., Renckens, B., Seto, Y., Kaida, A., and Koopmans, M. (2008b). Genetic analysis of the capsid gene of genotype GII.2 noroviruses. J. Virol. *82*, 7336–7345.

Iturriza Gomara, M., Simpson, R., Perault, A.M., Redpath, C., Lorgelly, P., Joshi, D., Mugford, M., Hughes, C.A., Dalrymple, J., Desselberger, U., *et al.* (2008). Structured surveillance of infantile gastroenteritis in East Anglia, UK: incidence of infection with common viral gastroenteric pathogens. Epidemiol. Infect. *136*, 23–33.

Jiang, X., Graham, D.Y., Wang, K.N., and Estes, M.K. (1990). Norwalk virus genome cloning and characterization. Science *250*, 1580–1583.

Jiang, X., Wang, M., Wang, K., and Estes, M.K. (1993). Sequence and genomic organization of Norwalk virus. Virology *195*, 51–61.

Johnson, P.C., Mathewson, J.J., DuPont, H.L., and Greenberg, H.B. (1990). Multiple-challenge study of host susceptibility to Norwalk gastroenteritis in US adults. J. Infect. Dis. *161*, 18–21.

Johnston, C.P., Qiu, H., Ticehurst, J.R., Dickson, C., Rosenbaum, P., Lawson, P., Stokes, A.B., Lowenstein, C.J., Kaminsky, M., Cosgrove, S.E., *et al.* (2007). Outbreak management and implications of a nosocomial norovirus outbreak. Clin. Infect. Dis. *45*, 534–540.

Kageyama, T., Shinohara, M., Uchida, K., Fukushi, S., Hoshino, F.B., Kojima, S., Takai, R., Oka, T., Takeda, N., and Katayama, K. (2004). Coexistence of multiple genotypes, including newly identified genotypes, in outbreaks of gastroenteritis due to Norovirus in Japan. J. Clin. Microbiol. *42*, 2988–2995.

Kanerva, M., Maunula, L., Lappalainen, M., Mannonen, L., von Bonsdorff, C.H., and Anttila, V.J. (2009). Prolonged norovirus outbreak in a Finnish tertiary care hospital caused by GII.4–2006b subvariants. J. Hosp. Infect. *71*, 206–213.

Kapikian, A.Z., Wyatt, R.G., Dolin, R., Thornhill, T.S., Kalica, A.R., and Chanock, R.M. (1972). Visualization by immune electron microscopy of a 27-nm particle associated with acute infectious nonbacterial gastroenteritis. J. Virol. *10*, 1075–1081.

Kaplan, J.E., Feldman, R., Campbell, D.S., Lookabaugh, C., and Gary, G.W. (1982). The frequency of a Norwalk-like pattern of illness in outbreaks of acute gastroenteritis. Am. J. Public Health *72*, 1329–1332.

Karst, S.M., Wobus, C.E., Lay, M., Davidson, J., and Virgin, H.W.t. (2003). STAT1-dependent innate immunity to a Norwalk-like virus. Science *299*, 1575–1578.

Kelly, R.J., Ernst, L.K., Larsen, R.D., Bryant, J.G., Robinson, J.S., and Lowe, J.B. (1994). Molecular basis for H blood group deficiency in Bombay (Oh) and para-Bombay individuals. Proc. Natl. Acad. Sci. U.S.A. *91*, 5843–5847.

Khamrin, P., Takanashi, S., Chan-It, W., Kobayashi, M., Nishimura, S., Katsumata, N., Okitsu, S., Maneekarn, N., Nishio, O., and Ushijima, H. (2009). Immunochromatography test for rapid detection of norovirus in fecal specimens. J. Virol. Methods *157*, 219–222.

Kirkwood, C.D., and Streitberg, R. (2008). Calicivirus shedding in children after recovery from diarrhoeal disease. J. Clin. Virol. *43*, 346–348.

Koelle, K., Cobey, S., Grenfell, B., and Pascual, M. (2006). Epochal evolution shapes the phylodynamics of inter-pandemic influenza A (H3N2) in humans. Science *314*, 1898–1903.

Koopmans, M. (2005). Food-borne norovirus outbreaks: a nuisance or more than that? Wien Klin. Wochenschr. *117*, 789–791.

Koopmans, M., and Duizer, E. (2004). Foodborne viruses: an emerging problem. Int. J. Food Microbiol. *90*, 23–41.

Koopmans, M., Harris, J., Verhoef, L., Depoortere, E., Takkinen, J., and Coulombier, D. (2006). European investigation into recent norovirus outbreaks on cruise ships: update. Euro. Surveill. *11*, E060706, 060705.

Koopmans, M., Vennema, H., Heersma, H., van Strien, E., van Duynhoven, Y., Brown, D., Reacher, M., and Lopman, B. (2003). Early identification of common-source foodborne virus outbreaks in Europe. Emerg. Infect. Dis. *9*, 1136–1142.

Koopmans, M., von Bonsdorff, C.H., Vinje, J., de Medici, D., and Monroe, S. (2002). Foodborne viruses. FEMS Microbiol. Rev. *26*, 187–205.

Krisztalovics, K., Reuter, G., Szucs, G., Csohan, A., and Borocz, K. (2006). Increase in norovirus circulation in Hungary in October-November 2006. Euro. Surveill. *11*, E061214, 061212.

Kroneman, A., Harris, J., Vennema, H., Duizer, E., van Duynhoven, Y., Gray, J., Iturriza, M., Bottiger, B., Falkenhorst, G., Johnsen, C., *et al.* (2008a). Data quality of 5 years of central norovirus outbreak reporting in the European Network for food-borne viruses. J. Public Health (Oxf.) *30*, 82–90.

Kroneman, A., Verhoef, L., Harris, J., Vennema, H., Duizer, E., van Duynhoven, Y., Gray, J., Iturriza, M., Bottiger, B., Falkenhorst, G., *et al.* (2008b). Analysis of integrated virological and epidemiological reports of norovirus outbreaks collected within the foodborne viruses in Europe Network from 1 July 2001 to 30 June 2006. J. Clin. Microbiol. *46*, 2959–2965.

Kuritsky, J.N., Osterholm, M.T., Greenberg, H.B., Korlath, J.A., Godes, J.R., Hedberg, C.W., Forfang, J.C.,

Kapikian, A.Z., McCullough, J.C., and White, K.E. (1984). Norwalk gastroenteritis: a community outbreak associated with bakery product consumption. Ann. Intern. Med. *100*, 519–521.

Kuusi, M., Nuorti, J.P., Maunula, L., Miettinen, I., Pesonen, H., and von Bonsdorff, C.H. (2004). Internet use and epidemiologic investigation of gastroenteritis outbreak. Emerg. Infect. Dis. *10*, 447–450.

Kuusi, M., Nuorti, J.P., Maunula, L., Minh, N.N., Ratia, M., Karlsson, J., and von Bonsdorff, C.H. (2002). A prolonged outbreak of Norwalk-like calicivirus (NLV) gastroenteritis in a rehabilitation centre due to environmental contamination. Epidemiol. Infect. *129*, 133–138.

Le Guyader, F., Loisy, F., Atmar, R.L., Hutson, A.M., Estes, M.K., Ruvoen-Clouet, N., Pommepuy, M., and Le Pendu, J. (2006a). Norwalk virus-specific binding to oyster digestive tissues. Emerg. Infect. Dis. *12*, 931–936.

Le Guyader, F.S., Bon, F., DeMedici, D., Parnaudeau, S., Bertone, A., Crudeli, S., Doyle, A., Zidane, M., Suffredini, E., Kohli, E., *et al.* (2006b). Detection of multiple noroviruses associated with an international gastroenteritis outbreak linked to oyster consumption. J. Clin. Microbiol. *44*, 3878–3882.

Lee, B.E., Pang, X.L., Robinson, J.L., Bigam, D., Monroe, S.S., and Preiksaitis, J.K. (2008). Chronic norovirus and adenovirus infection in a solid organ transplant recipient. Pediatr. Infect. Dis. J. *27*, 360–362.

Lee, J.I., Chung, J.Y., Han, T.H., Song, M.O., and Hwang, E.S. (2007a). Detection of human bocavirus in children hospitalized because of acute gastroenteritis. J. Infect. Dis. *196*, 994–997.

Lee, N., Chan, M.C., Wong, B., Choi, K.W., Sin, W., Lui, G., Chan, P.K., Lai, R.W., Cockram, C.S., Sung, J.J., *et al.* (2007b). Fecal viral concentration and diarrhea in norovirus gastroenteritis. Emerg. Infect. Dis. *13*, 1399–1401.

Lindell, A.T., Grillner, L., Svensson, L., and Wirgart, B.Z. (2005). Molecular epidemiology of norovirus infections in Stockholm, Sweden, during the years 2000 to 2003: association of the GGIIb genetic cluster with infection in children. J. Clin. Microbiol. *43*, 1086–1092.

Lindesmith, L., Moe, C., Lependu, J., Frelinger, J.A., Treanor, J., and Baric, R.S. (2005). Cellular and humoral immunity following Snow Mountain Virus challenge. J. Virol. *79*, 2900–2909.

Lindesmith, L.C., Donaldson, E.F., Lobue, A.D., Cannon, J.L., Zheng, D.P., Vinje, J., and Baric, R.S. (2008). Mechanisms of GII.4 norovirus persistence in human populations. PLoS Med. *5*, e31.

Liu, G.Q., Ni, Z., Yun, T., Yu, B., Zhu, J.M., Hua, J.G., and Chen, J.P. (2008). Rabbit hemorrhagic disease virus poly(A) tail is not essential for the infectivity of the virus and can be restored in vivo. Arch. Virol. *153*, 939–944.

Lopman, B., Vennema, H., Kohli, E., Pothier, P., Sanchez, A., Negredo, A., Buesa, J., Schreier, E., Reacher, M., Brown, D., *et al.* (2004a). Increase in viral gastroenteritis outbreaks in Europe and epidemic spread of new norovirus variant. Lancet *363*, 682–688.

Lopman, B.A., Gallimore, C., Gray, J.J., Vipond, I.B., Andrews, N., Sarangi, J., Reacher, M.H., and Brown, D.W. (2006). Linking healthcare associated norovirus outbreaks: a molecular epidemiologic method for investigating transmission. BMC Infect. Dis. *6*, 108.

Lopman, B.A., Reacher, M.H., Vipond, I.B., Sarangi, J., and Brown, D.W. (2004b). Clinical manifestation of norovirus gastroenteritis in health care settings. Clin. Infect. Dis. *39*, 318–324.

Ludwig, A., Adams, O., Laws, H.J., Schroten, H., and Tenenbaum, T. (2008). Quantitative detection of norovirus excretion in pediatric patients with cancer and prolonged gastroenteritis and shedding of norovirus. J. Med. Virol. *80*, 1461–1467.

Lyman, W.H., Walsh, J.F., Kotch, J.B., Weber, D.J., Gunn, E., and Vinje, J. (2009). Prospective study of etiologic agents of acute gastroenteritis outbreaks in child care centers. J. Pediatr. *154*, 253–257.

Lynch, M., Painter, J., Woodruff, R., and Braden, C. (2006). Surveillance for foodborne-disease outbreaks – United States, 1998–2002. MMWR Surveill. Summ. *55*, 1–42.

Makary, P., Maunula, L., Niskanen, T., Kuusi, M., Virtanen, M., Pajunen, S., Ollgren, J., and Tran Minh, N.N. (2009). Multiple norovirus outbreaks among workplace canteen users in Finland, July 2006. Epidemiol. Infect. *137*, 402–407.

Malek, M., Barzilay, E., Kramer, A., Camp, B., Jaykus, L.A., Escudero-Abarca, B., Derrick, G., White, P., Gerba, C., Higgins, C., *et al.* (2009). Outbreak of norovirus infection among river rafters associated with packaged delicatessen meat, Grand Canyon, 2005. Clin. Infect. Dis. *48*, 31–37.

Marionneau, S., Ruvoen, N., Le Moullac-Vaidye, B., Clement, M., Cailleau-Thomas, A., Ruiz-Palacois, G., Huang, P., Jiang, X., and Le Pendu, J. (2002). Norwalk virus binds to histo-blood group antigens present on gastroduodenal epithelial cells of secretor individuals. Gastroenterology *122*, 1967–1977.

Marks, P.J., Vipond, I.B., Carlisle, D., Deakin, D., Fey, R.E., and Caul, E.O. (2000). Evidence for airborne transmission of Norwalk-like virus (NLV) in a hotel restaurant. Epidemiol. Infect. *124*, 481–487.

Martella, V., Campolo, M., Lorusso, E., Cavicchio, P., Camero, M., Bellacicco, A.L., Decaro, N., Elia, G., Greco, G., Corrente, M., *et al.* (2007). Norovirus in captive lion cub (Panthera leo). Emerg. Infect. Dis. *13*, 1071–1073.

Martella, V., Lorusso, E., Decaro, N., Elia, G., Radogna, A., D'Abramo, M., Desario, C., Cavalli, A., Corrente, M., Camero, M., *et al.* (2008). Detection and molecular characterization of a canine norovirus. Emerg. Infect. Dis. *14*, 1306–1308.

Martinez, N., Espul, C., Cuello, H., Zhong, W., Jiang, X., Matson, D.O., and Berke, T. (2002). Sequence diversity of human caliciviruses recovered from children with diarrhea in Mendoza, Argentina, 1995–1998. J. Med. Virol. *67*, 289–298.

Mattison, K., Shukla, A., Cook, A., Pollari, F., Friendship, R., Kelton, D., Bidawid, S., and Farber, J.M. (2007). Human noroviruses in swine and cattle. Emerg. Infect. Dis. *13*, 1184–1188.

Mattner, F., Mattner, L., Borck, H.U., and Gastmeier, P. (2005). Evaluation of the impact of the source (patient versus staff) on nosocomial norovirus outbreak severity. Infect. Control Hosp. Epidemiol. *26*, 268–272.

Mattner, F., Sohr, D., Heim, A., Gastmeier, P., Vennema, H., and Koopmans, M. (2006). Risk groups for clinical complications of norovirus infections: an outbreak investigation. Clin. Microbiol. Infect. *12*, 69–74.

Maunula, L., Kalso, S., Von Bonsdorff, C.H., and Ponka, A. (2004). Wading pool water contaminated with both noroviruses and astroviruses as the source of a gastroenteritis outbreak. Epidemiol. Infect. *132*, 737–743.

Mead, P.S., Slutsker, L., Dietz, V., McCaig, L.F., Bresee, J.S., Shapiro, C., Griffin, P.M., and Tauxe, R.V. (1999). Food-related illness and death in the United States. Emerg. Infect. Dis. *5*, 607–625.

Mills, C.E., Robins, J.M., and Lipsitch, M. (2004). Transmissibility of 1918 pandemic influenza. Nature *432*, 904–906.

Muller, B., Klemm, U., Mas Marques, A., and Schreier, E. (2007). Genetic diversity and recombination of murine noroviruses in immunocompromised mice. Arch. Virol. *152*, 1709–1719.

Murata, T., Katsushima, N., Mizuta, K., Muraki, Y., Hongo, S., and Matsuzaki, Y. (2007). Prolonged norovirus shedding in infants <or=6 months of age with gastroenteritis. Pediatr. Infect. Dis. J. *26*, 46–49.

Nayak, M.K., Balasubramanian, G., Sahoo, G.C., Bhattacharya, R., Vinje, J., Kobayashi, N., Sarkar, M.C., Bhattacharya, M.K., and Krishnan, T. (2008). Detection of a novel intergenogroup recombinant Norovirus from Kolkata, India. Virology *377*, 117–123.

Ng, T.L., Chan, P.P., Phua, T.H., Loh, J.P., Yip, R., Wong, C., Liaw, C.W., Tan, B.H., Chiew, K.T., Chua, S.B., *et al.* (2005). Oyster-associated outbreaks of Norovirus gastroenteritis in Singapore. J. Infect. *51*, 413–418.

Nilsson, M., Hedlund, K.O., Thorhagen, M., Larson, G., Johansen, K., Ekspong, A., and Svensson, L. (2003). Evolution of human calicivirus RNA in vivo: accumulation of mutations in the protruding P2 domain of the capsid leads to structural changes and possibly a new phenotype. J. Virol. *77*, 13117–13124.

Noda, M., Fukuda, S., and Nishio, O. (2008). Statistical analysis of attack rate in norovirus foodborne outbreaks. Int. J. Food Microbiol. *122*, 216–220.

Noel, J.S., Fankhauser, R.L., Ando, T., Monroe, S.S., and Glass, R.I. (1999). Identification of a distinct common strain of 'Norwalk-like viruses' having a global distribution. J. Infect. Dis. *179*, 1334–1344.

Nuanualsuwan, S., Mariam, T., Himathongkham, S., and Cliver, D.O. (2002). Ultraviolet inactivation of feline calicivirus, human enteric viruses and coliphages. Photochem. Photobiol. *76*, 406–410.

Nygard, K., Torven, M., Ancker, C., Knauth, S.B., Hedlund, K.O., Giesecke, J., Andersson, Y., and Svensson, L. (2003). Emerging genotype (GGIIb) of norovirus in drinking water, Sweden. Emerg. Infect. Dis. *9*, 1548–1552.

O'Reilly, C.E., Bowen, A.B., Perez, N.E., Sarisky, J.P., Shepherd, C.A., Miller, M.D., Hubbard, B.C., Herring,

M., Buchanan, S.D., Fitzgerald, C.C., *et al.* (2007). A waterborne outbreak of gastroenteritis with multiple etiologies among resort island visitors and residents: Ohio, 2004. Clin. Infect. Dis. *44*, 506–512.

Okada, M., Tanaka, T., Oseto, M., Takeda, N., and Shinozaki, K. (2006). Genetic analysis of noroviruses associated with fatalities in healthcare facilities. Arch. Virol. *151*, 1635–1641.

Oliver, S.L., Asobayire, E., Charpilienne, A., Cohen, J., and Bridger, J.C. (2007). Complete genomic characterization and antigenic relatedness of genogroup III, genotype 2 bovine noroviruses. Arch. Virol. *152*, 257–272.

Oliver, S.L., Asobayire, E., Dastjerdi, A.M., and Bridger, J.C. (2006). Genomic characterization of the unclassified bovine enteric virus Newbury agent-1 (Newbury1) endorses a new genus in the family Caliciviridae. Virology *350*, 240–250.

Parrino, T.A., Schreiber, D.S., Trier, J.S., Kapikian, A.Z., and Blacklow, N.R. (1977). Clinical immunity in acute gastroenteritis caused by Norwalk agent. N. Engl. J. Med. *297*, 86–89.

Patel, M.M., Hall, A.J., Vinje, J., and Parashar, U.D. (2009). Noroviruses: a comprehensive review. J. Clin. Virol. *44*, 1–8.

Patel, M.M., Widdowson, M.A., Glass, R.I., Akazawa, K., Vinje, J., and Parashar, U.D. (2008). Systematic literature review of role of noroviruses in sporadic gastroenteritis. Emerg. Infect. Dis. *14*, 1224–1231.

Reimerink, J., Stelma, F., Rockx, B., Brouwer, D., Stobberingh, E., van Ree, R., Dompeling, E., Mommers, M., Thijs, C., and Koopmans, M. (2009). Early-life rotavirus and norovirus infections in relation to development of atopic manifestation in infants. Clin. Exp. Allergy *39*, 254–260.

Rockx, B., De Wit, M., Vennema, H., Vinje, J., De Bruin, E., Van Duynhoven, Y., and Koopmans, M. (2002). Natural history of human calicivirus infection: a prospective cohort study. Clin. Infect. Dis. *35*, 246–253.

Scott, F.W. (1980). Virucidal disinfectants and feline viruses. Am. J. Vet. Res. *41*, 410–414.

Sdiri-Loulizi, K., Gharbi-Khelifi, H., de Rougemont, A., Chouchane, S., Sakly, N., Ambert-Balay, K., Hassine, M., Guediche, M.N., Aouni, M., and Pothier, P. (2008). Acute infantile gastroenteritis associated with human enteric viruses in Tunisia. J. Clin. Microbiol. *46*, 1349–1355.

Siebenga, J.J., Beersma, M.F., Vennema, H., van Biezen, P., Hartwig, N.J., and Koopmans, M. (2008). High prevalence of prolonged norovirus shedding and illness among hospitalized patients: a model for in vivo molecular evolution. J. Infect. Dis. *198*, 994–1001.

Siebenga, J.J., Vennema, H., Renckens, B., de Bruin, E., van der Veer, B., Siezen, R.J., and Koopmans, M. (2007). Epochal evolution of GGII.4 norovirus capsid proteins from 1995 to 2006. J. Virol. *81*, 9932–9941.

Siebenga, J.J., Vennema, H., Zheng, D.P., Vinje, J., Lee, B.E., Pang, X.L., Ho, E.C., Lim, W., Choudekar, A., Broor, S., *et al.* (2009). Norovirus Illness Is a Global Problem: Emergence and Spread of Norovirus GII.4 Variants, 2001–2007. J. Infect. Dis. *200*, 802–812.

Simmons, G., Garbutt, C., Hewitt, J., and Greening, G. (2007). A New Zealand outbreak of norovirus gastroenteritis linked to the consumption of imported raw Korean oysters. N.Z. Med. J. *120*, U2773.

Sosnovtsev, S.V., Belliot, G., Chang, K.O., Onwudiwe, O., and Green, K.Y. (2005). Feline calicivirus VP2 is essential for the production of infectious virions. J. Virol. *79*, 4012–4024.

Souza, M., Azevedo, M.S., Jung, K., Cheetham, S., and Saif, L.J. (2008). Pathogenesis and immune responses in gnotobiotic calves after infection with the genogroup II.4-HS66 strain of human norovirus. J. Virol. *82*, 1777–1786.

Suffredini, E., Corrain, C., Arcangeli, G., Fasolato, L., Manfrin, A., Rossetti, E., Biazzi, E., Mioni, R., Pavoni, E., Losio, M.N., *et al.* (2008). Occurrence of enteric viruses in shellfish and relation to climatic-environmental factors. Lett. Appl. Microbiol. *47*, 467–474.

Sugieda, M., Nagaoka, H., Kakishima, Y., Ohshita, T., Nakamura, S., and Nakajima, S. (1998). Detection of Norwalk-like virus genes in the caecum contents of pigs. Arch. Virol. *143*, 1215–1221.

Sugieda, M., and Nakajima, S. (2002). Viruses detected in the caecum contents of healthy pigs representing a new genetic cluster in genogroup II of the genus 'Norwalk-like viruses'. Virus Res. *87*, 165–172.

Tajiri, H., Kiyohara, Y., Tanaka, T., Etani, Y., and Mushiake, S. (2008). Abnormal computed tomography findings among children with viral gastroenteritis and symptoms mimicking acute appendicitis. Pediatr. Emerg. Care *24*, 601–604.

Tan, M., and Jiang, X. (2005). The *p* domain of norovirus capsid protein forms a subviral particle that binds to histo-blood group antigen receptors. J. Virol. *79*, 14017–14030.

Tan, M., and Jiang, X. (2008). Association of histo-blood group antigens with susceptibility to norovirus infection may be strain-specific rather than genogroup dependent. J. Infect. Dis. *198*, 940–941; author reply 942–943.

Teunis, P.F., Moe, C.L., Liu, P., Miller, S. E., Lindesmith, L., Baric, R.S., Le Pendu, J., and Calderon, R.L. (2008). Norwalk virus: How infectious is it? J. Med. Virol. *80*, 1468–1476.

Thurston-Enriquez, J.A., Haas, C.N., Jacangelo, J., Riley, K., and Gerba, C.P. (2003). Inactivation of feline calicivirus and adenovirus type 40 by UV radiation. Appl. Environ. Microbiol. *69*, 577–582.

Tian, P., Bates, A.H., Jensen, H.M., and Mandrell, R.E. (2006). Norovirus binds to blood group A-like antigens in oyster gastrointestinal cells. Lett. Appl. Microbiol. *43*, 645–651.

Tian, P., Engelbrektson, A.L., Jiang, X., Zhong, W., and Mandrell, R.E. (2007). Norovirus recognizes histo-blood group antigens on gastrointestinal cells of clams, mussels, and oysters: a possible mechanism of bioaccumulation. J. Food Prot. *70*, 2140–2147.

Torner, N., Dominguez, A., Ruiz, L., Martinez, A., Bartolome, R., Buesa, J., and Ferrer, M.D. (2008). Acute gastroenteritis outbreaks in Catalonia, Spain: norovirus versus Salmonella. Scand. J. Gastroenterol. *43*, 567–573.

Troeger, H., Loddenkemper, C., Schneider, T., Schreier, E., Epple, H.J., Zeitz, M., Fromm, M., and Schulzke, J.D. (2008). Structural And Functional Changes Of The Duodenum In Human Norovirus Infection. Gut *58*, 1070–1077.

Tsang, O.T., Wong, A.T., Chow, C.B., Yung, R.W., Lim, W.W., and Liu, S.H. (2008). Clinical characteristics of nosocomial norovirus outbreaks in Hong Kong. J. Hosp. Infect. *69*, 135–140.

Tu, E.T., Bull, R.A., Greening, G.E., Hewitt, J., Lyon, M.J., Marshall, J.A., McIver, C.J., Rawlinson, W.D., and White, P.A. (2008a). Epidemics of gastroenteritis during 2006 were associated with the spread of norovirus GII.4 variants 2006a and 2006b. Clin. Infect. Dis. *46*, 413–420.

Tu, E.T., Bull, R.A., Kim, M.J., McIver, C.J., Heron, L., Rawlinson, W.D., and White, P.A. (2008b). Norovirus excretion in an aged-care setting. J. Clin. Microbiol. *46*, 2119–2121.

Tu, E.T., Nguyen, T., Lee, P., Bull, R.A., Musto, J., Hansman, G., White, P.A., Rawlinson, W.D., and McIver, C.J. (2007). Norovirus GII.4 strains and outbreaks, Australia. Emerg. Infect. Dis. *13*, 1128–1130.

Turcios, R.M., Widdowson, M.A., Sulka, A.C., Mead, P.S., and Glass, R.I. (2006). Reevaluation of epidemiological criteria for identifying outbreaks of acute gastroenteritis due to norovirus: United States, 1998–2000. Clin. Infect. Dis. *42*, 964–969.

Turcios-Ruiz, R.M., Axelrod, P., St John, K., Bullitt, E., Donahue, J., Robinson, N., and Friss, H.E. (2008). Outbreak of necrotizing enterocolitis caused by norovirus in a neonatal intensive care unit. J. Pediatr. *153*, 339–344.

van Duynhoven, Y.T., de Jager, C.M., Kortbeek, L.M., Vennema, H., Koopmans, M.P., van Leusden, F., van der Poel, W.H., and van den Broek, M.J. (2005). A one-year intensified study of outbreaks of gastroenteritis in The Netherlands. Epidemiol. Infect. *133*, 9–21.

Van Houtven, G., Honeycutt, A., Gilman, B., McCall, N., and Throneburg, W. (2005). Costs of Illness for Environmentally Related Health Effects in Older Americans. Research Triangle Institute Report Project Number 08687.029.001.

Vardy, J., Love, A.J., and Dignon, N. (2007). Outbreak of acute gastroenteritis among emergency department staff. Emerg. Med. J. *24*, 699–702.

Verhoef, L., Boxman, I.L., Duizer, E., Rutjes, S.A., Vennema, H., Friesema, I.H., de Roda Husman, A.M., and Koopmans, M. (2008a). Multiple exposures during a norovirus outbreak on a river-cruise sailing through Europe, 2006. Euro. Surveill. *13*.

Verhoef, L., Depoortere, E., Boxman, I., Duizer, E., van Duynhoven, Y., Harris, J., Johnsen, C., Kroneman, A., Le Guyader, S., Lim, W., *et al.* (2008b). Emergence of new norovirus variants on spring cruise ships and prediction of winter epidemics. Emerg. Infect. Dis. *14*, 238–243.

Verhoef, L., Duizer, E., Vennema, H., Siebenga, J., Swaan, C., Isken, L., Koopmans, M., Balay, K., Pothier, P., McKeown, P., *et al.* (2008c). Import of norovirus infections in the Netherlands and Ireland following pilgrimages to Lourdes, 2008 – preliminary report. Euro. Surveill. *13*, pii: 19025.

Verhoef, L.P., Kroneman, A., van Duynhoven, Y., Boshuizen, H., van Pelt, W., and Koopmans, M. (2009). Selection tool for foodborne norovirus outbreaks. Emerg. Infect. Dis. *15*, 31–38.

Vinje, J., and Koopmans, M.P. (1996). Molecular detection and epidemiology of small round-structured viruses in outbreaks of gastroenteritis in the Netherlands. J. Infect. Dis. *174*, 610–615.

Vinje, J., Altena, S.A., and Koopmans, M.P. (1997). The incidence and genetic variability of small round-structured viruses in outbreaks of gastroenteritis in The Netherlands. J. Infect. Dis. *176*, 1374–1378.

Vinje, J., Vennema, H., Maunula, L., von Bonsdorff, C.H., Hoehne, M., Schreier, E., Richards, A., Green, J., Brown, D., Beard, S.S., *et al.* (2003). International collaborative study to compare reverse transcriptase PCR assays for detection and genotyping of noroviruses. J. Clin. Microbiol. *41*, 1423–1433.

Vinje, J.J., Hamidjaja, R.A., and Sobsey, M.D. (2004). Developmental and application of a capsid VP1 (region D) based reverse transcription PCR assay for genotyping of genogroup I and II noroviruses. J. Virol. Methods *116*, 109–117.

Wang, Q.H., Han, M.G., Cheetham, S., Souza, M., Funk, J.A., and Saif, L.J. (2005a). Porcine noroviruses related to human noroviruses. Emerg. Infect. Dis. *11*, 1874–1881.

Wang, Q.H., Han, M.G., Funk, J.A., Bowman, G., Janies, D.A., and Saif, L.J. (2005b). Genetic diversity and recombination of porcine sapoviruses. J. Clin. Microbiol. *43*, 5963–5972.

Wang, Q.H., Souza, M., Funk, J.A., Zhang, W., and Saif, L.J. (2006). Prevalence of noroviruses and sapoviruses in swine of various ages determined by reverse transcription-PCR and microwell hybridization assays. J. Clin. Microbiol. *44*, 2057–2062.

Wei, C., Farkas, T., Sestak, K., and Jiang, X. (2008). Recovery of infectious virus by transfection of in vitro-generated RNA from tulane calicivirus cDNA. J. Virol. *82*, 11429–11436.

Wheeler, J.G., Sethi, D., Cowden, J.M., Wall, P.G., Rodrigues, L.C., Tompkins, D.S., Hudson, M.J., and Roderick, P.J. (1999). Study of infectious intestinal disease in England: rates in the community, presenting to general practice, and reported to national surveillance. The Infectious Intestinal Disease Study Executive. BMJ *318*, 1046–1050.

Widdowson, M.A., Cramer, E.H., Hadley, L., Bresee, J.S., Beard, R.S., Bulens, S.N., Charles, M., Chege, W., Isakbaeva, E., Wright, J.G., *et al.* (2004). Outbreaks of acute gastroenteritis on cruise ships and on land: identification of a predominant circulating strain of norovirus – United States, 2002. J. Infect. Dis. *190*, 27–36.

Wolf, S., Williamson, W., Hewitt, J., Lin, S., Rivera-Aban, M., Ball, A., Scholes, P., Savill, M., and Greening, G.E. (2009). Molecular detection of norovirus in sheep and pigs in New Zealand farms. Vet. Microbiol. *133*, 184–189.

Wood, D.J., David, T.J., Chrystie, I.L., and Totterdell, B. (1988). Chronic enteric virus infection in two T-cell immunodeficient children. J. Med. Virol. *24*, 435–444.

Wyatt, R.G., Dolin, R., Blacklow, N.R., DuPont, H.L., Buscho, R.F., Thornhill, T.S., Kapikian, A.Z., and Chanock, R.M. (1974). Comparison of three agents of acute infectious nonbacterial gastroenteritis by cross-challenge in volunteers. J. Infect. Dis. *129*, 709–714.

Xerry, J., Gallimore, C.I., Iturriza-Gomara, M., Allen, D.J., and Gray, J.J. (2008). Transmission events within outbreaks of gastroenteritis determined through analysis of nucleotide sequences of the P2 domain of genogroup II noroviruses. J. Clin. Microbiol. *46*, 947–953.

Yee, E.L., Palacio, H., Atmar, R.L., Shah, U., Kilborn, C., Faul, M., Gavagan, T.E., Feigin, R.D., Versalovic, J., Neill, F.H., *et al.* (2007). Widespread outbreak of norovirus gastroenteritis among evacuees of Hurricane Katrina residing in a large 'megashelter' in Houston, Texas: lessons learned for prevention. Clin. Infect. Dis. *44*, 1032–1039.

Zheng, D.P., Ando, T., Fankhauser, R.L., Beard, R.S., Glass, R.I., and Monroe, S.S. (2006). Norovirus classification and proposed strain nomenclature. Virology *346*, 312–323.

Zingg, W., Colombo, C., Jucker, T., Bossart, W., and Ruef, C. (2005). Impact of an outbreak of norovirus infection on hospital resources. Infect. Control Hosp. Epidemiol. *26*, 263–267.

Calicivirus Environmental Contamination

2

Gail E. Greening and Sandro Wolf

Abstract

The virus family *Caliciviridae* contains four genera: *Norovirus*, *Sapovirus*, *Lagovirus* and *Vesivirus*. Norovirus and sapovirus cause gastroenteritis in humans, whereas lagoviruses and vesiviruses mostly infect animals and cause a variety of diseases. Noroviruses and sapoviruses can also infect animals, including cow and pig, respectively. Noroviruses are the dominant cause of human gastroenteritis around the world, infecting all age groups. Their low infectious dose and stability in the natural environment allows noroviruses to be easily spread. Contamination in food and water destined for human consumption has lead to numerous outbreaks of gastroenteritis. Noroviruses have been detected in shellfish, sandwiches, fruit, ice, drinking water and treated wastewater. Direct transmission from food and water to humans is well documented. Increased monitoring and improvements in detection methods may help to reduce the number of infections but regulations and standards need to be addressed in order to reduce viral contamination in the natural environment.

Introduction

The family *Caliciviridae* consists of four recognized genera: *Norovirus*, *Sapovirus*, *Lagovirus* and *Vesivirus* (Fig. 2.1) (Green, 2007; Green *et al.*, 2000). Caliciviruses have a worldwide distribution, are considered environmentally stable and resist inactivation by environmental stressors. This stability is a key attribute because it enables them to survive in the environment until transmission to a suitable host. Many caliciviruses are able to withstand harsh conditions such as drying, high and low pH, and high and low temperatures. Shellfish, water, sewage, fomites, air and other environmental matrices are common distribution routes for human and animal calicivirus infections. Studies on the occurrence and behaviour of caliciviruses in different environmental settings have provided important information on their properties and pathogenicity and have assisted in developing control strategies. This chapter describes the occurrence and distribution of caliciviruses in the environment and their impact on humans and animals.

Human norovirus

Based on genetic analysis, human noroviruses can be classified into three distinct genogroups (GI, II and IV), and these can be further subdivided into numerous genotypes. Human noroviruses are believed to be the most common aetiological agent of non-bacterial gastroenteritis worldwide (Blanton *et al.*, 2006; Koopmans, 2008; Lopman *et al.*, 2003, 2004a; Mead *et al.*, 1999). Human noroviruses frequently cause outbreaks of gastroenteritis in retirement homes, hospitals, institutions, schools and cruise ships (Blanton *et al.*, 2006; Lopman *et al.*, 2003). All age groups are affected, but the young and elderly are particularly at risk. Noroviruses also cause epidemic gastroenteritis outbreaks and is believed to be a major aetiological agent of foodborne and waterborne infections (Hafliger *et al.*, 2000; Hewitt *et al.*, 2007; Maunula *et al.*, 2005; Turcios *et al.*, 2006). Prolonged shedding of virus in faecal material and projectile vomiting greatly facilitate effective transmission (Atmar *et al.*, 2008; Rockx *et al.*,

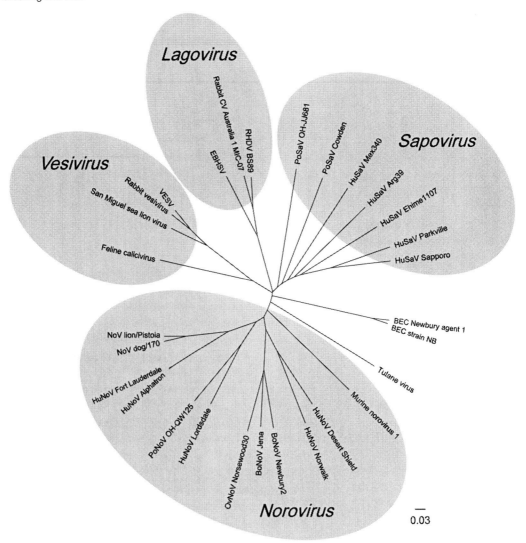

Figure 2.1 Classification of caliciviruses based on the nucleotide sequences of the complete capsid protein, VP1.

2002). In a recent study, norovirus was detected in faeces from over 50% of patients 4 weeks after infection and some patients were still shedding norovirus 56 days after infection (Atmar *et al.*, 2008). In other studies, noroviruses were shed for up to 3 weeks following infection and up to 6 weeks in infected infants (Murata *et al.*, 2007; Rockx *et al.*, 2002). The economic burden of norovirus outbreaks in terms of management expenses, lost work days, and closure of hospital wards, operating theatres, retirement homes, schools and cruise ships is substantial (Hafliger *et al.*, 2000; Lopman *et al.*, 2004b,c). GII genotype 4 (GII.4) norovirus strains are the most common genotype implicated in outbreaks in healthcare settings and these strains have been responsible for the majority of the institutional, healthcare and cruise ship outbreaks in recent years (Verhoef *et al.*, 2008). In contrast, many foodborne outbreaks are caused by genotypes other than GII.4 (Blanton *et al.*, 2006; Schmid *et al.*, 2007; Verhoef *et al.*, 2009). Multiple norovirus strains belonging to both GI and GII genotypes are commonly detected in shellfish-related outbreaks (Costantini *et al.*, 2006; Gallimore *et al.*, 2005b; Le Guyader *et al.*, 2006b). Waterborne outbreaks have been frequently associated with GI strains, which have been proposed as more environmentally stable than GII norovirus strains (Hewitt *et al.*, 2007; Maunula *et al.*, 2005).

Animal norovirus

Animal noroviruses can infect a number of different animals, including pig, cow, lion, dog and mouse. Based on genetic analysis these have been divided into different genogroups and genotypes. Porcine and bovine noroviruses belong to GII and GIII, respectively. These viruses have been identified in many countries worldwide including Japan, US, UK, Hungary, Germany, the Netherlands, South Korea and New Zealand (Ike et al., 2007; Oliver et al., 2007; Reuter et al., 2007; Sugieda et al., 1998; van Der Poel et al., 2000; Wang et al., 2005a; Wolf et al., 2007, 2009). Porcine GII norovirus infections are usually mild or asymptomatic in pigs (Reuter et al., 2007; Sugieda et al., 1998; Wang et al., 2005a, 2006; Wolf et al., 2009). Porcine GII norovirus are genetically distinct from the human norovirus genotypes and are subdivided into three genotypes, GII.11, GII.18 and GII.19. Bovine GIII noroviruses have been associated with diarrhoeal disease and asymptomatic infection in calves and cattle (Smiley et al., 2003; van der Poel et al., 2003; Wolf et al., 2007). Genetic analysis of bovine noroviruses suggests two distinct genotypes [GIII.1 (Jena-like) and GIII.2 (Newbury 2-like)] (Oliver et al., 2003; Oliver et al., 2007). Recently, norovirus strains closely related to the bovine GIII noroviruses were identified in faecal material from asymptomatic sheep in New Zealand, possibly representing a third GIII genotype (Wolf et al., 2009). GIV noroviruses were identified in a dead lion cub and a dog (Martella et al., 2007, 2008b). Genetic analysis revealed >90% amino acid similarity in the capsid protein and these two strains were designated as a new norovirus GIV genotype, GIV.2 (Patel et al., 2009). Murine norovirus belonging to GV was isolated from immunocompromised mice in 2003 (Karst et al., 2003). It has since been identified in laboratory mice across the US and Canada and is now considered to be one of the most prevalent pathogens in colonies of laboratory mice (Hsu et al., 2005; Perdue et al., 2007).

Human and animal sapovirus

Sapoviruses can infect both humans and animals. Sapoviruses can be divided into five genogroups (GI–V), of which GI, GII, GIV and GV infect humans and GIII infects pigs (Hansman et al., 2007a; Jeong et al., 2007). Recent studies have described several novel sapovirus genogroups in pigs (VI–VIII), which are genetically distinct from GIII sapoviruses (Barry et al., 2008; Wang et al., 2005b, 2006). Sapovirus outbreaks are less common than norovirus outbreaks, although the number of outbreaks appears to be increasing (Farkas et al., 2004; Hansman et al., 2007c; Moreno-Espinosa et al., 2004; Pang et al., 2008; Rockx et al., 2002). Human sapoviruses have also been identified in wastewater, river water and clams (Hansman et al., 2007b,d). Porcine sapoviruses have been found to infect both juvenile and adult pigs and are associated with diarrhoea and subclinical disease patterns (Jeong et al., 2007; Martella et al., 2008a). Porcine sapoviruses appear to have a worldwide distribution and have been identified in the US, Brazil, Venezuela, Japan, the Netherlands, Belgium, Hungary and South Korea (Barry et al., 2008; Guo et al., 1999; Jeong et al., 2007; Mauroy et al., 2008; Reuter et al., 2007; Wang et al., 2006). The prevalence of porcine sapoviruses in these countries ranged from 9% to 62%. Although most porcine sapovirus strains belonged to GIII (Cowden-like), a number of unclassified genetically distinct sapoviruses, which were previously described human and porcine sapovirus strains, were detected in pigs from various countries (L'Homme et al., 2009). Their genetic characteristics suggest they belonged to new sapovirus genogroups: VI (JJ681-like), VII (K7/JP or LL26-like) and VIII (K15/JP-like).

Vesivirus

Vesiviruses appear to be the exception in the family *Caliciviridae* because they have a broad host range. Moreover, vesiviruses have been reported to infect across species (Smith et al., 1998b). Two species of *Vesivirus* are officially recognized by the International Committee on Taxonomy of Viruses (ICTV): feline calicivirus (FCV) and vesicular exanthema of swine virus (VESV), although other potential species, including canine calicivirus (CaCV) and San Miguel sea lion virus (SMSV) exist. Many vesiviruses, such as SMSV, are of marine origin but are genetically related to terrestrial animal vesiviruses (Kurth et al., 2006a; Martín-Alonso et al., 2005; Smith et al., 1973). FCV is a common pathogen of cats, causing a range of diseases from mild oral and upper respiratory infections to systemic disease. Although rarely fatal, cats may develop chronic disease and remain persistently infected. FCV is genetically diverse and is able to adapt to environmental pressures (Radford *et al.*, 2007). Recently, a new variant of FCV has emerged which causes severe systemic disease in cats (Hurley *et al.*, 2004). CaCV was first isolated from dog faeces in 1985 (Schaffer *et al.*, 1985). Since then, it has been associated with watery diarrhoea in dogs and was prevalent among dogs in Japan (Matsuura *et al.*, 2002; Mochizuki *et al.*, 1993, 2002). The presence of antibodies to vesiviruses has been reported in association with abortion in horses and cattle (Kurth *et al.*, 2006b; Smith *et al.*, 2002). Although vesiviruses are not regarded as human pathogens, there is a report of an infected laboratory worker with SMSV (Smith *et al.*, 1998b). Serological evidence of vesivirus infections and viraemia in humans has also been reported (Smith *et al.*, 2006), but the findings were inconsistent among laboratories. In a recent study of vesivirus infection in humans, a generic vesivirus PCR assay was used to investigate the presence of vesiviruses in samples collected from patients with gastroenteritis, hepatitis, rash illnesses and acute respiratory disease (Svraka *et al.*, 2009). No evidence of vesiviruses was obtained from these human samples and the study concluded that vesiviruses were an unlikely cause of human illness.

Lagovirus

Lagoviruses have been detected only in rabbits and hares. The most significant member of the *Lagovirus* genus is rabbit haemorrhagic disease virus (RHDV), which causes a severe acute systemic viral haemorrhagic disease in rabbits (Cooke, 2002; Nowotny *et al.*, 1997; Yang *et al.*, 2008). Non-pathogenic rabbit calicivirus strains were probably predominant before the pathogenic RHDV strains emerged (Forrester *et al.*, 2006). RHDV was first recognized in rabbits in China, in 1984, and spread rapidly through the domestic rabbit population, killing 14 million of rabbits in nine months (Cooke, 2002; Forrester *et al.*, 2006; Yang *et al.*, 2008). RHDV has a widespread distribution across Europe, Mexico, Australia and New Zealand. Recent genetic analysis of RHDV strains has led to belief that epidemic RHDV strains of distinct lineage have emerged twice since 1984 and are co-circulating in both wild and domestic rabbits (Forrester *et al.*, 2006). RHDV was introduced into Australia in the 1990s as a method for control of wild rabbits and was then illegally imported into New Zealand. The mortality of RHDV-infected rabbits has resulted in the death of millions of free living and domestic rabbits worldwide. The other species in the *Lagovirus* genus is European brown hare syndrome virus (EBHSV), which is closely related to RHDV but causes disease in European brown hares (*Lepus europaeus*) and other species of hares (*Lepus timidus*). The mortality rate of EBHSV is significantly lower than that of RHDV (Nowotny *et al.*, 1997; Wirblich *et al.*, 1994).

Occurrence, environmental stability and transmission of calicivirus in the environment

Enteric caliciviruses share several characteristics that allow them to maintain a widespread distribution in the environment. These viruses are reported to have a low infectious dose, possibly as low as 10 viral particles, prolonged shedding, environmental stability and strain diversity (Atmar *et al.*, 2008; Caul, 1994; Green *et al.*, 2001; Koopmans, 2008; Rockx *et al.*, 2002; Rzezutka and Cook, 2004; Siebenga *et al.*, 2007; Teunis *et al.*, 2008). These factors facilitate virus transmission by a number of routes, including faecal-oral, aerosol and direct contact. One of the major factors influencing the widespread presence of enteric caliciviruses in and on matrices outside their natural hosts lies, as with most other enteric viruses, in their environmental stability. They are believed to be resistant to heat, disinfection, high pressure and pH changes (Duizer *et al.*, 2004; Hewitt and Greening, 2004, 2006; Kingsley *et al.*, 2007; Rzezutka and Cook, 2004). However, as many caliciviruses cannot be cultured, their true survival in the environment can only be estimated and is frequently based on molecular detection methods and/or surrogate virus studies. Although molecular detection methods, including reverse transcription (RT)-PCR and real-time RT-PCR, are powerful in terms of sensitivity and specificity, they are unable to discriminate between infectious and non-infectious virus particles.

In a comprehensive report, Rzeuztka and Cook reviewed survival studies carried out on a

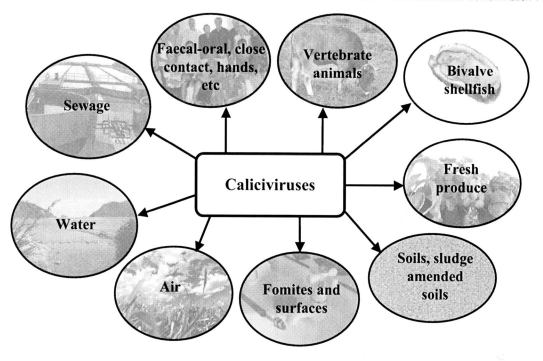

Figure 2.2 Transmission of caliciviruses in the environment.

range of enteric viruses, but were unable to report information on the survival of noroviruses and sapoviruses because of the absence of an infectivity assay (Rzezutka and Cook, 2004). Our knowledge of human norovirus survival is therefore based mainly on studies using virus surrogates, such as poliovirus, FCV, murine norovirus and hepatitis A virus (HAV) under restricted conditions (Duizer et al., 2004; Hewitt and Greening, 2004, 2006; Roda de Husman et al., 2004). Inactivation studies of FCV and CaCV demonstrated that, when assessed by PCR methods, human noroviruses always showed more resistance to environmental stressors, including heat, high and low pH, ultra-violet light and disinfectants, than FCV and CaCV (Duizer et al., 2004). FCV was found to be problematic as a norovirus surrogate because it was sensitive to pH and heat (Duizer et al., 2004; Hewitt and Greening, 2004). These data indicate that FCV and CaCV may not be suitable surrogates for noroviruses. Recent studies have shown that murine norovirus and HAV are environmentally stable when exposed to a range of environmental stressors, indicating that these viruses may be more appropriate surrogates than FCV or enteroviruses

(Baert et al., 2008; Cannon et al., 2006; Hewitt et al., 2009; Kingsley et al., 2007).

Apart from FCV, few studies have been undertaken to elucidate transmission routes of the non-enteric caliciviruses. These viruses are believed to be less environmentally stable than the enteric caliciviruses and it is likely that airborne infection and direct contact with animal secretions and excretions play major roles in their transmission.

Caliciviruses are found in and on many environmental matrices including sewage and wastewaters, soils and sludge-amended soils, fresh and marine waters, and cultivated foods such as shellfish and fresh produce. In most cases, enteric caliciviruses are derived from excreted human or animal faecal material in which subsequently contaminates the environment and food products. The primary reservoirs of the non-enteric caliciviruses are unknown.

Sewage discharges and wastewater

Caliciviruses, especially human noroviruses, are usually present in untreated influent sewage (Iwai et al., 2009; Lodder and de Roda Husman,

2005; Wyn-Jones *et al.*, 2000) and often in treated sewage and wastewater (Myrmel *et al.*, 2006; van den Berg *et al.*, 2005). Levels can reach 10^7 norovirus RT-PCR units per litre. Evidence of the ability of noroviruses to resist sewage treatment was reported by Myrmel and coworkers (Myrmel *et al.*, 2006). They detected noroviruses in 43%, 53% and 24% of influent sewage samples and in 26%, 40% and 21% of effluent samples from three different sewage treatment plants (one primary and two secondary treatment plants). They demonstrated that, despite the different sewage treatment processes, noroviruses travelled through the treatment plants and were discharged into the environment. The secondary treatment plants were more efficient at virus removal than the primary plant. Van den Berg and coworkers examined both raw and treated sewage and found that 10^2–10^3 norovirus RT-PCR units per litre were present in effluent prior to discharge into surface waters (van den Berg *et al.*, 2005). Multiple norovirus strains were identified in both raw and treated sewage. In another evaluation of norovirus removal through different sewage treatment processes, da Silva and coworkers found that 53% and 88% of influent samples and 30% and 14% of effluent samples were positive for GI and GII noroviruses, respectively (da Silva *et al.*, 2008). The prevalence of GI noroviruses in both raw and treated sewage was unexpected given that the majority of community outbreaks are caused by GII norovirus strains. The treatment processes included a waste stabilization pond, activated sludge treatment and a state-of-the-art submerged membrane bioreactor. Their findings indicated that all treatment processes reduced viral loads in receiving waters. Iwai and coworkers reported a continuous co-existence of many different human norovirus genotypes and human sapoviruses in raw sewage in a Japanese city (Iwai *et al.*, 2009). Some of these genotypes were present in clinical cases and, to a lesser extent, in healthy children. They concluded that, as would be expected, the genotypes detected in sewage reflected those circulating in the community, and that subclinical infections were also common. However, as noted above, the infectivity of these viruses following sewage treatment processes cannot be determined. The transmission of animal noroviruses and sapoviruses via domestic and animal waste is

feasible but there are few reports in the literature. Costantini and coworkers investigated the effect of waste treatment processes on the detection and infectivity of porcine sapoviruses in animal waste and manure (Costantini *et al.*, 2007). They found that although viral RNA could still be detected after waste treatment processes, when inoculated into permissive gnotobiotic pigs no infectious virus was detected and no seroconversion was observed.

Soil and sludge-amended soils

Sludge is a complex mixture of organic and inorganic compounds of both biological and chemical origin that is derived from wastewater and sewage during sewage treatment processes. It usually contains a range of microorganisms, including viral, bacterial, protozoan, fungal and helminth pathogens. However, treatment of sludge is not always effective in reducing pathogen numbers, and so may create a public health risk because the infectious dose for some viral pathogens, including noroviruses, is very low. Soils can be contaminated with enteric viruses following treatment with sewage sludge and other waste materials. Naturally occurring human enteric viruses including noroviruses have been detected in soils and sludge. Viruses may reach concentrations of up to 1000 virus particles per litre of sludge (Straub *et al.*, 1993). Following irrigation or long periods of rainfall, these viruses may be transported into groundwater and surface water. This can lead to a potential disease risk in communities that use the groundwater aquifers for general domestic purposes.

Temperature, soil moisture content and virus adsorption to soil have all been shown to be important factors influencing virus survival in soils. Research on poliovirus showed that during winter, when temperatures were low and moisture content was generally higher, poliovirus survival was greater (Hurst *et al.*, 1980; Straub *et al.*, 1993). High soil temperature and low soil moisture content ($<5\%$) were major factors in reducing virus numbers in sludge-amended soils. Hurst and coworkers observed that the presence of aerobic soil microorganisms was also an important factor for virus survival (Hurst *et al.*, 1980). It is probable that, given their similar structural and morphological properties, the enteric caliciviruses show similar survival characteristics to poliovirus in soils (Carter, 2005; Rzezutka and Cook, 2004).

Water and foods

Preharvest contamination of fresh produce, shellfish and water have caused many outbreaks of viral gastroenteritis and hepatitis A disease, and more recently poor food handling practices have been recognized as a major cause of foodborne viral disease. Indirect transmission of noroviruses via water or food can rapidly spread through populations to reach epidemic levels thus creating large outbreaks (Koopmans and Duizer, 2004). The main routes for viral contamination of foods are shown in Table 2.1.

Water

The main sources of caliciviruses in surface and groundwater are sewage discharges and faecal waste from humans and animals. Caliciviruses have been frequently detected in environmental water, including wells, rivers, lakes, groundwater and marine water in a number of countries (Fout *et al.*, 2003; Haramoto *et al.*, 2005; Lodder and de Roda Husman, 2005; Pusch *et al.*, 2005; Wyn-Jones *et al.*, 2000). These findings present a health risk when these waters are used for drinking water, recreational use and shellfish growing. Numerous waterborne norovirus outbreaks have occurred following exposure to contaminated drinking water (Boccia *et al.*, 2002; Hafliger *et al.*, 2000; Hewitt *et al.*, 2007), recreational water used for swimming (Maunula *et al.*, 2004; Podewils *et al.*, 2007), canoeing (Gray *et al.*, 1997), fountain water (Hoebe *et al.*, 2004), well water (Anderson *et al.*, 2001; Beller *et al.*, 1997; Kukkula *et al.*, 1997; Parshionikar *et al.*, 2003) and ice on cruise ships (Khan *et al.*, 1994). The use of molecular

methods for subsequent identification of norovirus strains and tracing contamination sources in these outbreaks has greatly assisted epidemiological investigations.

Contamination of well water is of major concern given that these waters are commonly used for drinking in many parts of the world. A study of private household wells in rural areas near septic tank and septage land application sites in Wisconsin, US, showed that four of 50 wells were transiently contaminated with enteric viruses, including noroviruses (Borchardt *et al.*, 2003). Wells can contaminate groundwater, but groundwater contamination may also occur directly from sewage discharge and septic tank leachates. A study of the microbial quality of urban groundwater in British sandstone aquifers showed that contamination with noroviruses and enteroviruses occurred regularly over the 15-month study period and corresponded with seasonal variation in virus discharge to the sewage system (Powell *et al.*, 2003).

There is limited information on the occurrence of other caliciviruses in various waters. Hansman and coworkers detected sapoviruses in untreated wastewater, treated wastewater, and river water in Japan (Hansman *et al.*, 2007d). A total of 7 of 69 water samples were positive by RT-PCR. Smith and coworkers hypothesized that glacial ice, ice sheets and lake ice may be potential reservoirs for human viral pathogens including caliciviruses, but there is no evidence to support the existence of caliciviruses in these ice repositories (Smith *et al.*, 2004).

The factors influencing virus survival in water are similar to those influencing survival in soil. These factors include temperature, light, pH, microbial flora, and the presence of salts, suspended solids and organic matter. It is interesting to note

Table 2.1 Transmission routes for viral contamination of water and foods

Water and food types	Contamination source
Bivalve shellfish (oysters, mussels, clams)	Pre-harvest, growing waters
Fresh produce, berry fruits, leafy greens, herbs	Pre-harvest, irrigation waters, soil
Drinking water, ice, processing water	Primary (direct use)
Ready to eat foods (RTE): bakery goods, pre-prepared foods, salads and delicatessen foods	Post-harvest, poor hygiene, food processing and handling

that although in most countries, the majority of human norovirus outbreaks are caused by GII strains, GI strains may exhibit greater stability in water than the GII strains. Maunula and co-workers found that while only 13% of norovirus outbreaks in Finland between 1998 and 2003 were caused by GI strains, approximately 50% of the waterborne outbreaks were attributable to GI strains and suggested that there may be differences in virus stability among the different genotypes (Maunula et al., 2005). GI norovirus strains have also been implicated in other waterborne outbreaks (Greening et al., 2001, 2002; Hafliger et al., 2000; Hewitt et al., 2007; Hoebe et al., 2004).

Shellfish

One of the most common routes of foodborne norovirus infections is via bivalve molluscan shellfish, especially oysters, mussels and clams. These filter-feeders bioaccumulate and concentrate microorganisms, including enteric viruses derived from faecal material, which may be present in their growing waters. Sewage, septic tank and boat discharges are usually the main sources of faecal contamination in their growing waters. These sources of pollution generally contain many different enteric viruses so that contaminated shellfish frequently bioaccumulate a mixture of viruses. Although bacteria are generally eliminated within 2–3 days, viruses are known to persist for several weeks or months (Greening et al., 2003; Nappier et al., 2008). Noroviruses mainly concentrate in their digestive gland tissue but have also been detected in the gill and other tissues (McLeod et al., 2009; Wang et al., 2008). Although the infectivity of these viruses is unclear, the consumption of contaminated raw oysters or undercooked contaminated shellfish presents a public health risk potentially leading to viral gastroenteritis outbreaks.

Recent studies have shown that noroviruses bind to human histo-blood group antigens (HBGAs), and that, in common with humans, shellfish possess some of these carbohydrate antigens (Le Guyader et al., 2006a; Tian et al., 2006). Noroviruses have been observed to bind to the HBGAs in shellfish digestive tissue, which may facilitate virus persistence inside the shellfish for several weeks following uptake and accumulation

(Le Guyader et al., 2006a). Bivalve shellfish naturally depurate or purge themselves of microbial contaminants and this process is used to purify shellfish for retail markets. Although the process is successful for the removal of bacteria, it has proved inefficient for the elimination of enteric viruses, especially noroviruses.

Detection of noroviruses in naturally contaminated shellfish was not achieved until 1995 (Lees et al., 1995). Since then, naturally occurring caliciviruses, mainly human noroviruses, have been detected in shellfish in many countries, including Canada, France, Greece, Hong Kong, Italy, New Zealand, Spain, Sweden, UK and US (Berg et al., 2000; Boxman et al., 2006; Cheng et al., 2005; Costantini et al., 2006; Croci et al., 2007; David et al., 2007; Elamri et al., 2006; Formiga-Cruz et al., 2002; Gray et al., 2009; Kohn et al., 1995; Le Guyader et al., 2000, 2006b, 2008; Murphy et al., 1979; Myrmel et al., 2004; Prato et al., 2004; Roda de Husman et al., 2007; Simmons et al., 2001). Sapoviruses have also been detected in Japanese clams that were sold at local markets (Hansman et al., 2007b). Multiple norovirus genotypes have been identified in human faecal specimens from cases associated with gastroenteritis outbreaks following consumption of contaminated shellfish (Gallimore et al., 2005a; Le Guyader et al., 2006b, 2008; Symes et al., 2007). In Japan, multiple norovirus and sapovirus strains were detected in patient faeces from oyster-associated outbreaks including one outbreak having GI.1 and GII.2 sapoviruses and GI.2 and GI.4 noroviruses (Nakagawa-Okamoto et al., 2009). In France, a large outbreak of gastroenteritis associated with oyster consumption occurred following a flood event near a shellfish farming area (Le Guyader et al., 2008). Up to five different enteric viruses were identified in shellfish samples, and seven different viruses, including Aichi virus, enterovirus, astrovirus, rotavirus and multiple norovirus strains, were detected in one faecal sample (Le Guyader et al., 2008).

The presence of both human and animal caliciviruses in shellfish has also been described (Costantini et al., 2006; Le Guyader et al., 2008). In a comprehensive study of shellfish in US coastal waters, Costantini and co-workers identified a range of animal and human enteric caliciviruses in shellfish, including human, porcine and bovine noroviruses and porcine sapoviruses (Costantini *et al.*, 2006). In this study, several shellfish samples contained both animal and human noroviruses and sapoviruses. The presence of animal caliciviruses occurred in states where there was high livestock production, indicating that the source of contamination was probably animal waste.

Many reports indicate a seasonal effect for norovirus prevalence in shellfish, with norovirus es observed to be more prevalent during winter months (Croci *et al.*, 2007; Myrmel *et al.*, 2004). This also correlates with a peak prevalence of norovirus outbreaks in most countries. Noroviruses are believed to persist longer in the environment and shellfish in winter when temperatures are colder and UV exposure is shorter. At this time, shellfish also are less physiologically active and their feeding and depuration activities are slower, which facilitates longer retention times in the shellfish.

Foods and fresh produce

Fresh produce, including leafy salad greens, fresh herbs, soft berry fruits and green onions, which are generally consumed raw, is now grown on large scale in many countries and transported globally. Faecal contamination on the produce may occur at the pre-harvest stage through contaminated irrigation water and soil, or post-harvest through poor hygiene practices at the handling and processing stages. Irrigation water derived from sewage effluent or recycled waters may be contaminated since sewage treatment does not always render the effluent free from infectious viruses. Currently there are no international guidelines for the quality of irrigation water; consequently the extent of irrigation carried out using faecally-contaminated water is unknown. Moreover, there is little information on the microbial quality or the occurrence of enteric viruses in irrigation water, and so the potential risks from the contaminated water are still poorly understood. Several large viral outbreaks have occurred in recent years in Europe, New Zealand and the US following consumption of soft fruits, salad vegetables and green onions contaminated with noroviruses or HAV (Calder *et al.*, 2003; Cotterelle *et al.*, 2005; Le Guyader *et al.*, 2004; Ponka *et al.*, 1999; Wheeler, 2005).

Fomites and environmental surfaces

Enteric viruses are known to persist and retain infectivity for long periods on environmental surfaces and objects (Carter, 2005; Rzezutka and Cook, 2004). Recent studies have shown that contaminated fomites and surfaces can be important factors in the spread of viruses. Food can become contaminated if placed onto a virus-contaminated surface. The overall contribution of fomites and surfaces to the high number of norovirus infections is unknown (Boone and Gerba, 2007) but there are studies showing the importance of fomites in prolonging outbreaks in institutional settings (Kuusi *et al.*, 2002; Wu *et al.*, 2005). During one hospital outbreak of gastroenteritis, noroviruses were detected in 11 of 36 environmental swabs collected from lockers, curtains and commodes and the immediate environment of patients (Green *et al.*, 1998). In a prolonged outbreak in a Finnish rehabilitation centre, noroviruses were detected in three environmental samples but not in food or water. These norovirus strains were indistinguishable from those in the faeces from patients (Kuusi *et al.*, 2002). In a similar study of environmental contamination, Wu and coworkers identified the mode of transmission and risk factors for acute gastroenteritis illness in a veterans' long-term care facility where 52% of residents and 46% of staff were ill (Wu *et al.*, 2005). Environmental surface swabs were taken from various surfaces and noroviruses were detected from a dining room table and an elevator button used only by employees. The norovirus sequences from both environmental and clinical samples were indistinguishable and the researchers concluded that extensive contamination of environmental surfaces may play a role in prolonged norovirus outbreaks and therefore should be addressed in control interventions.

There is evidence to suggest that noroviruses can remain infectious on carpets for long periods. Twelve days after a retirement home outbreak, two carpet layers removed a soiled carpet which had been previously cleaned. The carpet layers, who had no other known exposure to viral infection, subsequently became ill with norovirus infections (Cheesbrough *et al.*, 1997). In an investigation of environmental contamination during a prolonged norovirus outbreak in a hotel, the same researchers detected noroviruses in carpets nine weeks after the initial outbreak. Noroviruses were also detected on hard horizontal surfaces above 1.5 m high (e.g. light fittings), which were unlikely to be directly contaminated (Cheesbrough et al., 2000).

In a study of potential calicivirus transmission via commonly used fomites, Clay and coworkers used FCV as a norovirus surrogate to investigate calicivirus survival on a computer mouse, keyboard keys, telephone receiver and buttons, and brass disks to simulate door handles (Clay et al., 2006). FCV survival varied according to the contaminated surface, with the longest survival period being 3 days on telephones, and the shortest being 8–12 hours on keyboards and brass disks.

In a longitudinal study of environmental contamination in a paediatric institution, Gallimore and coworkers examined environmental surfaces and equipment at 11 sites twice per week over six months to determine whether enteric viruses (noroviruses, rotaviruses and astroviruses) were present (Gallimore et al., 2006). Noroviruses were detected on 50% of the swabbing dates, were the most commonly identified viruses, and were found mostly on toilet taps. Following these findings, new cleaning schedules were introduced as a means of reducing environmental contamination. In a follow-up study 1 year later, norovirus contamination was found to have decreased and the improved cleaning regimes were deemed to have reduced the level of environmental contamination (Gallimore et al., 2008).

Intervention strategies and infection control measures for reducing norovirus infections in healthcare settings include closure of wards, visitor restrictions in retirement homes, containment of infected individuals, improved hand washing hygiene among staff and patients, environmental decontamination and stand-down of staff until they are asymptomatic. These strategies have proven to be effective control measures in reducing the spread of norovirus outbreaks.

Food handling

Poor food handling practices are now believed to be a major route of transmission for foodborne viral disease, although accurate data to prove the significance of the food handling route are not readily available. At the post-harvest stage, food can become contaminated by poor hygiene and food handling practices leading to direct contamination with faecal material. In South and Latin America, Eastern Europe and Asia (especially China), large areas of land are now being cultivated to produce fresh produce on a large scale for the global marketplace. Many of these countries follow international Good Agricultural Practices (GAP), but these practices are not in place in several countries; consequently, the growing areas may have contaminated irrigation water and poor hygiene practices.

Poor food handling has now been recognized as a major cause of norovirus outbreaks. One of the first reports of a major outbreak that was related to poor food handling occurred in 1984, when a baker manually mixed a large batch of icing and at the same time contaminated it with norovirus. He subsequently caused an outbreak of an estimated 3000 cases in a US city (Kuritsky et al., 1984). Other more recent outbreaks in the Netherlands and the US have also implicated bakery products contaminated with noroviruses by those preparing foods (de Wit et al., 2007; Friedman et al., 2005). Infected food handlers were implicated in norovirus outbreaks associated with consumption of contaminated ham sandwiches (Daniels et al., 2000) and salads (Anderson et al., 2001). In the latter outbreak, the contaminated salads were distributed as part of a banquet to groups across the US. Illness was reported in 13 states following consumption of the salads and the same norovirus strain was identified in cases from eight states. Other outbreaks linked to poor food handling practices have been well documented (Malek et al., 2009; Marshall et al., 2001; Schmid et al., 2007).

Cruise ships

In recent years there has been an increased incidence in norovirus outbreaks on cruise ships in Europe and North America, which has attracted international publicity and also had a substantial economic influence on the cruise ship industry. Outbreaks have been associated with norovirus-contaminated ice and foods followed by ongoing person-to-person and airborne transmission, which have prolonged outbreaks for several weeks on successive cruises (Ho *et al.*, 1989; Khan *et al.*, 1994; Koo *et al.*, 1996; McEvoy *et al.*, 1996; Widdowson *et al.*, 2004). Multiple transmission routes and high attack rates (>10%) have been recorded in many of these outbreaks (Widdowson *et al.*, 2004). It is believed that environmental contamination in cabins and through the ships contributed to the longevity of these outbreaks. Environmental swabbing has proven useful in detecting noroviruses in such situations (Boxman *et al.*, 2009). In conjunction with improved handwashing and food handling procedures during cruises, rigorous cleaning and environmental disinfection of vessels between cruises have provided effective disruption of virus transmission (Lawrence, 2004). Many countries have prepared guidelines for control of outbreaks on ships, but to date there are no international guidelines on prevention and control of norovirus outbreaks on cruise ships.

Airborne infection

Although noroviruses are primarily spread via the faecal–oral route, secondary spread also occurs following projectile vomiting, a common symptom of human norovirus infection. Vomitus contains high numbers of norovirus particles, which can be aerosolized and potentially infectious. An estimated 30 million virus particles may be expelled in a single vomiting episode (Caul, 1994). This is a main secondary transmission route of noroviruses and is a major factor in the rapid spread of noroviruses through institutions, especially retirement homes and hospitals. Transmission of norovirus by aerosolization of vomit in hospitals and in a hotel dining room has been reported (Chadwick and McCann, 1994; Marks *et al.*, 2000). In the latter situation, an outbreak of norovirus gastroenteritis occurred after a diner vomited in a UK hotel dining room. Of the diners at the same table as the vomiter, 91% subsequently became ill. The attack rates in diners at other tables reduced inversely to their proximity to the affected table, ranging from 71% at the next table to 25% at a table in an adjoining annex. Ceiling fans close to the index case's table probably contributed to the spread of the virus. Food was not implicated in the outbreak. Probable airborne transmission of norovirus from eight ill crew members to five passengers on an airplane

was also reported (Widdowson *et al.*, 2005a). No specimens were collected but the epidemiological findings indicated airborne transmission as the probable route.

Other caliciviruses, such as FCV, are spread by respiratory routes. Following clinical FCV infection, many cats remain persistent carriers. Highly virulent strains of FCV associated with a high mortality rate and systemic disease have recently been identified. These strains appear to be transmitted via fomites as well as the respiratory route and virus can be shed for up to 16 weeks following recovery (Hurley *et al.*, 2004). Murine norovirus may also be transmitted by both the airborne route and direct contact as infection occurs between cages in colonies of laboratory mice (Perdue *et al.*, 2007).

Mechanical vectors

RHDV is excreted in all secretions and excretions of diseased rabbits. Aside from direct contact of infected animals it has been speculated that insects may play a role as mechanical vectors in RHDV transmission (McColl *et al.*, 2002). Insects that fed on diseased rabbits under laboratory conditions were able to transfer the virus to previously uninfected animals (Lenghaus *et al.*, 1994). In New Zealand, blowflies and flesh flies were shown to be likely vectors for RHDV (Henning *et al.*, 2005). In an Australian study a strong association between flea infestation, low temperatures, breeding in rabbits and rabbit haemorrhagic disease was found (Lugton, 1999).

Zoonotic calicivirus infection

The potential for calicivirus zoonotic transmission from animal to human or vice versa has been discussed in several papers (Costantini *et al.*, 2006; Koopmans, 2008; Radford *et al.*, 2004; Scipioni *et al.*, 2008), but convincing evidence to support this idea is limited to a few studies. Smith and coworkers reported hand lesions and infection with a vesivirus (SMSV) in a laboratory worker but there are no other reports of this type of infection (Smith *et al.*, 1998a). Although GIII and GV norovirus strains have been identified only in animals, GII and GIV norovirus strains are now known to infect both humans and animals. Mattison and coworkers investigated the presence of noroviruses in animal faeces, manure and meat and identified GIII (bovine), GII.18 (swine), and GII.4 (human) norovirus sequences in faeces (Mattison *et al.*, 2007). This was the first report of human noroviruses in livestock. In addition, they also detected a GII.4-like norovirus on a retail meat sample. These findings indicated a possible route for indirect zoonotic transmission of noroviruses through the food chain. In another study, antibodies to bovine noroviruses were found in 28% of veterinarians without accompanying symptoms of infection compared with 20% of matched population controls (Pratelli *et al.*, 2000; Widdowson *et al.*, 2005b). Overall exposure to human norovirus as determined by antibody presence was between 76–90% across both groups, depending on norovirus genotype, compared with 22% for bovine norovirus. Using these data, the authors postulated that bovine norovirus strains may infect humans, though less frequently than human norovirus strains. There are reports of cross species transmission of caliciviruses among different animals, with most reports associated with vesiviruses. A recent example was the isolation of FCV from infected dogs (Pratelli *et al.*, 2000). Smith and coworkers have presented evidence that marine animals may be a reservoir for the sporadic reintroduction of vesiviruses to terrestrial animals (Smith *et al.*, 1976, 1998b). These authors hypothesized that outbreaks of VESV, which affected the swine industry in the US in the first half of the last century, were caused by marine vesivirus serotypes. They speculated that these viruses were transmitted to domestic pig populations through the food chain by the feeding of uncooked waste, which included fish and marine mammal tissues. Recently a GIV norovirus was identified in faeces of a dead lion cub in Italy (Martella *et al.*, 2007). It was genetically distinct from human GIV noroviruses and classified as GIV.2. Two months later the same laboratory identified a similar GIV norovirus in a dog with diarrhoea, indicating possible cross-species transmission (Martella *et al.*, 2008b). This is the first evidence of a norovirus genotype naturally infecting two animal species.

The potential for zoonotic transmission of porcine norovirus to humans was also raised following the detection of GII noroviruses in pigs from Japan and Europe and GII norovirus antibodies in US pigs. Wang and coworkers identified calicivirus strains in pigs that were closely related to human GII noroviruses (Wang *et al.*, 2005a). Among these, a potential recombinant strain was identified. One porcine norovirus was characterized as genetically and antigenically related to human norovirus and was able to replicate in gnotobiotic pigs. These findings raised concerns that pigs may be reservoirs of new human norovirus strains and that porcine/human GII recombinant norovirus strains could emerge. Further work by Wang and coworkers showed close genetic relatedness between some human and porcine sapoviruses, which could also pose a potential risk of zoonotic infection with the emergence of new human and animal recombinant strains (Wang *et al.*, 2007).

Emergence of recombinants and new variants

Environmental selection pressures on RNA viruses are known to lead to the emergence of new variants and recombinants (Worobey and Holmes, 1999). The genetically diverse caliciviruses are no exception, and it is likely that environmental pressures have had a major influence on the emergence of new calicivirus strains. Noroviruses in particular are genetically very diverse and the recognition of new strains is a frequent occurrence. Recombinant GII.b norovirus strains were first recognized in France in 2000, and then in other regions of the world (Ambert-Balay *et al.*, 2005; Buesa *et al.*, 2002; Bull *et al.*, 2007; Reuter *et al.*, 2006). The emergence of the first GII.b sequences was believed to be associated with a multinational oyster-associated gastroenteritis outbreak (Reuter *et al.*, 2006). Reuter *et al.* identified four recombinant GII.b (pol) norovirus strains in Hungary, which had capsids belonging to GII.1, GII.2, GII.3 and GII.4.

Bivalve molluscan shellfish provide an ideal environment for concentrating a mix of enteric viruses, including multiple norovirus strains. Subsequent consumption of the shellfish containing this virus cocktail may cause infection in a human host. Co-infection in the human gut with multiple norovirus strains may lead to genetic recombination and the emergence of recombinant strains. Infection with multiple norovirus strains has been frequently reported in conjunction with shellfish-related norovirus outbreaks (Costantini *et al.*, 2006; Gallimore *et al.*, 2005a; Le Guyader *et al.*, 2006b, 2008; Symes *et al.*, 2007). Symes and coworkers identified three different GII norovirus strains in the faeces of a case linked to a shellfish outbreak and found that each strain showed evidence of a potential recombination event when the sequences were compared with other norovirus sequences (Symes *et al.*, 2007).

Gallimore and coworkers identified a recombinant GII norovirus, rGII-3a (recombinant Harrow/Mexico), in cases from an outbreak associated with consumption of oysters. Recently, another new recombinant strain, GIIc-GII.12, was identified in cases from several New Zealand outbreaks associated with consumption of contaminated oysters, and also in a sample of the implicated oysters (Grey *et al.*, 2009). This new strain has since been detected in other norovirus non-shellfish related outbreaks throughout New Zealand (Anonymous, NZPHSR, March 2009).

Since 2002, several new GII.4 variant strains have emerged at regular intervals and have been responsible for many large outbreaks across Europe, the US and on cruise ships (Bull *et al.*, 2006; Siebenga *et al.*, 2007; Tu *et al.*, 2007, 2008). These GII.4 variants include the 2002 Farmington Hills cruise ship variant, the 2004 Hunter variant, and more recently the 2006a (Laurens) and 2006b (Minerva) variants, which have been the predominant causes of outbreaks in Europe, North America and Australasia since 2006.

Conclusion

Caliciviruses are ubiquitous in the environment and are major causes of disease in humans and many animals. Outbreaks of calicivirus infection, especially human norovirus outbreaks, and the accompanying economic burden will continue until intervention and prevention strategies can be greatly improved.

References

Ambert-Balay, K., Bon, F., Le Guyader, F., Pothier, P., and Kohli, E. (2005). Characterization of new recombinant noroviruses. J. Clin. Microbiol. *43*, 5179–5186.

Anderson, A.D., Garrett, V.D., Sobel, J., Monroe, S.S., Fankhauser, R.L., Schwab, K.J., Bresee, J.S., Mead, P.S., Higgins, C., Campana, J., *et al.* (2001). Multistate outbreak of Norwalk-like virus gastroenteritis associated with a common caterer. Am. J. Epidemiol. *154*, 1013–1019.

Atmar, R.L., Opekun, A.R., Gilger, M.A., Estes, M.K., Crawford, S.E., Neill, F.H., and Graham, D.Y. (2008). Norwalk virus shedding after experimental human infection. Emerg. Infect. Dis. *14*, 1553–1557.

Baert, L., Uyttendaele, M., Vermeersch, M., Van Coillie, E., and Debevere, J. (2008). Survival and transfer of murine norovirus 1, a surrogate for human noroviruses, during the production process of deep-frozen onions and spinach. J. Food Prot. *71*, 1590–1597.

Barry, A.F., Alfieri, A.F., and Alfieri, A.A. (2008). High genetic diversity in RdRp gene of Brazilian porcine sapovirus strains. Vet. Microbiol. *131*, 185–191.

Beller, M., Ellis, A., Lee, S.H., Drebot, M.A., Jenkerson, S.A., Funk, E., Sobsey, M.D., III, O.D.S., Monroe, S.S., Ando, T., *et al.* (1997). Outbreak of viral gastroenteritis due to a contaminated well – International consequences. J. Am. Med. Assoc. *278*, 563–568.

Berg, D.E., Kohn, M.A., Farley, T.A., and McFarland, L.M. (2000). Multi-state outbreaks of acute gastroenteritis traced to fecal-contaminated oysters harvested in Louisiana. J. Infect. Dis. *181* Suppl 2, S381–386.

Blanton, L.H., Adams, S.M., Beard, R.S., Wei, G., Bulens, S.N., Widdowson, M.A., Glass, R.I., and Monroe, S.S. (2006). Molecular and Epidemiologic Trends of Caliciviruses Associated with Outbreaks of Acute Gastroenteritis in the United States, 2000–2004. J. Infect. Dis. *193*, 413–421.

Boccia, D., Tozzi, A.E., Cotter, B., Rizzo, C., Russo, T., Buttinelli, G., Caprioli, A., Marziano, M.L., and Ruggeri, F.M. (2002). Waterborne outbreak of Norwalk-like virus gastroenteritis at a tourist resort, Italy. Emerg. Infect. Dis. *8*, 563–568.

Boone, S.A., and Gerba, C.P. (2007). Significance of fomites in the spread of respiratory and enteric viral disease. Appl. Environ. Microbiol. *73*, 1687–1696.

Borchardt, M.A., Bertz, P.D., Spencer, S.K., and Battigelli, D.A. (2003). Incidence of enteric viruses in groundwater from household wells in Wisconsin. Appl. Environ. Microbiol. *69*, 1172–1180.

Boxman, I.L., Dijkman, R., te Loeke, N.A., Hagele, G., Tilburg, J.J., Vennema, H., and Koopmans, M. (2009). Environmental swabs as a tool in norovirus outbreak investigation, including outbreaks on cruise ships. J. Food Prot. *72*, 111–119.

Boxman, I.L., Tilburg, J.J., Te Loeke, N.A., Vennema, H., Jonker, K., de Boer, E., and Koopmans, M. (2006). Detection of noroviruses in shellfish in the Netherlands. Int. J. Food Microbiol. *108*, 391–396.

Buesa, J., Collado, B., Lopez-Andujar, P., Abu-Mallouh, R., Rodriguez Diaz, J., Garcia Diaz, A., Prat, J., Guix, S., Llovet, T., Prats, G., *et al.* (2002). Molecular epidemiology of caliciviruses causing outbreaks and sporadic cases of acute gastroenteritis in Spain. J. Clin. Microbiol. *40*, 2854–2859.

Bull, R.A., Tanaka, M.M., and White, P.A. (2007). Norovirus recombination. J. Gen. Virol. *88*, 3347–3359.

Bull, R.A., Tu, E.T., McIver, C.J., Rawlinson, W.D., and White, P.A. (2006). Emergence of a new norovirus genotype II.4 variant associated with global outbreaks of gastroenteritis. J. Clin. Microbiol. *44*, 327–333.

Calder, L., Simmons, G., Thornley, C., Taylor, P., Pritchard, K., Greening, G., and Bishop, J. (2003). An outbreak of hepatitis A associated with consumption of raw blueberries. Epidemiol. Infect. 131, 745–751.

Cannon, J.L., Papafragkou, E., Park, G.W., Osborne, J., Jaykus, L.A., and Vinje, J. (2006). Surrogates for the study of norovirus stability and inactivation in the environment: aA comparison of murine norovirus and feline calicivirus. J. Food Prot. 69, 2761–2765.

Carter, M. (2005). Enterically infecting viruses: pathogenicity, transmission and significance for food and waterborne infection. J. Appl. Microbiol. 98, 1354–1380.

Caul, E.O. (1994). Small round structured viruses: airborne transmission and hospital control. Lancet 343, 1240–1242.

Chadwick, P.R., and McCann, R. (1994). Transmission of a small round structured virus by vomiting during a hospital outbreak of gastroenteritis. J. Hosp. Infect. 26, 251–259.

Cheesbrough, J.S., Barkess-Jones, L., and Brown, D.W. (1997). Possible prolonged environmental survival of small round structured viruses. J. Hosp. Infect. 35, 325–326.

Cheesbrough, J.S., Green, J., Gallimore, C.I., Wright, P.A., and Brown, D.W. (2000). Widespread environmental contamination with Norwalk-like viruses (NLV) detected in a prolonged hotel outbreak of gastroenteritis. Epidemiol. Infect. 125, 93–98.

Cheng, P.K., Wong, D.K., Chung, T.W., and Lim, W.W. (2005). Norovirus contamination found in oysters worldwide. J. Med. Virol. 76, 593–597.

Clay, S., Maherchandani, S., Malik, Y.S., and Goyal, S.M. (2006). Survival on uncommon fomites of feline calicivirus, a surrogate of noroviruses. Am. J. Infect. Control 34, 41–43.

Cooke, B.D. (2002). Rabbit haemorrhagic disease: field epidemiology and the management of wild rabbit populations. Rev. Sci. Tech. 21, 347–358.

Costantini, V., Loisy, F., Joens, L., Le Guyader, F.S., and Saif, L.J. (2006). Human and animal enteric caliciviruses in oysters from different coastal regions of the United States. Appl. Environ. Microbiol. 72, 1800–1809.

Costantini, V.P., Azevedo, A.C., Li, X., Williams, M.C., Michel, F.C., Jr., and Saif, L.J. (2007). Effects of different animal waste treatment technologies on detection and viability of porcine enteric viruses. Appl. Environ. Microbiol. 73, 5284–5291.

Cotterelle, B., Drougard, C., Rolland, J., Becamel, M., Boudon, M., Pinede, S., Traore, O., Balay, K., Pothier, P., and Espie, E. (2005). Outbreak of norovirus infection associated with the consumption of frozen raspberries, France, March 2005. Euro. Surveill. 10, E050428, 050421.

Croci, L., Losio, M.N., Suffredini, E., Pavoni, E., Di Pasquale, S., Fallacara, F., and Arcangeli, G. (2007). Assessment of human enteric viruses in shellfish from the northern Adriatic sea. Int. J. Food Microbiol. 114, 252–257.

da Silva, A.K., Le Guyader, F.S., Le Saux, J.C., Pommepuy, M., Montgomery, M.A., and Elimelech, M. (2008). Norovirus removal and particle association in a waste stabilization pond. Environ. Sci. Technol. 42, 9151–9157.

Daniels, N.A., Bergmire-Sweat, D.A., Schwab, K.J., Hendricks, K.A., Reddy, S., Rowe, S.M., Fankhauser, R.L., Monroe, S.S., Atmar, R.L., Glass, R.I., et al. (2000). A foodborne outbreak of gastroenteritis associated with Norwalk-like viruses: first molecular traceback to deli sandwiches contaminated during preparation. J. Infect. Dis. 181, 1467–1470.

David, S.T., McIntyre, L., MacDougall, L., Kelly, D., Liem, S., Schallie, K., McNabb, A., Houde, A., Mueller, P., Ward, P., et al. (2007). An outbreak of norovirus caused by consumption of oysters from geographically dispersed harvest sites, British Columbia, Canada, 2004. Foodborne Pathog. Dis. 4, 349–358.

de Wit, M.A., Widdowson, M.A., Vennema, H., de Bruin, E., Fernandes, T., and Koopmans, M. (2007). Large outbreak of norovirus: the baker who should have known better. J. Infect. 55, 188–193.

Duizer, E., Bijkerk, P., Rockx, B., De Groot, A., Twisk, F., and Koopmans, M. (2004). Inactivation of caliciviruses. Appl. Environ. Microbiol. 70, 4538–4543.

Elamri, D.E., Aouni, M., Parnaudeau, S., and Le Guyader, F.S. (2006). Detection of human enteric viruses in shellfish collected in Tunisia. Lett. Appl. Microbiol. 43, 399–404.

Farkas, T., Zhong, W.M., Jing, Y., Huang, P.W., Espinosa, S.M., Martinez, N., Morrow, A.L., Ruiz-Palacios, G.M., Pickering, L.K., and Jiang, X. (2004). Genetic diversity among sapoviruses. Arch. Virol. 149, 1309–1323.

Formiga-Cruz, M., Tofino-Quesada, G., Bofill-Mas, S., Lees, D.N., Henshilwood, K., Allard, A.K., Conden-Hansson, A.C., Hernroth, B.E., Vantarakis, A., Tsibouxi, A., et al. (2002). Distribution of human virus contamination in shellfish from different growing areas in Greece, Spain, Sweden, and the United Kingdom. Appl. Environ. Microbiol. 68, 5990–5998.

Forrester, N.L., Abubakr, M.I., Abu Elzein, E.M., Al-Afaleq, A.I., Housawi, F.M., Moss, S.R., Turner, S.L., and Gould, E.A. (2006). Phylogenetic analysis of rabbit haemorrhagic disease virus strains from the Arabian Peninsula: did RHDV emerge simultaneously in Europe and Asia? Virology 344, 277–282.

Fout, G.S., Martinson, B.C., Moyer, M.W., and Dahling, D.R. (2003). A Multiplex Reverse Transcription-PCR Method for Detection of Human Enteric Viruses in Groundwater. Appl. Environ. Microbiol. 69, 3158–3164.

Friedman, D.S., Heisey-Grove, D., Argyros, F., Berl, E., Nsubuga, J., Stiles, T., Fontana, J., Beard, R.S., Monroe, S., McGrath, M.E., et al. (2005). An outbreak of norovirus gastroenteritis associated with wedding cakes. Epidemiol. Infect. 133, 1057–1063.

Gallimore, C.I., Cheesbrough, J.S., Lamden, K., Bingham, C., and Gray, J.J. (2005a). Multiple norovirus genotypes characterised from an oyster-associated outbreak of gastroenteritis. Int. J. Food Microbiol. 103, 323–330.

Gallimore, C.I., Pipkin, C., Shrimpton, H., Green, A.D., Pickford, Y., McCartney, C., Sutherland, G., Brown, D.W., and Gray, J.J. (2005b). Detection of multiple enteric virus strains within a foodborne outbreak of gastroenteritis: an indication of the source of contamination. Epidemiol. Infect. 133, 41–47.

Gallimore, C.I., Taylor, C., Gennery, A.R., Cant, A.J., Galloway, A., Iturriza-Gomara, M., and Gray, J.J. (2006). Environmental monitoring for gastroenteric viruses in a pediatric primary immunodeficiency unit. J. Clin. Microbiol. 44, 395–399.

Gallimore, C.I., Taylor, C., Gennery, A.R., Cant, A.J., Galloway, A., Xerry, J., Adigwe, J., and Gray, J.J. (2008). Contamination of the hospital environment with gastroenteric viruses: comparison of two pediatric wards over a winter season. J. Clin. Microbiol. 46, 3112–3115.

Gray, J.J., Green, J., Cunliffe, C., Gallimore, C., Lee, J.V., Neal, K., and Brown, D.W. (1997). Mixed genogroup SRSV infections among a party of canoeists exposed to contaminated recreational water. J. Med. Virol. 52, 425–429.

Green, J., Wright, P.A., Gallimore, C.I., Mitchell, O., Morgan-Capner, P., and Brown, D.W. (1998). The role of environmental contamination with small round structured viruses in a hospital outbreak investigated by reverse-transcriptase polymerase chain reaction assay. J. Hosp. Infect. 39, 39–45.

Green, K.Y. (2007). Caliciviridae: The Noroviruses. In Fields virology, H.P. Knipe, and P.M. Howley, eds. (Philadelphia, Pa, Lippincott Williams & Wilkins), pp. 949–979.

Green, K.Y., Ando, T., Balayan, M.S., Berke, T., Clarke, I.N., Estes, M.K., Matson, D.O., Nakata, S., Neill, J.D., Studdert, M.J., et al. (2000). Taxonomy of the caliciviruses. J. Infect. Dis. 181 Suppl 2, S322–330.

Green, K.Y., Kapikian, A., and Chanock, R.M. (2001). Human caliciviruses. In Fields virology, H.P. Knipe, ed. (Philadelphia, PA: Lippincott Williams and Wilkins) pp. 841–874.

Greening, G., Hewitt, J., Hay, B., and Grant, C. (2002). Persistence of Norwalk-like viruses over time in Pacific oysters grown in the natural environment. Paper presented at: 4th International Conference on Molluscan Shellfish Safety (Santiago de Compostela: Consellería de Pesca e Asuntos Marítimos da Xunta de Galicia & Intergovernmental Oceanographic Commission of UNESCO).

Greening, G., Hewitt, J., Hay, B., and Grant, C., eds. (2003). Persistence of Norwalk-like viruses over time in Pacific oysters grown in the natural environment. (Santiago de Compostela: Consellería de Pesca e Asuntos Marítimos da Xunta de Galicia & Intergovernmental Oceanographic Commission of UNESCO).

Greening, G.E., Mirams, M., and Berke, T. (2001). Molecular epidemiology of 'Norwalk-like viruses' associated with gastroenteritis outbreaks in New Zealand. J. Med. Virol. 64, 58–66.

Grey, C., Simmons, G., Ormsby, C., Hewitt, J., and Greening, G. (2009). Novel recombinant strain of norovirus identified from an oyster-borne outbreak in Auckland. New Zealand Public Health Surveillance Report 7, 6–7.

Guo, M., Chang, K.O., Hardy, M.E., Zhang, Q., Parwani, A.V., and Saif, L.J. (1999). Molecular characterization of a porcine enteric calicivirus genetically related to Sapporo-like human caliciviruses. J. Virol. 73, 9625–9631.

Hafliger, D., Hubner, P., and Luthy, J. (2000). Outbreak of viral gastroenteritis due to sewage-contaminated drinking water. Int. J. Food Microbiol. 54, 123–126.

Hansman, G.S., Oka, T., Katayama, K., and Takeda, N. (2007a). Human sapoviruses: genetic diversity, recombination, and classification. Rev. Med. Virol. 17, 133–141.

Hansman, G.S., Oka, T., Okamoto, R., Nishida, T., Toda, S., Noda, M., Sano, D., Ueki, Y., Imai, T., Omura, T., et al. (2007b). Human sapovirus in clams, Japan. Emerg. Infect. Dis. 13, 620–622.

Hansman, G.S., Saito, H., Shibata, C., Ishizuka, S., Oseto, M., Oka, T., and Takeda, N. (2007c). An outbreak of gastroenteritis due to Sapovirus. J. Clin. Microbiol. 45, 1347–1349.

Hansman, G.S., Sano, D., Ueki, Y., Imai, T., Oka, T., Katayama, K., Takeda, N., and Omura, T. (2007d). Sapovirus in water, Japan. Emerg. Infect. Dis. 13, 133–135.

Haramoto, E., Katayama, H., Oguma, K., and Ohgaki, S. (2005). Application of cation-coated filter method to detection of noroviruses, enteroviruses, adenoviruses, and torque teno viruses in the Tamagawa River in Japan. Appl. Environ. Microbiol. 71, 2403–2411.

Henning, J., Schnitzler, F.R., Pfeiffer, D.U., and Davies, P. (2005). Influence of weather conditions on fly abundance and its implications for transmission of rabbit haemorrhagic disease virus in the North Island of New Zealand. Med. Vet. Entomol. 19, 251–262.

Hewitt, J., Bell, D., Simmons, G.C., Rivera-Aban, M., Wolf, S., and Greening, G.E. (2007). Gastroenteritis outbreak caused by waterborne norovirus at a New Zealand ski resort. Appl. Environ. Microbiol. 73, 7853–7857.

Hewitt, J., and Greening, G.E. (2004). Survival and persistence of norovirus, hepatitis A virus, and feline calicivirus in marinated mussels. J. Food Prot. 67, 1743–1750.

Hewitt, J., and Greening, G.E. (2006). Effect of heat treatment on hepatitis A virus and norovirus in New Zealand greenshell mussels (Perna canaliculus) by quantitative real-time reverse transcription PCR and cell culture. J. Food Prot. 69, 2217–2223.

Hewitt, J., Rivera-Aban, M., and Greening, G. (2009). Evaluation of murine norovirus as a surrogate for human norovirus and hepatitis A virus in heat inactivation studies. J. Appl. Microbiol. 107, 65–71.

Ho, M.S., Glass, R.I., Monroe, S.S., Madore, H.P., Stine, S., Pinsky, P.F., Cubitt, D., Ashley, C., and Caul, E.O. (1989). Viral gastroenteritis aboard a cruise ship. Lancet 2, 961–965.

Hoebe, C.J., Vennema, H., de Roda Husman, A.M., and van Duynhoven, Y.T. (2004). Norovirus outbreak among primary schoolchildren who had played in a recreational water fountain. J. Infect. Dis. 189, 699–705.

Hsu, C.C., Wobus, C.E., Steffen, E.K., Riley, L.K., and Livingston, R.S. (2005). Development of a microsphere-based serologic multiplexed fluorescent immunoassay and a reverse transcriptase PCR assay to detect murine norovirus 1 infection in mice. Clin. Diagn. Lab. Immunol. 12, 1145–1151.

Hurley, K.E., Pesavento, P.A., Pedersen, N.C., Poland, A.M., Wilson, E., and Foley, J.E. (2004). An outbreak of virulent systemic feline calicivirus disease. J. Am. Vet. Med. Assoc. 224, 241–249.

Hurst, C.J., Gerba, C.P., and Cech, I. (1980). Effects of environmental variables and soil characteristics on virus survival in soil. Appl. Environ. Microbiol. 40, 1067–1079.

Ike, A.C., Roth, B.N., Bohm, R., Pfitzner, A.J., and Marschang, R.E. (2007). Identification of bovine enteric Caliciviruses (BEC) from cattle in Baden-Wurttemberg. Dtsch Tierarztl Wochenschr 114, 12–15.

Iwai, M., Hasegawa, S., Obara, M., Nakamura, K., Horimoto, E., Takizawa, T., Kurata, T., Sogen, S.I., and Shiraki, K. (2009). Continuous existence of noroviruses and sapoviruses in raw sewage reveals infection among inhabitants in Toyama, Japan (2006–2008). Appl. Environ. Microbiol. 75, 1264–1270.

Jeong, C., Park, S.I., Park, S.H., Kim, H.H., Park, S.J., Jeong, J.H., Choy, H.E., Saif, L.J., Kim, S.K., Kang, M.I., et al. (2007). Genetic diversity of porcine sapoviruses. Vet. Microbiol. 122, 246–257.

Karst, S.M., Wobus, C.E., Lay, M., Davidson, J., and Virgin, H.W.t. (2003). STAT1-dependent innate immunity to a Norwalk-like virus. Science 299, 1575–1578.

Khan, A.S., Moe, C.L., Glass, R.I., Monroe, S.S., Estes, M.K., Chapman, L.E., Jiang, X., Humphrey, C., Pon, E., Iskander, J.K., et al. (1994). Norwalk virus-associated gastroenteritis traced to ice consumption aboard a cruise ship in Hawaii: comparison and application of molecular method-based assays. J. Clin. Microbiol. 32, 318–322.

Kingsley, D.H., Holliman, D.R., Calci, K.R., Chen, H., and Flick, G.J. (2007). Inactivation of a norovirus by high-pressure processing. Appl. Environ. Microbiol. 73, 581–585.

Kohn, M.A., Farley, T.A., Ando, T., Curtis, M., Wilson, S.A., Jin, Q., Monroe, S.S., Baron, R.C., McFarland, L.M., and Glass, R.I. (1995). An outbreak of Norwalk virus gastroenteritis associated with eating raw oysters. Implications for maintaining safe oyster beds. JAMA 273, 466–471.

Koo, D., Maloney, K., and Tauxe, R. (1996). Epidemiology of diarrheal disease outbreaks on cruise ships, 1986 through 1993. JAMA 275, 545–547.

Koopmans, M. (2008). Progress in understanding norovirus epidemiology. Curr. Opin. Infect. Dis. 21, 544–552.

Koopmans, M., and Duizer, E. (2004). Foodborne viruses: an emerging problem. Int. J. Food Microbiol. 90, 23–41.

Kukkula, M., Arstila, P., Klossner, M.L., Maunula, L., Bonsdorff, C.H., and Jaatinen, P. (1997). Waterborne outbreak of viral gastroenteritis. Scand. J. Infect. Dis. 29, 415–418.

Kuritsky, J.N., Osterholm, M.T., Greenberg, H.B., Korlath, J.A., Godes, J.R., Hodberg, C.W., Forfang, J.C., Kapikian, A.Z., McCullough, J.C., and White, K.E. (1984). Norwalk gastroenteritis: a community outbreak associated with bakery product consumption. Ann. Intern. Med. 100, 519–521.

Kurth, A., Evermann, J.F., Skilling, D.E., Matson, D.O., and Smith, A.W. (2006a). Prevalence of vesivirus in a laboratory-based set of serum samples obtained from dairy and beef cattle. Am. J. Vet. Res. 67, 114–119.

Kurth, A., Skilling, D.E., and Smith, A.W. (2006b). Serologic evidence of vesivirus-specific antibodies associated with abortion in horses. Am. J. Vet. Res. 67, 1033–1039.

Kuusi, M., Nuorti, J.P., Maunula, L., Minh, N.N., Ratia, M., Karlsson, J., and von Bonsdorff, C.H. (2002). A prolonged outbreak of Norwalk-like calicivirus (NLV) gastroenteritis in a rehabilitation centre due to environmental contamination. Epidemiol. Infect. 129, 133–138.

L'Homme, Y., Sansregret, R., Plante-Fortier, E., Lamontagne, A.M., Lacroix, G., Ouardani, M., Deschamps, J., Simard, G., and Simard, C. (2009). Genetic diversity of porcine Norovirus and Sapovirus: Canada, 2005–2007. Arch. Virol. 154, 581–593.

Lawrence, D.N. (2004). Outbreaks of gastrointestinal diseases on cruise ships: lessons from three decades of progress. Curr. Infect. Dis. Rep. 6, 115–123.

Le Guyader, F., Haugarreau, L., Miossec, L., Dubois, E., and Pommepuy, M. (2000). Three-year study to assess human enteric viruses in shellfish. Appl. Environ. Microbiol. 66, 3241–3248.

Le Guyader, F., Loisy, F., Atmar, R.L., Hutson, A.M., Estes, M.K., Ruvoen-Clouet, N., Pommepuy, M., and Le Pendu, J. (2006a). Norwalk virus-specific binding to oyster digestive tissues. Emerg. Infect. Dis. 12, 931–936.

Le Guyader, F.S., Bon, F., DeMedici, D., Parnaudeau, S., Bertone, A., Crudeli, S., Doyle, A., Zidane, M., Suffredini, E., Kohli, E., et al. (2006b). Detection of multiple noroviruses associated with an international gastroenteritis outbreak linked to oyster consumption. J. Clin. Microbiol. 44, 3878–3882.

Le Guyader, F.S., Le Saux, J.C., Ambert-Balay, K., Krol, J., Serais, O., Parnaudeau, S., Giraudon, H., Delmas, G., Pommepuy, M., Pothier, P., et al. (2008). Aichi virus, norovirus, astrovirus, enterovirus, and rotavirus involved in clinical cases from a French oyster-related gastroenteritis outbreak. J. Clin. Microbiol. 46, 4011–4017.

Le Guyader, F.S., Mittelholzer, C., Haugarreau, L., Hedlund, K.O., Alsterlund, R., Pommepuy, M., and Svensson, L. (2004). Detection of noroviruses in raspberries associated with a gastroenteritis outbreak. Int. J. Food Microbiol. 97, 179–186.

Lees, D.N., Henshilwood, K., Green, J., Gallimore, C.I., and Brown, D.W. (1995). Detection of small round structured viruses in shellfish by reverse transcription-PCR. Appl. Environ. Microbiol. 61, 4418–4424.

Lenghaus, C., Westbury, H., Collins, B., Ratnamoban, N., and Morrissy, C. (1994). Overview of the RHD project in Australia. In R K Munro; R T Williams (ed),

Rabbit haemorrhagic disease: issues in assessment for biological control Canberra, Bureau of Resource Sciences, 104–129.

Lodder, W.J., and de Roda Husman, A.M. (2005). Presence of noroviruses and other enteric viruses in sewage and surface waters in The Netherlands. Appl. Environ. Microbiol. 71, 1453–1461.

Lopman, B., Vennema, H., Kohli, E., Pothier, P., Sanchez, A., Negredo, A., Buesa, J., Schreier, E., Reacher, M., Brown, D., et al. (2004a). Increase in viral gastroenteritis outbreaks in Europe and epidemic spread of new norovirus variant. Lancet 363, 682–688.

Lopman, B.A., Reacher, M.H., Van Duijnhoven, Y., Hanon, F.X., Brown, D., and Koopmans, M. (2003). Viral gastroenteritis outbreaks in Europe, 1995–2000. Emerg. Infect. Dis. 9, 90–96.

Lopman, B.A., Reacher, M.H., Vipond, I.B., Hill, D., Perry, C., Halladay, T., Brown, D.W., Edmunds, W.J., and Sarangi, J. (2004b). Epidemiology and cost of nosocomial gastroenteritis, Avon, England, 2002–2003. Emerg. Infect. Dis. 10, 1827–1834.

Lopman, B.A., Reacher, M.H., Vipond, I.B., Sarangi, J., and Brown, D.W. (2004c). Clinical manifestation of norovirus gastroenteritis in health care settings. Clin. Infect. Dis. 39, 318–324.

Lugton, I.W. (1999). A cross-sectional study of risk factors affecting the outcome of rabbit haemorrhagic disease virus releases in New South Wales. Aust. Vet. J. 77, 322–328.

Malek, M., Barzilay, E., Kramer, A., Camp, B., Jaykus, L.A., Escudero-Abarca, B., Derrick, G., White, P., Gerba, C., Higgins, C., et al. (2009). Outbreak of norovirus infection among river rafters associated with packaged delicatessen meat, Grand Canyon, 2005. Clin. Infect. Dis. 48, 31–37.

Marks, P.J., Vipond, I.B., Carlisle, D., Deakin, D., Fey, R.E., and Caul, E.O. (2000). Evidence for airborne transmission of Norwalk-like virus (NLV) in a hotel restaurant. Epidemiol. Infect. 124, 481–487.

Marshall, J.A., Yuen, L.K., Catton, M.G., Gunesekere, I.C., Wright, P.J., Bettelheim, K.A., Griffith, J.M., Lightfoot, D., Hogg, G.G., Gregory, J., et al. (2001). Multiple outbreaks of Norwalk-like virus gastro-enteritis associated with a Mediterranean-style restaurant. J. Med. Microbiol. 50, 143–151.

Martella, V., Banyai, K., Lorusso, E., Bellacicco, A.L., Decaro, N., Mari, V., Saif, L., Costantini, V., De Grazia, S., Pezzotti, G., et al. (2008a). Genetic heterogeneity of porcine enteric caliciviruses identified from diarrhoeic piglets. Virus Genes 36, 365–373.

Martella, V., Campolo, M., Lorusso, E., Cavicchio, P., Camero, M., Bellacicco, A.L., Decaro, N., Elia, G., Greco, G., Corrente, M., et al. (2007). Norovirus in captive lion cub (Panthera leo). Emerg. Infect. Dis. 13, 1071–1073.

Martella, V., Lorusso, E., Decaro, N., Elia, G., Radogna, A., D'Abramo, M., Desario, C., Cavalli, A., Corrente, M., Camero, M., et al. (2008b). Detection and molecular characterization of a canine norovirus. Emerg. Infect. Dis. 14, 1306–1308.

Martín-Alonso, J.M., Skilling, D.E., González-Molleda, L., del Barrio, G., Machín, Á., Keefer, N.K., Matson, D.O., Iversen, P.L., Smith, A.W., and Parra, F. (2005). Isolation and characterization of a new Vesivirus from rabbits. Virology 337, 373–383.

Matsuura, Y., Tohya, Y., Nakamura, K., Shimojima, M., Roerink, F., Mochizuki, M., Takase, K., Akashi, H., and Sugimura, T. (2002). Complete nucleotide sequence, genome organization and phylogenic analysis of the canine calicivirus. Virus Genes 25, 67–73.

Mattison, K., Shukla, A., Cook, A., Pollari, F., Friendship, R., Kelton, D., Bidawid, S., and Farber, J.M. (2007). Human noroviruses in swine and cattle. Emerg. Infect. Dis. 13, 1184–1188.

Maunula, L., Kalso, S., Von Bonsdorff, C.H., and Ponka, A. (2004). Wading pool water contaminated with both noroviruses and astroviruses as the source of a gastroenteritis outbreak. Epidemiol. Infect. 132, 737–743.

Maunula, L., Miettinen, I.T., and von Bonsdorff, C.H. (2005). Norovirus outbreaks from drinking water. Emerg. Infect. Dis. 11, 1716–1721.

Mauroy, A., Scipioni, A., Mathijs, E., Miry, C., Ziant, D., Thys, C., and Thiry, E. (2008). Noroviruses and sapoviruses in pigs in Belgium. Arch. Virol. 153, 1927–1931.

McColl, K.A., Merchant, J.C., Hardy, J., Cooke, B.D., Robinson, A., and Westbury, H.A. (2002). Evidence for insect transmission of rabbit haemorrhagic disease virus. Epidemiol. Infect. 129, 655–663.

McEvoy, M., Blake, W., Brown, D., Green, J., and Cartwright, R. (1996). An outbreak of viral gastroenteritis on a cruise ship. Commun. Dis. Rep. CDR Rev. 6, R188–192.

McLeod, C., Hay, B., Grant, C., Greening, G., and Day, D. (2009). Localization of norovirus and poliovirus in Pacific oysters. J. Appl. Microbiol. 106, 1220–1230.

Mead, P.S., Slutsker, L., Dietz, V., McCaig, L.F., Bresee, J.S., Shapiro, C., Griffin, P.M., and Tauxe, R.V. (1999). Food-related illness and death in the United States. Emerg. Infect. Dis. 5, 607–625.

Mochizuki, M., Hashimoto, M., Roerink, F., Tohya, Y., Matsuura, Y., and Sasaki, N. (2002). Molecular and seroepidemiological evidence of canine calicivirus infections in Japan. J. Clin. Microbiol. 40, 2629–2631.

Mochizuki, M., Kawanishi, A., Sakamoto, H., Tashiro, S., Fujimoto, R., and Ohwaki, M. (1993). A calicivirus isolated from a dog with fatal diarrhoea. Vet. Rec. 132, 221–222.

Moreno-Espinosa, S., Farkas, T., and Jiang, X. (2004). Human caliciviruses and pediatric gastroenteritis. Semin. Pediatr. Infect. Dis. 15, 237–245.

Murata, T., Katsushima, N., Mizuta, K., Muraki, Y., Hongo, S., and Matsuzaki, Y. (2007). Prolonged norovirus shedding in infants <or=6 months of age with gastroenteritis. Pediatr. Infect. Dis. J. 26, 46–49.

Murphy, A.M., Grohmann, G.S., Christopher, P.J., Lopez, W.A., Davey, G.R., and Millsom, R.H. (1979). An Australia-wide outbreak of gastroenteritis from oysters caused by Norwalk virus. Med. J. Austral. 2, 329–333.

Myrmel, M., Berg, E.M., Grinde, B., and Rimstad, E. (2006). Enteric viruses in inlet and outlet samples from sewage treatment plants. J. Water Health 4, 197–209.

Myrmel, M., Berg, E.M., Rimstad, E., and Grinde, B. (2004). Detection of enteric viruses in shellfish from the Norwegian coast. Appl. Environ. Microbiol. 70, 2678–2684.

Nakagawa-Okamoto, R., Arita-Nishida, T., Toda, S., Kato, H., Iwata, H., Akiyama, M., Nishio, O., Kimura, H., Noda, M., Takeda, N., et al. (2009). Detection of multiple sapovirus genotypes and genogroups in oyster-associated outbreaks. Jpn J. Infect. Dis. 62, 63–66.

Nappier, S.P., Graczyk, T.K., and Schwab, K.J. (2008). Bioaccumulation, retention, and depuration of enteric viruses by Crassostrea virginica and Crassostrea ariakensis oysters. Appl. Environ. Microbiol. 74, 6825–6831.

Nowotny, N., Bascunana, C.R., Ballagi-Pordany, A., Gavier-Widen, D., Uhlen, M., and Belak, S. (1997). Phylogenetic analysis of rabbit haemorrhagic disease and European brown hare syndrome viruses by comparison of sequences from the capsid protein gene. Arch. Virol. 142, 657–673.

Oliver, S.L., Dastjerdi, A.M., Wong, S., El-Attar, L., Gallimore, C., Brown, D.W., Green, J., and Bridger, J.C. (2003). Molecular characterization of bovine enteric caliciviruses: a distinct third genogroup of noroviruses (Norwalk-like viruses) unlikely to be of risk to humans. J. Virol. 77, 2789–2798.

Oliver, S.L., Wood, E., Asobayire, E., Wathes, D.C., Brickell, J.S., Elschner, M., Otto, P., Lambden, P.R., Clarke, I.N., and Bridger, J.C. (2007). Serotype 1 and 2 bovine noroviruses are endemic in cattle in the United kingdom and Germany. J. Clin. Microbiol. 45, 3050–3052.

Pang, X.L., Lee, B.E., Tyrrell, G.J., and Preiksaitis, J.K. (2008). Epidemiology and Genotype Analysis of Sapovirus Associated with Gastroenteritis Outbreaks in Alberta, Canada: 2004–2007. J. Infect. Dis. 199, 547–551.

Parshionikar, S.U., Willian-True, S., Fout, G.S., Robbins, D.E., Seys, S.A., Cassady, J.D., and Harris, R. (2003). Waterborne outbreak of gastroenteritis associated with a norovirus. Appl. Environ. Microbiol. 69, 5263–5268.

Patel, M.M., Hall, A.J., Vinje, J., and Parashar, U.D. (2009). Noroviruses: a comprehensive review. J. Clin. Virol. 44, 1–8.

Perdue, K.A., Green, K.Y., Copeland, M., Barron, E., Mandel, M., Faucette, L.J., Williams, E.M., Sosnovtsev, S.V., Elkins, W.R., and Ward, J.M. (2007). Naturally occurring murine norovirus infection in a large research institution. J. Am. Assoc. Lab. Anim. Sci. 46, 39–45.

Podewils, L.J., Zanardi Blevins, L., Hagenbuch, M., Itani, D., Burns, A., Otto, C., Blanton, L., Adams, S., Monroe, S.S., Beach, M.J., et al. (2007). Outbreak of norovirus illness associated with a swimming pool. Epidemiol. Infect. 135, 827–833.

Ponka, A., Maunula, L., von Bonsdorff, C.H., and Lyytikainen, O. (1999). An outbreak of calicivirus associated with consumption of frozen raspberries. Epidemiol. Infect. 123, 469–474.

Powell, K.L., Taylor, R.G., Cronin, A.A., Barrett, M.H., Pedley, S., Sellwood, J., Trowsdale, S.A., and Lerner, D.N. (2003). Microbial contamination of two urban sandstone aquifers in the UK. Water Res. 37, 339–352.

Pratelli, A., Greco, G., Camero, M., Corrente, M., Normanno, G., and Buonavoglia, C. (2000). Isolation and identification of a calicivirus from a dog with diarrhoea. New Microbiol. 23, 257–260.

Prato, R., Lopalco, P.L., Chironna, M., Barbuti, G., Germinario, C., and Quarto, M. (2004). Norovirus gastroenteritis general outbreak associated with raw shellfish consumption in south Italy. BMC Infect. Dis. 4, 37.

Pusch, D., Oh, D.Y., Wolf, S., Dumke, R., Schroter-Bobsin, U., Hohne, M., Roske, I., and Schreier, E. (2005). Detection of enteric viruses and bacterial indicators in German environmental waters. Arch. Virol. 150, 929–947.

Radford, A.D., Coyne, K.P., Dawson, S., Porter, C.J., and Gaskell, R.M. (2007). Feline calicivirus. Vet. Res. 38, 319–335.

Radford, A.D., Gaskell, R.M., and Hart, C.A. (2004). Human norovirus infection and the lessons from animal caliciviruses. Curr. Opin. Infect. Dis. 17, 471–478.

Reuter, G., Biro, H., and Szucs, G. (2007). Enteric caliciviruses in domestic pigs in Hungary. Arch. Virol. 152, 611–614.

Reuter, G., Vennema, H., Koopmans, M., and Szucs, G. (2006). Epidemic spread of recombinant noroviruses with four capsid types in Hungary. J. Clin. Virol. 35, 84–88.

Rockx, B., De Wit, M., Vennema, H., Vinje, J., De Bruin, E., Van Duynhoven, Y., and Koopmans, M. (2002). Natural history of human calicivirus infection: a prospective cohort study. Clin. Infect. Dis. 35, 246–253.

Roda de Husman, A.M., Bijkerk, P., Lodder, W., Van Den Berg, H., Pribil, W., Cabaj, A., Gehringer, P., Sommer, R., and Duizer, E. (2004). Calicivirus inactivation by nonionizing (253.7-nanometer-wavelength [UV]) and ionizing (gamma) radiation. Appl. Environ. Microbiol. 70, 5089–5093.

Roda de Husman, A.M., Lodder-Verschoor, F., van den Berg, H.H., Le Guyader, F.S., van Pelt, H., van der Poel, W.H., and Rutjes, S.A. (2007). Rapid virus detection procedure for molecular tracing of shellfish associated with disease outbreaks. J. Food Prot. 70, 967–974.

Rzezutka, A., and Cook, N. (2004). Survival of human enteric viruses in the environment and food. FEMS Microbiol. Rev. 28, 441–453.

Schaffer, F.L., Soergel, M.E., Black, J.W., Skilling, D.E., Smith, A.W., and Cubitt, W.D. (1985). Characterization of a new calicivirus isolated from feces of a dog. Arch. Virol. 84, 181–195.

Schmid, D., Stuger, H.P., Lederer, I., Pichler, A.M., Kainz-Arnfelser, G., Schreier, E., and Allerberger, F. (2007). A foodborne norovirus outbreak due to manually prepared salad, Austria 2006. Infection 35, 232–239.

Scipioni, A., Mauroy, A., Vinje, J., and Thiry, E. (2008). Animal noroviruses. Vet. J. 178, 32–45.

Siebenga, J.J., Vennema, H., Duizer, E., and Koopmans, M.P. (2007). Gastroenteritis caused by norovirus GGII.4, The Netherlands, 1994–2005. Emerg. Infect. Dis. 13, 144–146.

Simmons, G., Greening, G., Gao, W., and Campbell, D. (2001). Raw oyster consumption and outbreaks of viral gastroenteritis in New Zealand: evidence for risk to the public's health. Aust. N.Z. J. Publ. Health *25*, 234–240.

Smiley, J.R., Hoet, A.E., Traven, M., Tsunemitsu, H., and Saif, L.J. (2003). Reverse transcription-PCR assays for detection of bovine enteric caliciviruses (BEC) and analysis of the genetic relationships among BEC and human caliciviruses. J. Clin. Microbiol. *41*, 3089–3099.

Smith, A.W., Akers, T.G., Madin, S.H., and Vedros, N.A. (1973). San Miguel sea lion virus isolation, preliminary characterization and relationship to vesicular exanthema of swine virus. Nature *244*, 108–110.

Smith, A.W., Akers, T.G., Prato, C.M., and Bray, H. (1976). Prevalence and distribution of four serotypes of SMSV serum neutralizing antibodies in wild animal populations. J. Wildl. Dis. *12*, 326–334.

Smith, A.W., Berry, E.S., Skilling, D.E., Barlough, J.E., Poet, S.E., Berke, T., Mead, J., and Matson, D.O. (1998a). In vitro isolation and characterization of a calicivirus causing a vesicular disease of the hands and feet. Clin. Infect. Dis. *26*, 434–439.

Smith, A.W., Iversen, P.L., Skilling, D.E., Stein, D.A., Bok, K., and Matson, D.O. (2006). Vesivirus viremia and seroprevalence in humans. J. Med. Virol. *78*, 693–701.

Smith, A.W., Skilling, D.E., Castello, J.D., and Rogers, S.O. (2004). Ice as a reservoir for pathogenic human viruses: specifically, caliciviruses, influenza viruses, and enteroviruses. Med. Hypotheses *63*, 560–566.

Smith, A.W., Skilling, D.E., Cherry, N., Mead, J.H., and Matson, D.O. (1998b). Calicivirus emergence from ocean reservoirs: zoonotic and interspecies movements. Emerg. Infect. Dis. *4*, 13–20.

Smith, A.W., Skilling, D.E., Matson, D.O., Kroeker, A.D., Stein, D.A., Berke, T., and Iversen, P.L. (2002). Detection of vesicular exanthema of swine-like calicivirus in tissues from a naturally infected spontaneously aborted bovine fetus. J. Am. Vet. Med. Assoc. *220*, 455–458.

Straub, T.M., Pepper, I.L., and Gerba, C.P. (1993). Virus Survival in Sewage Sludge Amended Desert Soil. Water Science and Technology WSTED4, *27*, No. 3/4.

Sugieda, M., Nagaoka, H., Kakishima, Y., Ohshita, T., Nakamura, S., and Nakajima, S. (1998). Detection of Norwalk-like virus genes in the caecum contents of pigs. Arch. Virol. *143*, 1215–1221.

Svraka, S., Duizer, E., Egberink, H., Dekkers, J., Vennema, H., and Koopmans, M. (2009). A new generic real-time reverse transcription polymerase chain reaction assay for vesiviruses; vesiviruses were not detected in human samples. J. Virol. Methods *157*, 1–7.

Symes, S.J., Gunesekere, I.C., Marshall, J.A., and Wright, P.J. (2007). Norovirus mixed infection in an oyster-associated outbreak: an opportunity for recombination. Arch. Virol. *152*, 1075–1086.

Teunis, P.F., Moe, C.L., Liu, P., Miller, S. E. , Lindesmith, L., Baric, R.S., Le Pendu, J., and Calderon, R.L. (2008). Norwalk virus: How infectious is it? J. Med. Virol. *80*, 1468–1476.

Tian, P., Bates, A.H., Jensen, H.M., and Mandrell, R.E. (2006). Norovirus binds to blood group A-like antigens in oyster gastrointestinal cells. Lett. Appl. Microbiol. *43*, 645–651.

Tu, E.T., Bull, R.A., Kim, M.J., McIver, C.J., Heron, L., Rawlinson, W.D., and White, P.A. (2008). Norovirus excretion in an aged-care setting. J. Clin. Microbiol. *46*, 2119–2121.

Tu, E.T., Nguyen, T., Lee, P., Bull, R.A., Musto, J., Hansman, G., White, P.A., Rawlinson, W.D., and McIver, C.J. (2007). Norovirus GII.4 strains and outbreaks, Australia. Emerg. Infect. Dis. *13*, 1128–1130.

Turcios, R.M., Widdowson, M.A., Sulka, A.C., Mead, P.S., and Glass, R.I. (2006). Reevaluation of epidemiological criteria for identifying outbreaks of acute gastroenteritis due to norovirus: United States, 1998–2000. Clin. Infect. Dis. *42*, 964–969.

van den Berg, H., Lodder, W., van der Poel, W., Vennema, H., and de Roda Husman, A.M. (2005). Genetic diversity of noroviruses in raw and treated sewage water. Res. Microbiol. *156*, 532–540.

van der Poel, W.H., van der Heide, R., Verschoor, F., Gelderblom, H., Vinje, J., and Koopmans, M.P. (2003). Epidemiology of Norwalk-like virus infections in cattle in The Netherlands. Vet. Microbiol. *92*, 297–309.

van Der Poel, W.H., Vinje, J., van Der Heide, R., Herrera, M.I., Vivo, A., and Koopmans, M.P. (2000). Norwalk-like calicivirus genes in farm animals. Emerg. Infect. Dis. *6*, 36–41.

Verhoef, L., Depoortere, E., Boxman, I., Duizer, E., van Duynhoven, Y., Harris, J., Johnsen, C., Kroneman, A., Le Guyader, S., Lim, W., *et al.* (2008). Emergence of new norovirus variants on spring cruise ships and prediction of winter epidemics. Emerg. Infect. Dis. *14*, 238–243.

Verhoef, L.P., Kroneman, A., van Duynhoven, Y., Boshuizen, H., van Pelt, W., and Koopmans, M. (2009). Selection tool for foodborne norovirus outbreaks. Emerg. Infect. Dis. *15*, 31–38.

Wang, D., Wu, Q., Yao, L., Wei, M., Kou, X., and Zhang, J. (2008). New target tissue for food-borne virus detection in oysters. Lett. Appl. Microbiol. *47*, 405–409.

Wang, Q.H., Costantini, V., and Saif, L.J. (2007). Porcine enteric caliciviruses: genetic and antigenic relatedness to human caliciviruses, diagnosis and epidemiology. Vaccine *25*, 5453–5466.

Wang, Q.H., Han, M.G., Cheetham, S., Souza, M., Funk, J.A., and Saif, L.J. (2005a). Porcine noroviruses related to human noroviruses. Emerg. Infect. Dis. *11*, 1874–1881.

Wang, Q.H., Han, M.G., Funk, J.A., Bowman, G., Janies, D.A., and Saif, L.J. (2005b). Genetic diversity and recombination of porcine sapoviruses. J. Clin. Microbiol. *43*, 5963–5972.

Wang, Q.H., Souza, M., Funk, J.A., Zhang, W., and Saif, L.J. (2006). Prevalence of noroviruses and sapoviruses in swine of various ages determined by reverse transcription-PCR and microwell hybridization assays. J. Clin. Microbiol. *44*, 2057–2062.

Wheeler, M. (2005). Ethanol and HCV-induced cytotoxicity: the perfect storm. Gastroenterology *128*, 232–234.

Widdowson, M.A., Cramer, E.H., Hadley, L., Bresee, J.S., Beard, R.S., Bulens, S.N., Charles, M., Chege, W., Isakbaeva, E., Wright, J.G., *et al.* (2004). Outbreaks of acute gastroenteritis on cruise ships and on land: identification of a predominant circulating strain of norovirus – United States, 2002. J. Infect. Dis. *190*, 27–36.

Widdowson, M.A., Glass, R., Monroe, S., Beard, R.S., Bateman, J.W., Lurie, P., and Johnson, C. (2005a). Probable transmission of norovirus on an airplane. JAMA*293*, 1859–1860.

Widdowson, M.A., Rockx, B., Schepp, R., van der Poel, W.H., Vinje, J., van Duynhoven, Y.T., and Koopmans, M.P. (2005b). Detection of serum antibodies to bovine norovirus in veterinarians and the general population in the Netherlands. J. Med. Virol. *76*, 119–128.

Wirblich, C., Meyers, G., Ohlinger, V.F., Capucci, L., Eskens, U., Haas, B., and Thiel, H.J. (1994). European brown hare syndrome virus: relationship to rabbit hemorrhagic disease virus and other caliciviruses. J. Virol. *68*, 5164–5173.

Wolf, S., Williamson, W., Hewitt, J., Lin, S., Rivera-Aban, M., Ball, A., Scholes, P., Savill, M., and Greening, G.E. (2009). Molecular detection of norovirus in sheep and pigs in New Zealand farms. Vet. Microbiol. *133*, 184–189.

Wolf, S., Williamson, W.M., Hewitt, J., Rivera-Aban, M., Lin, S., Ball, A., Scholes, P., and Greening, G.E. (2007). Sensitive multiplex real-time reverse transcription-PCR assay for the detection of human and animal noroviruses in clinical and environmental samples. Appl. Environ. Microbiol. *73*, 5464–5470.

Worobey, M., and Holmes, E.C. (1999). Evolutionary aspects of recombination in RNA viruses. J. Gen. Virol. *80*, 2535–2543.

Wu, H.M., Fornek, M., Schwab, K.J., Chapin, A.R., Gibson, K., Schwab, E., Spencer, C., and Henning, K. (2005). A norovirus outbreak at a long-term-care facility: the role of environmental surface contamination. Infect. Control Hosp. Epidemiol. *26*, 802–810.

Wyn-Jones, A.P., Pallin, R., Dedoussis, C., Shore, J., and Sellwood, J. (2000). The detection of small round-structured viruses in water and environmental materials. J. Virol. Methods *87*, 99–107.

Yang, L., Wang, F., Hu, B., Xue, J., Hu, Y., Zhou, B., Wang, D., and Xu, W. (2008). Development of an RT-PCR for rabbit hemorrhagic disease virus (RHDV) and the epidemiology of RHDV in three eastern provinces of China. J. Virol. Methods *151*, 24–29.

Genome Organization and Recombination

3

Rowena Bull and Peter A. White

Abstract

Recombination was first described in the human caliciviruses in 1997 (Hardy et al., 1997). Since then naturally occurring recombinants have been detected for all four genera of the family Caliciviridae and has become an important mechanism in the emergence of new calicivirus variants. Owing to similarities in genome organization between the different genera, recombination predominantly occurs at the start of the major structural gene which encodes the capsid, VP1. Knowledge of the mechanisms of calicivirus recombination is important as new variants can emerge, with potentially different pathogenesis and virulence.

Introduction

Improved molecular techniques have lead to the characterization of full-length genomes for all four genera of the family Caliciviridae: Norovirus, Sapovirus, Lagovirus and Vesivirus. Despite the high sequence diversity between the caliciviruses their genomes share common traits and a common mechanism of evolution, recombination. This chapter examines primarily recombination and its importance in calicivirus evolution and also reviews the similarities in genomic structures of the caliciviruses.

Calicivirus genome organization

At present the positive-sense, polyadenylated, single-stranded RNA virus family Caliciviridae contains four genera, Norovirus, Sapovirus, Lagovirus and Vesivirus. Evidence supporting the creation of three additional genera in the family has been

reported (Farkas et al., 2008; L'Homme et al., 2009; Oliver et al., 2006a), but these have not yet been accepted by the International Committee on Taxonomy of Viruses (ICTV). All members of the Caliciviridae have a short nontranslated region (NTR) that starts with GU at the 5′ end of their genomes. The 5′ NTR varies between 4 and 75 nucleotides depending on the genus and part of which is generally repeated internally within the genome, near to the transcription start site of the subgenomic RNA (sgRNA) species (Farkas et al., 2008) (Table 3.1). The genome is covalently linked to virus protein genome (VPg) at the 5′ end and polyadenylated (pA) at the 3′ end.

The calicivirus genome is divided into two to four open reading frames (ORFs) depending on the genus (Table 3.1) (Fig. 3.1). Irrespective of the genus, the first ORF (ORF1), located at the 5′ end of the genome, encodes for the mature non-structural proteins. ORF2, in norovirus and vesivirus, encodes for the major structural protein, VP1 [commonly referred to as VP60 in rabbit haemorrhagic disease virus (RHDV)], and ORF3 encodes for a protein that has been identified as a minor structural protein in feline calicivirus (FCV), sapovirus and norovirus (commonly referred to as VP2 in norovirus, sapovirus and vesivirus but as VP10 in lagovirus).

The sapovirus and lagovirus genomes are organized in a slightly different way from those of noroviruses or vesiviruses, with the VP1 gene in frame with the non-structural genes and located near the 3′ end of ORF1 (Fig. 3.1). ORF2 in sapoviruses and lagoviruses is thought to encode a protein similar to the VP2 protein encoded by

Table 3.1 Calicivirus genome organization

Genus	Reference strain	Genogroup	GenBank accession no.	No. of ORFs	Proteins encoded				Genome length (bp)	NTR length	
					ORF1	ORF2	ORF3	ORF4		5'	3'
Norovirus	Norwalk	GI	M87661	3	non-structural	VP1	VP2	N/A	7654	4	66
	MNV-1[a]	GV	DQ285629	4	non-structural	VP1	VP2	unknown	7382	5	75
Sapovirus	Mc10	GII-III	NC_010624	2	non-structural, VP1	VP2	N/A	N/A	7458	14	108
	Mc114[a]	GI, GIV-V	AY237422	3	non-structural, VP1	VP2	unknown	N/A	7429	13	80
Vesivirus	FCV	N/A	NC_001481	3	non-structural	VP1	VP2	N/A	7683	19	46
Lagovirus	RHDV	N/A	AB300693	2	non-structural, VP1	VP2	N/A	N/A	7437	9	59

a An additional reference strain is included for *Norovirus* and *Sapovirus* as some genogroups from each have an extra ORF.

b N/A, not applicable, indicates the absence of that ORF for that genogroup.

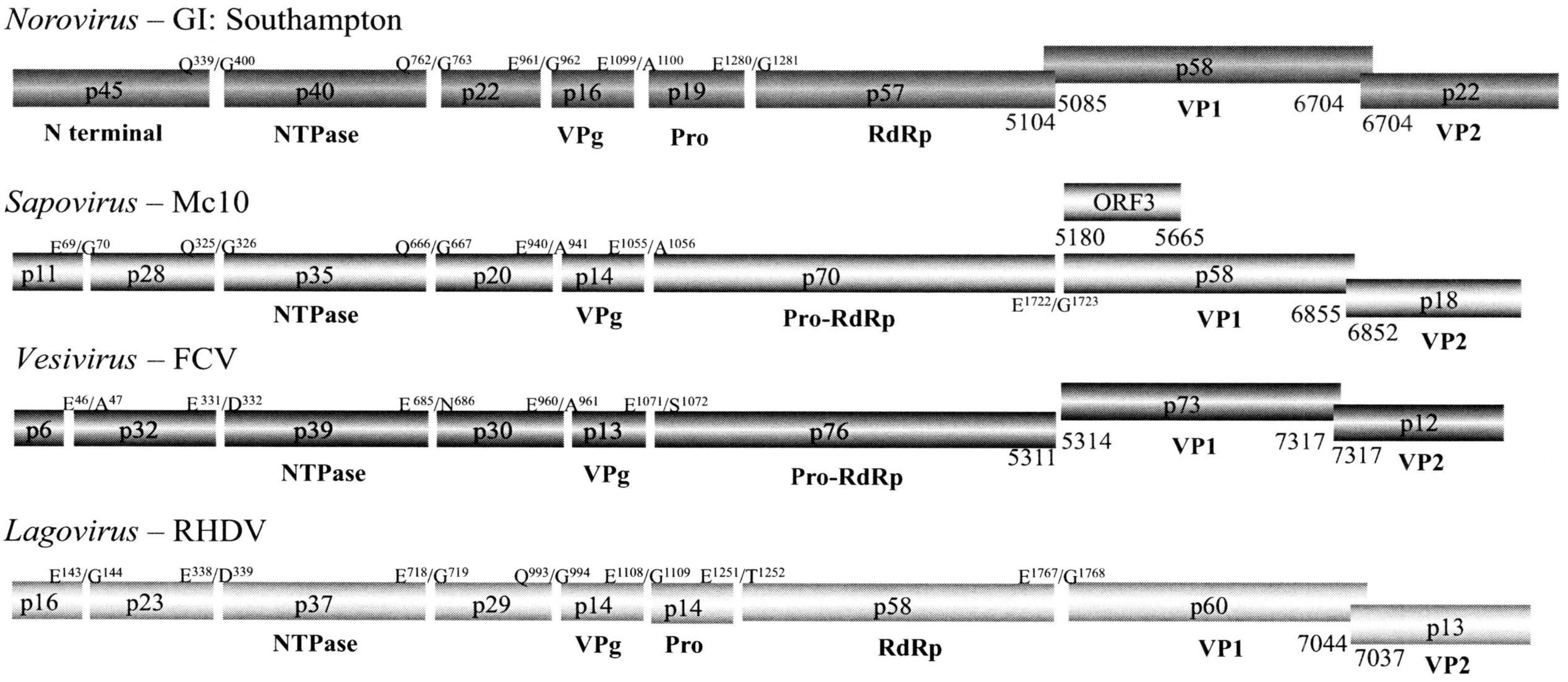

Figure 3.1 Genomic organization of the four genera of the family *Caliciviridae*. The genomic organization of the four major genera in the family *Caliciviridae* is diverse with between two to four ORFs depending on the genus. The expected viral specific protease cleavage sites within ORF1 are indicated and the function of each protein is listed for a representative of each genus [Southampton, L07418 (Liu *et al.*, 1999); Mc10, AY237420 (Oka *et al.*, 2006); FCV, D31836 (Oka *et al.*, 2007); RHDV, M67473 (Meyers *et al.*, 2000)]. The structural protein VP1 is either encoded for at the 3′ end of ORF1 or encoded for by ORF2, depending on the genus. The structural protein VP2 is translated from its own ORF in all the genera. In most genera, ORF1 overlaps ORF2 and ORF2 overlaps ORF3. For some sapovirus and norovirus strains (not shown) an additional ORF is found within the 5′ end of the VP1 coding sequence.

ORF3 in noroviruses and vesiviruses (Glass *et al.*, 2000; Sosnovtsev and Green, 2000).

In addition to the ORFs described above, some caliciviruses, i.e. sapovirus genogroup II and III (GII and GIII) and norovirus GV, have additional ORFs, ORF3 and ORF4 respectively. The additional ORFs are encoded from a separate reading frame located within the 5′ end of the major structural protein coding sequence (Thackray *et al.*, 2007; Hansman *et al.*, 2005; Katayama *et al.*, 2004) (Fig. 3.1). The functions of the post-translational product from these additional ORFs remain unknown.

A characteristic feature of the caliciviruses is that certain ORFs may overlap. In many caliciviruses, the 3′ end of ORF1 overlaps with the 5′ end of ORF2. The length of the overlap (1–20 nt) varies among strains. Other caliciviruses contain spacer sequences between ORF1 and ORF2. For example, canine caliciviruses have a 3-nt spacer region as opposed to an overlap (Martella *et al.*, 2008). The 5′ end of the small terminal ORF (present in all caliciviruses) characteristically overlaps with the 3′ end of the capsid encoding sequence.

Proteolytic processing of the viral polypeptide encoded by ORF1 is rapid, co-translational and leads to a number of intermediate proteins before further cleavage to produce at least six mature proteins. The mature proteins have been designated [from the amino (N) terminus to the carboxy (C) terminus] N-terminal protein (45 kDa), NTPase (40 kDa), a picornavirus 3A-like protein (22 kDa), VPg (16 kDa), proteinase (Pro) (20 kDa) and RNA-dependent RNA polymerase (RdRp) (57 kDa) (Blakeney *et al.*, 2003; Green *et al.*, 2001) (Fig. 3.1). A nomenclature system has been proposed for the non-structural proteins, NS1 through NS7 (N to C termini) (Sosnovtsev *et al.*, 2006). Proteolytic processing of ORF1 is mediated by the virus Pro, a '3C-like' cysteine proteinase (Belliot *et al.*, 2003; Boniotti *et al.*, 1994; Joubert *et al.*, 2000; Liu *et al.*, 1996, 1999; Oka *et al.*, 2005a,b; Someya *et al.*, 2002).

Initially, the calicivirus non-structural proteins were identified through amino acid sequence motif similarity to characterized proteins from the family *Picornaviridae* and included a '2C-like' helicase, a '3C-like' proteinase and a '3D-like' RdRp (Neill, 1990). Structure and function studies have confirmed the identities of the calicivirus Pro and RdRp (Belliot *et al.*, 2005; Ng *et al.*, 2002; Ng *et al.*, 2004; Oka *et al.*, 2007). The identities and functions of the individual non-structural proteins are discussed further in Chapter four.

RNA recombination

The high genetic variability found in RNA viruses is attributed to three major evolutionary forces: mutation, reassortment (for segmented viral genomes) and recombination (Roossinck, 1997). RNA-RNA recombination occurs less frequently than other methods of mutation like nucleotide substitution and reassortment (Lai, 1992). However, an increasing number of RNA viruses have been shown to undergo recombination (Aaziz and Tepfer, 1999). Furthermore, many emergent viruses have been shown to belong to families that are reportedly able to undergo recombination (Morse, 1994). A recombinant virus can be defined as one virus with genetic information from two or more separate sources. In caliciviruses, the genome of a recombinant virus clusters into two distinct phylogenetic groups when different regions (normally the structural and non-structural genes) of the genome are subjected to phylogenetic analysis. Recombination in viruses has the potential to increase the virulence of the strain and has major implications in viral vaccine design. In HIV recombinant viruses now account for greater than 20% of all viruses demonstrating the selective advantages that can be achieved by mixing their genomes with other viruses (Osmanov *et al.*, 2002).

Furthermore, it has been proposed that recombination is an important mechanism for viral survival as it enables viruses to replace deleterious mutations that would otherwise result in defective proteins (Muller, 1964). Viruses have evolved different mechanisms of recombination and these have been classified into three categories: (i) Homologous (legitimate) recombination: involves crossover events at precision matched sites between two related RNA molecules; (ii) Aberrant homologous recombination: involves RNA molecules with similar sequences but with crossover events not at corresponding sites, thereby leading to sequence duplication or deletions; (iii) Non-homologous (illegitimate) recombination: occurs between two unrelated RNA molecules at non-corresponding sites with no requirements for sequence homology (reviewed in Lai, 1992). To date only the first type of recombination, homologous recombination, has been identified in the caliciviruses.

Recombination methodology

The large genomic diversity of the noroviruses, the lack of a prototype for each recombinant type, and the use of different methods to define recombinants has confused their classification and resulted in the duplication of the reporting of norovirus recombinant types (Bull *et al.*, 2005). Many methods have been used for detecting recombinants with the following used commonly in the detection of calicivirus recombinants; phylogenetics, Simplot (Lole *et al.*, 1999) and maximum chi-squared analysis (Smith, 1992). Simplot analysis and the maximum chi-squared method are useful for identifying the breakpoint, but require sequence data for the parental strains and this is not always available. Phylogenetic analysis is useful when a parental strain cannot be identified for one of the regions being examined. However, phylogenetic analysis may not be solely relied upon to identify recombinants as the norovirus non-structural region (ORF1) does not cluster as definitively as the norovirus structural (ORF2) region (Vinje *et al.*, 2004). Due to the limitations of each method, analysis of recombinants should include a combination of different approaches. Online packages are available that contain a compilation of different recombination analysis models (Maydt and Lengauer, 2006).

Norovirus recombination

Naturally occurring recombination in norovirus was first described in 1997 when the nucleotide sequence of the capsid region of a norovirus GII prototype, Snow Mountain virus, was shown to have 94% identity to the capsid region of another norovirus GII prototype, Melksham virus, but only 79% identity in the RdRp region (Hardy *et al.*, 1997). Further phylogenetic analysis confirmed that Snow Mountain virus was a naturally occurring recombinant (Hardy *et al.*, 1997). Since then increased awareness and identification of multiple naturally occurring norovirus

recombinant viruses across the globe has lead to a rise in their reports (Ambert-Balay *et al.*, 2005; Bon *et al.*, 2005; Bull *et al.*, 2005; Etherington *et al.*, 2006; Gallimore *et al.*, 2004a,b, 2005; Han *et al.*, 2004; Hansman *et al.*, 2004a,b; Iritani *et al.*, 2003; Katayama *et al.*, 2002; Martinez *et al.*, 2002; Oliver *et al.*, 2006b; Phan *et al.*, 2006a,c; Reuter *et al.*, 2005, 2006; Sasaki *et al.*, 2006; Tsugawa *et al.*, 2006; van den Berg *et al.*, 2005; Wang *et al.*, 2005). To date at least 22 unique naturally occurring norovirus recombinant types have been identified; one in GI, 16 in GII, two in GIII, two in GV and one intergenogroup recombinant, GI/GII. As yet no recombinants have been identified in norovirus GIV (Table 3.2). The high number of norovirus recombinants suggests that it is an important mechanism in norovirus evolution.

Norovirus GI recombinant

The GI recombinant, WUGI/01/JP (GenBank accession number AB081723), was isolated in 2001 and has a GI genotype 2 (GI.2) RdRp region and a GI.6 capsid (Katayama *et al.*, 2002) (Table 3.2). The breakpoint lies between −11 and +1 nucleotides from the start of ORF2 overlap. WUGI/01/JP is the prototype for this recombinant type. Six other WUGI/01/JP-like strains have been identified in GenBank and were isolated in Japan and the US between 2000 and 2004 suggesting that this recombinant type virus is prevalent (Bull *et al.*, 2007) (Fig. 3.2).

Norovirus GII recombinants

Over the past five years there has been a significant increase in the number of norovirus GII recombinants detected. To date 16 norovirus GII recombinant types have been confirmed (Table 3.2) (Fig. 3.3). The 16 prototype GII recombinants are a combination of one of eight different RdRp genotypes and one of nine different capsid genotypes. Five of the 16 GII recombinants belong to GII.4 in the RdRp region and four belong to the norovirus GII.b RdRp cluster (Buesa *et al.*, 2002). The remaining seven RdRps belong to GII.2, GII.5, GII.6 and three novel genotypes, GII.a, GII.c, and GII.d (Bull *et al.*, 2007). The capsid genotypes associated with recombinant GII were more diverse with one GII.1, three GII.2, three GII.3, two GII.4, two GII.5, one GII.7, one GII.10, one GII.12 and two GII.15. The

Table 3.2 Viruses representing 19 norovirus recombinant types

Norovirus genogroup	Prototype strain	RdRp genotype	Capsid genotype	Average breakpoint[b]	First published
I	WUGI/01/JP	GI.2	GI.6	5	Katayama et al. (2002)
II	Picton/03/AU	GII.b	GII.1	−26	Bull et al. (2005)
	Snow Mountain/76/US	GII.c	GII.2	−55	Hardy et al. (1997)
	E3/97/Crete	GII.4	GII.2	−25	Bull et al. (2005)
	Pont de Roide 673/04/Fr	GII.b	GII.2	−21	Bon et al. (2005)
	SydneyC14/02/AU	GII.b	GII.3	+23	Buesa et al. (2002), Lole et al. (1999)
	Sydney2212/98/AU	GII.a	GII.3	−59	Jiang et al. (1999)
	Chiba1/04/JP	GII.4	GII.3	−7	Vidal et al. (2006)
	771/05/IRL	GII.4/GII.d	GII.4	−595 − 212	Waters et al. (2007)
	Nyiregyhaza/1057/02/HUN[a]	GII.b	GII.4		Gallimore et al. (2004)
	S63/99/Fr	GII.2	GII.5	+35	Bull et al. (2005)
	Hokkaido133/03/JP	GII.d	GII.5	−18	Bull et al. (2007)
	Kunming/04/CH	GII.6	GII.7	−54	Phan et al. (2006b)
	Mc37/01/Th	GII.4	GII.10	+4	Hansman et al. (2004)
	SaitamaU1/02/JP	GII.4	GII.12	−38	Katayama et al. (2002)
	Minato14/99/JP	GII.6	GII.15	−53	Sasaki et al. (2006)
	VannesL23/99/US	GII.5	GII.15	−31	Bull et al. (2005)
III	B-1SVD/03/US	GIII.2	GIII.1	−20	Bull et al. (2007)
	CV521/02/US	GIII.1	GIII.2	−10	Han et al. (2004)
V	S28/06/DE	ND[c]	ND	+17	Muller et al. (2007)
	MNV-4/05/US	ND	ND	+55	Muller et al. (2007)
I/II	L8775/06/ID	GI.3	GII.4	+22	Nayak et al. (2008)

a Sequence data for the RdRp and capsid were not available in the database for this strain.

b Average breakpoint as determined by two methods: Simplot and maximum chi-squared method (Bull et al., 2007). Numbering is in relation to the start of ORF2.

c ND, genotype not defined.

breakpoint detected for all 16 GII recombinants ranged between −594 and +38 nucleotides from the start of ORF2. Interestingly, half of the GII recombinant types identified had either a GII.4 or GII.b RdRp. The GII.b RdRp was associated with four genotypically different capsids; GII.1, GII.3, GII.4 and GII.14, indicating four separate recombination events for viruses with a GII.b RdRp. The three GII.b recombinant type RdRps, associated with GII.1, GII.3 or GII.14 capsid, share greater than 95% identity across 420 nucleotides of the 3′ end of the RdRp region. The GII.4 RdRp was associated with five genotypically different capsids: GII.2, GII.3, GII.10, GII.12 and GII.14. The recombinant GII.4 RdRps clustered independently of the GII.4 RdRps that have been

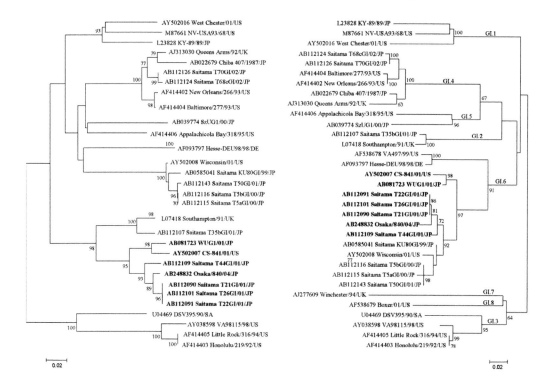

Figure 3.2 Phylogenetic analysis of norovirus GI recombinants. Phylogenetic analysis of the nucleotide sequences of RdRp and capsid regions from 30 norovirus GI strains. The left tree is obtained from a 277 bp region of the 3' end of the RdRp region. The right tree corresponds to 295 bp of the 5' end of the capsid sequence. Norovirus GI recombinants are represented in boldface. The percentage bootstrap values in which the major groupings were observed among 100 replicates are indicated. The branch lengths are proportional to the evolutionary distance between sequences and the distance scale, in nucleotide substitutions per position, is shown. The capsid clustering is shown in bold and is based on the classification of Zheng et al. (2006).

associated with four pandemics in the last decade, Farmington Hills, US95/95 (including Burwash Landing and Miami Beach 326), Hunter and 2006a (Bull et al., 2006; Noel et al., 1999; Tu et al., 2008; Widdowson et al., 2004). The GII.4 and GII.b RdRps are the two most prevalent RdRps circulating across the globe (Bull et al., 2006; Gallimore et al., 2004a; Phan et al., 2006b; Reuter et al., 2005). Therefore, the probability of finding a GII.4 or GII.b recombinant is higher as recombination requires coinfection with two norovirus strains. However, this is also true of the capsid and despite GII.4 being the predominant capsid genotype worldwide, there was only one recombinant type with a GII.4 capsid (not including recombinant 771/05/IRL). The lack of multiple RdRp genotypes associated with GII.4 capsids and the fact that the GII.b and GII.4 RdRps are

each associated with four and five different capsid types, respectively, suggests that the RdRp may be a driving factor in recombination.

Norovirus GIII recombinants

Two norovirus GIII recombinant types, GIII.1/ GIII.2 and GIII.2/GIII.1, have been identified in bovine in the US and UK (Bull et al., 2007; Han et al., 2004; Oliver et al., 2004) (Table 3.2) (Fig. 3.4). The recombinant type, GIII.1/GIII.2, includes both previously published strains, Thirsk10/00/UK and CV521-OH/02/US (Han et al., 2004; Oliver et al., 2004). Interestingly, Thirsk10/00/UK and CV521-OH/02/US were not initially identified as being the same recombinant type, probably due to low sequence homology (84% nucleotide identity over ORF1 and 2). The low sequence homology suggests

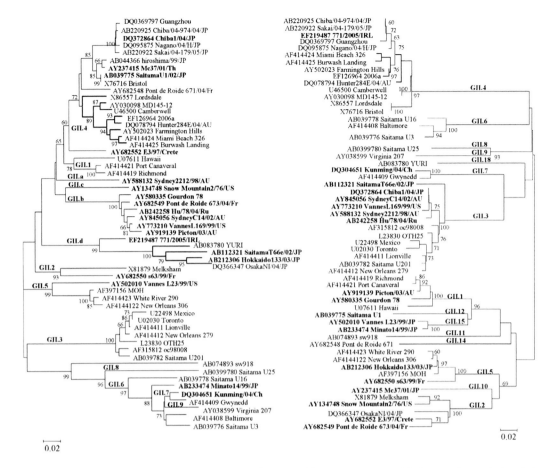

Figure 3.3 Phylogenetic analysis of norovirus GII recombinant sequences. Phylogenetic analysis of the nucleotide sequences of RdRp and capsid regions of 16 identified unique recombinant norovirus GII types. The left tree is based on a 420 bp region of the 3′ end of the RdRp region. The right tree is based on 550 bp of the 5′ end of the capsid sequence. Norovirus GII recombinants are represented in boldface. The percentage bootstrap values in which the major groupings were observed among 100 replicates are indicated. The branch lengths are proportional to the evolutionary distance between sequences and the distance scale, in nucleotide substitutions per position, is shown. The capsid clustering is shown in bold and is based on the classification of Zheng et al. (2006).

that recombination may have occurred some time ago and both isolates have since diverged. The breakpoint detected ranged between −7 and −27 nucleotides downstream of the start of ORF2 for all three GIII recombinants.

Norovirus GV recombinants

Two norovirus GV recombinant types have been identified in murine norovirus (Muller *et al.*, 2007) (Table 3.2). However, further subtyping of these recombinants is not possible as a genotype classification system has not yet been determined for norovirus GV. The breakpoint for these two

norovirus GV recombinants were +17 and +55 nucleotides upstream of the start of ORF2.

Norovirus intergenogroup recombinants

A norovirus intergenogroup recombinant has also been reported. Recombination occurred between a GI.3 virus and a GII.4 virus and the recombination breakpoint was identified as within the ORF1/ORF2 overlap (Nayak *et al.*, 2008). So far naturally occurring intergenogroup recombinants have only been identified in norovirus and sapovirus (see below) and only been between genogroups that infect humans. An incompatibility

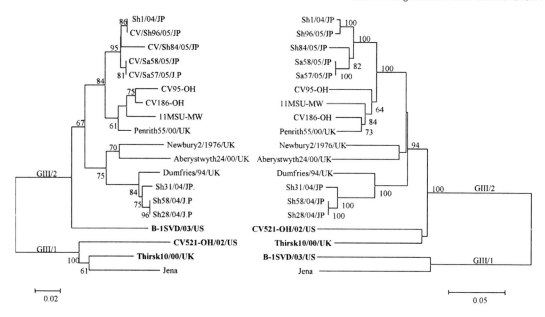

Figure 3.4 Phylogenetic analysis of norovirus GIII recombinants. Phylogenetic analysis of the nucleotide sequences of the RdRp and capsid regions illustrating two norovirus GIII recombinant types in relation to 19 known norovirus GIII strains and prototype strains. The left tree corresponds to a 210 bp region of the 3' end of the RdRp region. The right tree is based on 898 bp of the 5' end of the capsid sequence. Norovirus GIII recombinants are represented in boldface. The percentage bootstrap values in which the major groupings were observed among 100 replicates are indicated. The branch lengths are proportional to the evolutionary distance between sequences and the distance scale, in nucleotide substitutions per position, is shown. The capsid clustering is shown in bold and is based on the classification of Zheng et al. (2006).

between heterogeneous viral proteins may explain the rarity of homologous recombination between different viral species. However, norovirus has been detected in a wide range of mammals, including humans, mice, cows and pigs, with strong evidence of zoonotic transmission (Widdowson et al., 2005). Therefore understanding recombination in norovirus is important as recombination between a mammalian norovirus and a human norovirus may result in the emergence of new norovirus variants, with potentially different pathogenesis and virulence. Interspecies exchange has been reported for many RNA viruses with the most notable being reassortment (equivalent to recombination in segmented genomes) between the human and avian influenza strains to produce a highly virulent virus (de Jong et al., 1997).

Sapovirus recombinants

Recently, a number of intragenogroup and intergenogroup recombinant sapoviruses were identified by analyses of full-length genomes (Hansman et al., 2005; Katayama et al., 2004)

(Table 3.3). Simplot analysis showed a decrease in nucleotide identity after the RdRp region indicating a potential recombination site at the RdRp/capsid junction, consistent with the norovirus breakpoint. For the two-intragenogroup recombinant sapoviruses, Mc10 and C12 strains, there were 44 nucleotides at the RdRp/capsid junction that matched 100% between the two viruses (Katayama et al., 2004). This conserved site may represent either the break and rejoin site or the site for copy choice for Mc10 and C12 during viral replication (described in details below). Phylogenetic analysis grouped Mc10 and C12 non-structural sequences (i.e. between genome start and VP1 start) into the same GII genotype, but their capsid sequences grouped into two different GII genotypes, GII.2 and GII.3, respectively. For the intergenogroup recombinant sapoviruses, SW278 and Ehime1107 strains, their non-structural region clustered into the same GII genotype, but their complete capsid sequences grouped into the same GIV genotype, indicating a recombination event occurred between GII and

Table 3.3 Sapovirus recombinants

Recombinant type	Prototype strain	RdRp genotype	Capsid genotype	Breakpoint site	GenBank accession no.	First published
Intragenotype	6728/Maizuru/JP	GI.1a	GI.1b	RdRp/Capsid	DQ395300	Phan et al. (2006a)
Intergenotype	Mc10	GII	GII.2	RdRp/Capsid	NC_010624	Katayama et al. (2004)
	C12	GII	GII.3	RdRp/Capsid	NC_006554	Katayama et al. (2004)
	QW270	GIII	GIII	RdRp/Capsid	AY826426	Wang et al. (2005)
	MM280	GIII	GIII	RdRp/Capsid	AY823308	Wang et al. (2005)
Intergenogroup	Ehime1107	GII	GIV.1	RdRp/Capsid	AY237459	Hansman et al. (2005)

GIV. The parental sequences for the non-structural region of SW278 and Ehime1107 strains, both GIV strains, have not yet been determined.

Vesivirus recombinants

To date only one vesivirus recombinant has been reported. Recombination within vesivirus was reported in 2006 between two FCV strains (Coyne et al., 2006). The recombinant FCV strain was isolated from a cat colony that was endemically infected with two FCV variants, termed major variant and minor variant (>20% divergence) (Coyne et al., 2006). Longitudinal sampling of the cat colony revealed that the cats were initially only infected with the major variant and then six months later the minor variant was introduced. A recombinant strain was then detected 27 months later. The recombination breakpoint was located between nt 5227 and 5373, within the highly conserved region spanning the ORF1/ORF2 junction and includes the start of the sgRNA (Coyne et al., 2006). ORF1 was derived from the major strain and ORF2/3 was obtained from the minor strain. The discovery of this recombinant is significant for calicivirus recombination studies as it is the first report where both parental strains can be clearly identified. Coyne et al. also provided strong evidence indicating that recombination is an important survival mechanism, as following the introduction of the recombinant the minor variant was replaced, suggesting that despite the

similar antigenic properties of the recombinant and the minor variant the recombinant strain had a fitness benefit over the minor variant.

Lagovirus recombinants

In 2008, Forrester et al. analysed 26 near full-length RHDV strains for recombination breakpoints (Forrester et al., 2008). Of the 26 sequences ten were identified as naturally occurring recombinants (Abrantes et al., 2008; Forrester et al., 2008) (Table 3.4). Recombination breakpoints were located across the genome and were not isolated to the polymerase/capsid junction or to the ORF1/ORF2 overlap (in sapovirus and lagovirus the ORF1/ORF2 junction lies between the major and minor capsid proteins, whereas, for norovirus and vesivirus ORF1/ORF2 lies between the polymerase and capsid) as had predominantly been the case for other calicivirus recombinants. In fact, of the ten RHDV recombinants, only two had a single crossover site with a recombination breakpoint near the polymerase/capsid junction. The remaining eight recombinants had a minimum of two crossover sites, which resulted in at least a double recombination event. Three recombinants had a double recombination breakpoint within the polymerase, similar to 771/05/IRL, a norovirus recombinant. One had a double breakpoint spanning the helicase and VPg region of ORF1 and two had a double breakpoint spanning a 250-nt region within the capsid (Table 3.4).

Table 3.4 Recombinant RHDV strains

Recombinant strain	Crossover event	Crossover type	Breakpoint	Gene	GenBank accession no.
German FRG	1	Double	2000/2500	Helicase/VPg	NC_001543
Czech V351	1	Double	2000/2500	Helicase/VPg	U54983
	2	Double	6125/6375	Capsid	
Frankfurt5	1	Double	4125/4500	RdRp	EF558573
Frankfurt12	1	Double	4125/4500	RdRp	EF558572
Wika	1	Double	4125/4500	RdRp	EF558574
NZ54	1	Double	6125/6375	Capsid	EF558579
NZ61	1	Double	6125/6375	Capsid	EF558580
Hagenow	1	Single	5300	Capsid	EF558585
Mexico	1	Single	5500	Capsid	AF295785
Hartmannsdorf	1	Double	5750/5875	Capsid	EF558586
	2	Single	6875	Capsid	

Interestingly two recombinants, Hartmannsdorf and Czech V351, appeared to have more than two crossover events and appeared to be derived from three parental strains (Forrester *et al.*, 2008). The RHDV recombinant strain, Hartmannsdorf, had three crossover sites in the capsid gene. RHDV strain, Czech V351, also had a double crossover event within the capsid gene. However, the second double recombination event in Czech V351 spanned the helicase and VPg (Table 3.4). The high percentage (35.8%) of recombinant RHDV to wild-type RHDV strains and the identification of recombinant breakpoints across the genome indicate that recombination is also an important mechanism in RHDV evolution.

Recombination breakpoint

All sapovirus, vesivirus and norovirus recombinants, with the exception of one norovirus recombinant had a crossover point either within or close to the RdRp/capsid junction (also equivalent to the ORF1/ORF2 overlap in norovirus and vesivirus, but not sapovirus) (Fig. 3.5). Whereas,

lagovirus recombinants had breakpoints located across the genome (Fig. 3.5). However, the precise locations of these recombination breakpoints at the RdRp/capsid junction is difficult to determine as the junction is almost 100% conserved within each calicivirus genus and therefore masks the breakpoint. This fact and the identification of the average breakpoint located 16–20 nt upstream of the start of ORF2 in norovirus, strongly suggests that there is a recombination hotspot at the start of the capsid gene.

Recombination hotspots have also been identified in other viral species closely related to the caliciviruses, such as the picornaviruses and coronaviruses and recombination events have been seen to occur throughout the entire genome for both in *in vitro* studies. Though, interestingly, recombination has not been identified in the capsid regions encoding for VP1 and VP3 in picornaviruses (reviewed in Lai, 1992). While recombination within the capsid gene has been suggested for norovirus GII and RHDV (Phan *et al.*, 2006a; Rohayem *et al.*, 2005), further analysis

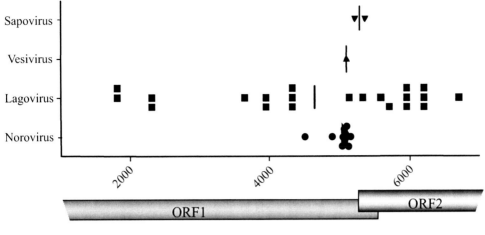

Figure 3.5 Comparison of the breakpoints for all calicivirus recombinants. The breakpoint for each calicivirus recombinant was plotted in relation to the genome of norovirus GII prototype strain, Lordsdale (GenBank accession number X86557). The average breakpoint site was determined for each genus and is represented by a vertical line.

of the putative norovirus recombinants was not able to identify any statistically significant breakpoints in the capsid (Bull *et al.*, 2007). It has been speculated for the picornaviruses that recombination in the capsid gene is not favourable because it can lead to non-functional or non-stable products (reviewed in Lai, 1992).

Recently calicivirus recombinants with breakpoints outside of the RdRp/capsid junction have been detected (Forrester *et al.*, 2008; Waters *et al.*, 2007), most likely due to double crossover events either in the RdRp or capsid. Given the fact that for most norovirus strains only partial sequence is available, the identification of recombination sites remote from the polymerase/capsid junction is under represented. In the norovirus and RHDV recombinants with double crossover sites there was a lack of visible RNA promoters or secondary structure at the crossover sites thereby suggesting that these recombination events may have arisen by other mechanisms compared to those that induce a breakpoint in or around the ORF1/2 overlap (discussed below). However, the lack of insertions or deletions at the putative crossover sites in the norovirus double recombinant suggests that it is a result of homologous recombination and lends support to the template

switching model proposed for norovirus (Bull *et al.*, 2005). *In vitro* recombination experiments using poliovirus have predominantly isolated recombinants with single crossover events, but recombinants with multiple crossover sites have also been generated (reviewed in Lai, 1992). Whether these multiple crossover sites are from a single recombination event or from multiple rounds of replication/recombination is unknown.

Recombination model

There are two proposed mechanisms responsible for RNA recombination: (i) a cleavage–ligation model (Chetverin *et al.*, 1997), which functions by cleavage of two RNA molecules, followed by ligation of fragments from different RNA molecules and (ii) an RdRp-mediated copy choice model (Cooper *et al.*, 1974). In the 'copy choice' model, recombinant RNA is formed when the viral RdRp complex switches, mid replication, from one RNA template to another. This results in homologous recombination if the replicase continues to copy the new strand precisely where it left the initial strand and aberrant or non-homologous recombination if it does not (Worobey and Holmes, 1999). Nearly all reports on RNA recombination support the 'copy choice' mechanism (Aaziz and Tepfer, 1999). Many of the RNA viruses with sgRNA have been shown to undergo recombination and in most the recombination crossover point has been localized to the internal initiation site of the sgRNA (reviewed in Carpenter et al., 1995). Therefore it is suspected that the sgRNA promoter sequence located at the start of ORF2 may direct recombination to occur there. The primary mechanism involved in recombination in RNA viruses is the copy-choice model (Cooper et al., 1974). In this model homologous recombination is driven by the viral encoded RdRp when pausing occurs during the transcription of a region of complex secondary structure. The RdRp then loses processivity and switches between RNA templates (reviewed in Lai, 1992; Worobey and Holmes, 1999). A number of models of sgRNA synthesis have been proposed, but the most accepted is the internal initiation mechanism (Miller and Koev, 2000). Here the replicase initiates positive strand sgRNA

transcription internally on a negative strand of genomic RNA (Miller et al., 1985). Using these two well-supported models, copy-choice and internal initiation model, a simple mechanism for recombination in norovirus has been proposed (Bull et al., 2005) (Fig. 3.6). Replication and internal sgRNA synthesis generates two positive RNA species. These templates direct RNA synthesis that leads to the generation of both a full-length negative genomic RNA and a negative sgRNA species, in the second round of replication. The negative sgRNA is the key player in our proposed model and such species have been identified in viruses that produce sgRNAs (Ishikawa et al., 1997). Recombination then occurs when the enzyme initiates positive strand synthesis at the 3′ end of the full-length negative strand and then after pausing due to the complex secondary structure of the RNA promoter sequence, the RdRp template switches to an available negative sgRNA species generated by a

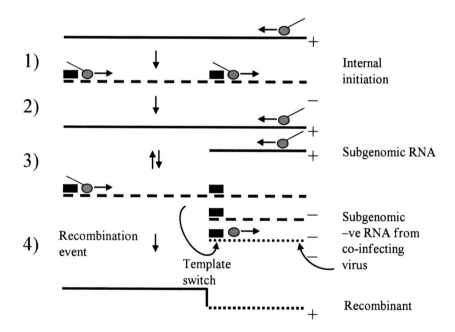

Figure 3.6 A simple mechanism for recombination in norovirus. (1) RNA transcription by the RdRp (grey circle) generates a negative strand intermediate (dashed line). (2) Binding of the RdRp to the almost identical RNA promoter sequences (filled boxes) generates positive stranded (straight line) genomes and subgenomic RNA. (3) These templates direct RNA synthesis from the 3′ end that leads to the generation of both a full-length negative genome and a negative subgenomic RNA species. (4) Recombination occurs when the enzyme initiates positive strand synthesis at the 3′ end of the full-length negative strand, stalls at the subgenomic promoter, and then template switches to an available negative subgenomic RNA species generated by a co-infecting virus. The net result is a recombinant virus that has acquired new ORF2 and ORF3 sequences. Reproduced with permission from Bull et al. (2005).

co-infecting virus. Alternatively, the RdRp could also template switch directly from one genomic RNA to another genomic RNA in the highly conserved ORF1/ORF2 overlap. The net result is a recombinant virus that has acquired new ORF2 and ORF3 sequences.

More recombinant noroviruses have been proposed but only statistically confirmed recombinants were discussed in this chapter (Bull *et al.*, 2007). Confirmation of a recombination event in some strains has been complicated by difficulties in genotyping the RdRp. In particular GII.6, GII.7, GII.8, GII.9, GII.13 and GII.14 polymerases cluster together (Fankhauser *et al.*, 2002; Jin *et al.*, 2008). Furthermore, sequence data of the putative non-recombinant parental strains is not always known.

The high diversity of norovirus has also complicated the detection of intragenotypic recombinants (recombination between strains from the same genotype). To date, evidence for nine intragenotypic norovirus GII recombinants has been published (Etherington *et al.*, 2006; Phan *et al.*, 2006a; Rohayem *et al.*, 2005). But further analysis of these recombinants failed to identify the recombination sites as statistically significant (Bull *et al.*, 2007). The inability to prove the presence of intragenotypic recombinants is more likely to be due to the lack of a sensitive data analysis method.

Recombination and its role in viral persistence

The large number of recombinants in circulation suggests that it could be important survival mechanism in calicivirus evolution. For 27 of 36 confirmed calicivirus recombinants, a common recombination breakpoint site near the RdRp/capsid junction was identified. Most likely this has important evolutionary implications because the capsid is responsible for host cell receptor binding and for the antigenic recognition (Chen *et al.*, 2004). As a result, the capsid gene is under strong selection by structural and functional constraints from the host. Alternatively the capsid is forced to evolve structurally in order to escape host immune defences. The ability of caliciviruses to swap their structural genes is analogous to antigenic shift in influenza virus. Antigenic shift occurs in influenza virus when it swaps genomic

segments encoding antigenic determinants, the haemagglutinin (HA) and neuraminidase (NA) gene. This enables influenza virus to reinfect the same host due to an ability to avoid immune detection because of the new antigenic determinants inherited from reassortment, and alter virus host species specificity (reviewed in Hay *et al.*, 2001; Scholtissek, 1995). Theoretical explanations for the evolution of recombination fit into one of two standard classes: (i) enables the creation and spread of advantageous traits and (ii) permits the removal of deleterious genes (reviewed in Domingo *et al.*, 1996). Both norovirus and FCV recombinants support the first hypothesis. The recombinant FCV outcompeted its antigenically related ancestor suggesting that the non-structural genes provided the recombinant with a replicative advantage (Coyne *et al.*, 2006). In norovirus the two most prevalent ORF1s in the recombinant population, GII.b and GII.4 [associated with four and five different ORF2 (capsid) genotypes, respectively], were also the most prevalent RdRp type in the non-recombinant population. This suggests that the GII.b and GII.4 ORF1 traits may provide a fitness benefit to the viruses. Therefore, recombination has potentially improved the fitness of circulating strains.

Due to the larger distribution of GII.4 and GII.b ORF1 in the population there is a higher probability that recombination events involve a GII.4 or a GII.b virus. However, it is also possible that some mechanistic enhancement may be increasing their ability to recombine. For example, single amino acid substitutions in the HIV RT have been shown to result in an increase in the rate of template switching (Nikolenko *et al.*, 2004).

Recombination requires co-infection, therefore, if recombination was based solely on chance then the ratio of recombinant types with GII.4 RdRps and GII.4 capsids should be close to 1. However, the ratio is one GII.4 capsid to four GII.4 RdRps, suggesting that the GII.4 RdRp is driving recombination and has a higher rate of recombination than the RdRps from the other genotypes. Alternatively, recombination evolved in GII.4 RdRp-based norovirus, as a means of replacing defective capsid genes, as has been shown with several RNA viruses (reviewed in Worobey and Holmes, 1999). For example, weak or even non-replicative mutant strains of tombusviruses and Sindbis virus have undergone recombination to form viable, highly fit viruses (Raju *et al.*, 1995; Weiss and Schlesinger, 1991; White and Morris, 1994). Also plant viruses have been observed to repair their genomes by recombining with host transgene transcripts (Borja *et al.*, 1999; Gal *et al.*, 1992; Rubio *et al.*, 1999). The accumulation of a deleterious allele in a population is known as Muller's ratchet (Duarte *et al.*, 1992). Muller's ratchet states that asexual populations of organisms with a small population size and a high mutation rate will tend to incorporate deleterious mutations in an essentially irreversible manner that will lead to viral extinction unless compensatory mechanisms such as recombination can restore the initial, mutation-free class of genomes (Muller, 1964). Experiments with RNA viruses have generally supported Muller's ratchet and have shown decreased fitness for populations in which it occurs (Worobey and Holmes, 1999). Evidence for Muller's ratchet and the beneficial effects of recombination to replace less fit genes can also be seen from the norovirus recombinant data.

Four of the eight norovirus RdRp genotypes (GII.a, GII.b, GII.c and GII.d) were found only in association with recombinant strains, that is, their matching non-recombinant (native) capsid sequence could not be identified. This suggests that the capsid gene belonging to these four RdRp genotypes has become extinct (Bull *et al.*, 2007). The ability of norovirus to recombine at the ORF1/ORF2 overlap, therefore, enables the virus to keep its ORF1 genes but swap its viral coat and subsequently adopt a different antigenic profile, which is, particularly advantageous for viruses whose host range has diminished either due to the development of herd immunity or deleterious mutations in host binding epitopes.

Understanding the role of calicivirus recombination is especially important when considering vaccine development. The assumption that recombination in RNA viruses either does not happen or is unimportant has a history of being proved wrong (Worobey and Holmes, 1999). For example, several live attenuated vaccine strains from RNA viruses such as poliovirus and infectious bronchitis virus have been shown to undergo recombination both experimentally

and in the host to produce viable and highly fit viruses (Georgescu *et al.*, 1994; Holmes *et al.*, 1999; Jia *et al.*, 1995; Karakasiliotis *et al.*, 2005; Kew *et al.*, 2002; Kusters *et al.*, 1990). Therefore, the potential of recombination to produce new pathogenic hybrid strains needs to be carefully considered whenever live-attenuated vaccines are used to control RNA viruses (Worobey and Holmes, 1999).

Caliciviruses have been detected in a wide range of mammals, including humans, mice, cows and pigs (Wang *et al.*, 2005), with strong evidence of zoonotic transmission (Widdowson *et al.*, 2005). Interspecies exchange has been reported for many RNA viruses with the most notable being reassortment (equivalent to recombination in segmented genomes) between the human and avian influenza strains to produce a highly virulent virus (de Jong *et al.*, 1997). In caliciviruses, recombination between mammalian caliciviruses belonging to different genogroups might also yield new strains. Indeed, intergenogroup recombination within the sapoviruses has already been reported (Hansman *et al.*, 2005).

Conclusion

Knowledge of the mechanisms of recombination in the caliciviruses will continue to be important as new calicivirus variants emerge, with potentially different pathogenesis and virulence.

References

Aaziz, R., and Tepfer, M. (1999). Recombination in RNA viruses and in virus-resistant transgenic plants. J. Gen. Virol. *80*, 1339–1346.

Abrantes, J., Esteves, P.J., and van der Loo, W. (2008). Evidence for recombination in the major capsid gene VP60 of the rabbit haemorrhagic disease virus (RHDV). Arch. Virol. *153*, 329–335.

Ambert-Balay, K., Bon, F., Le Guyader, F., Pothier, P., and Kohli, E. (2005). Characterization of new recombinant noroviruses. J. Clin. Microbiol. *43*, 5179–5186.

Belliot, G., Sosnovtsev, S.V., Chang, K.O., Babu, V., Uche, U., Arnold, J.J., Cameron, C.E., and Green, K.Y. (2005). Norovirus proteinase-polymerase and polymerase are both active forms of RNA-dependent RNA polymerase. J. Virol. *79*, 2393–2403.

Belliot, G., Sosnovtsev, S.V., Mitra, T., Hammer, C., Garfield, M., and Green, K.Y. (2003). In vitro proteolytic processing of the MD145 norovirus ORF1 nonstructural polyprotein yields stable precursors and products similar to those detected in calicivirus-infected cells. J. Virol. *77*, 10957–10974.

Blakeney, S.J., Cahill, A., and Reilly, P.A. (2003). Processing of Norwalk virus nonstructural proteins by a 3C-like cysteine proteinase. Virology 308, 216–224.

Bon, F., Ambert-Balay, K., Giraudon, H., Kaplon, J., Le Guyader, S., Pommepuy, M., Gallay, A., Vaillant, V., de Valk, H., Chikhi-Brachet, R., et al. (2005). Molecular epidemiology of caliciviruses detected in sporadic and outbreak cases of gastroenteritis in France from December 1998 to February 2004. J. Clin. Microbiol. 43, 4659–4664.

Boniotti, B., Wirblich, C., Sibilia, M., Meyers, G., Thiel, H.J., and Rossi, C. (1994). Identification and characterization of a 3C-like protease from rabbit hemorrhagic disease virus, a calicivirus. J. Virol. 68, 6487–6495.

Borja, M., Rubio, T., Scholthof, H.B., and Jackson, A.O. (1999). Restoration of wild-type virus by double recombination of tombusvirus mutants with a host transgene. Mol. Plant Microbe Interact. 12, 153–162.

Buesa, J., Collado, B., Lopez-Andujar, P., Abu-Mallouh, R., Rodriguez Diaz, J., Garcia Diaz, A., Prat, J., Guix, S., Llovet, T., Prats, G., et al. (2002). Molecular epidemiology of caliciviruses causing outbreaks and sporadic cases of acute gastroenteritis in Spain. J. Clin. Microbiol. 40, 2854–2859.

Bull, R.A., Hansman, G.S., Clancy, L.E., Tanaka, M.M., Rawlinson, W.D., and White, P.A. (2005). Norovirus recombination in ORF1/ORF2 overlap. Emerg. Infect. Dis. 11, 1079–1085.

Bull, R.A., Tanaka, M.M., and White, P.A. (2007). Norovirus recombination. J. Gen. Virol. 88, 3347–3359.

Bull, R.A., Tu, E.T., McIver, C.J., Rawlinson, W.D., and White, P.A. (2006). Emergence of a new norovirus genotype II.4 variant associated with global outbreaks of gastroenteritis. J. Clin. Microbiol. 44, 327–333.

Carpenter, C.D., Oh, J.W., Zhang, C., and Simon, A.E. (1995). Involvement of a stem–loop structure in the location of junction sites in viral RNA recombination. J. Mol. Biol. 245, 608–622.

Chen, R., Neill, J.D., Noel, J.S., Hutson, A.M., Glass, R.I., Estes, M.K., and Prasad, B.V. (2004). Inter- and intra-genus structural variations in caliciviruses and their functional implications. J. Virol. 78, 6469–6479.

Chetverin, A.B., Chetverina, H.V., Demidenko, A.A., and Ugarov, V.I. (1997). Nonhomologous RNA recombination in a cell-free system: evidence for a transesterification mechanism guided by secondary structure. Cell 88, 503–513.

Cooper, P.D., Steiner-Pryor, A., Scotti, P.D., and Delong, D. (1974). On the nature of poliovirus genetic recombinants. J. Gen. Virol. 23, 41–49.

Coyne, K.P., Reed, F.C., Porter, C.J., Dawson, S., Gaskell, R.M., and Radford, A.D. (2006). Recombination of Feline calicivirus within an endemically infected cat colony. J. Gen. Virol. 87, 921–926.

de Jong, J.C., Claas, E.C., Osterhaus, A.D., Webster, R.G., and Lim, W.L. (1997). A pandemic warning? Nature 389, 554.

Domingo, E., Escarmis, C., Sevilla, N., Moya, A., Elena, S.F., Quer, J., Novella, I.S., and Holland, J.J. (1996). Basic concepts in RNA virus evolution. Faseb J. 10, 859–864.

Duarte, E., Clarke, D., Moya, A., Domingo, E., and Holland, J. (1992). Rapid fitness losses in mammalian RNA virus clones due to Muller's ratchet. Proc. Natl. Acad. Sci. U.S.A. 89, 6015–6019.

Etherington, G.J., Dicks, J., and Roberts, I.N. (2006). High throughput sequence analysis reveals hitherto unreported recombination in the genus Norovirus. Virology 345, 88–95.

Fankhauser, R.L., Monroe, S.S., Noel, J.S., Humphrey, C.D., Bresee, J.S., Parashar, U.D., Ando, T., and Glass, R.I. (2002). Epidemiologic and molecular trends of 'Norwalk-like viruses' associated with outbreaks of gastroenteritis in the United States. J. Infect. Dis. 186, 1–7.

Farkas, T., Sestak, K., Wei, C., and Jiang, X. (2008). Characterization of a rhesus monkey calicivirus representing a new genus of Caliciviridae. J. Virol. 82, 5408–5416.

Forrester, N.L., Moss, S.R., Turner, S.L., Schirrmeier, H., and Gould, E.A. (2008). Recombination in rabbit haemorrhagic disease virus: possible impact on evolution and epidemiology. Virology 376, 390–396.

Gal, S., Pisan, B., Hohn, T., Grimsley, N., and Hohn, B. (1992). Agroinfection of transgenic plants leads to viable cauliflower mosaic virus by intermolecular recombination. Virology 187, 525–533.

Gallimore, C.I., Cheesbrough, J.S., Lamden, K., Bingham, C., and Gray, J.J. (2005). Multiple norovirus genotypes characterised from an oyster-associated outbreak of gastroenteritis. Int. J. Food Microbiol. 103, 323–330.

Gallimore, C.I., Cubitt, D., du Plessis, N., and Gray, J.J. (2004a). Asymptomatic and symptomatic excretion of noroviruses during a hospital outbreak of gastroenteritis. J. Clin. Microbiol. 42, 2271–2274.

Gallimore, C.I., Lewis, D., Taylor, C., Cant, A., Gennery, A., and Gray, J.J. (2004b). Chronic excretion of a norovirus in a child with cartilage hair hypoplasia (CHH). J. Clin. Virol. 30, 196–204.

Georgescu, M.M., Delpeyroux, F., Tardy-Panit, M., Balanant, J., Combiescu, M., Combiescu, A.A., Guillot, S., and Crainic, R. (1994). High diversity of poliovirus strains isolated from the central nervous system from patients with vaccine-associated paralytic poliomyelitis. J. Virol. 68, 8089–8101.

Glass, P.J., White, L.J., Ball, J.M., Leparc-Goffart, I., Hardy, M.E., and Estes, M.K. (2000). Norwalk virus open reading frame 3 encodes a minor structural protein. J. Virol. 74, 6581–6591.

Green, K.Y., Chanock, R.M., and Kapikian, A.Z. (2001). Human calicivirus. In Fields virology, K.D.M., and H.P.M., eds. (Philadelphia, Lippincott Williams & Wilkins), pp. 841–874.

Han, M.G., Smiley, J.R., Thomas, C., and Saif, L.J. (2004). Genetic recombination between two genotypes of genogroup III bovine noroviruses (BoNVs) and capsid sequence diversity among BoNVs and Nebraska-like bovine enteric caliciviruses. J. Clin. Microbiol. 42, 5214–5224.

Hansman, G.S., Doan, L.T., Kguyen, T.A., Okitsu, S., Katayama, K., Ogawa, S., Natori, K., Takeda, N., Kato,

Y., Nishio, O., *et al.* (2004a). Detection of norovirus and sapovirus infection among children with gastroenteritis in Ho Chi Minh City, Vietnam. Arch. Virol. *149*, 1673–1688.

Hansman, G.S., Katayama, K., Maneekarn, N., Peerakome, S., Khamrin, P., Tonusin, S., Okitsu, S., Nishio, O., Takeda, N., and Ushijima, H. (2004b). Genetic diversity of norovirus and sapovirus in hospitalized infants with sporadic cases of acute gastroenteritis in Chiang Mai, Thailand. J. Clin. Microbiol. *42*, 1305–1307.

Hansman, G.S., Takeda, N., Oka, T., Oseto, M., Hedlund, K.O., and Katayama, K. (2005). Intergenogroup recombination in sapoviruses. Emerg. Infect. Dis. *11*, 1916–1920.

Hardy, M.E., Kramer, S.F., Treanor, J.J., and Estes, M.K. (1997). Human calicivirus genogroup II capsid sequence diversity revealed by analyses of the prototype Snow Mountain agent. Arch. Virol. *142*, 1469–1479.

Hay, A.J., Gregory, V., Douglas, A.R., and Lin, Y.P. (2001). The evolution of human influenza viruses. Phil. Trans. R. Soc. Lond., B, Biol. Sci *356*, 1861–1870.

Holmes, E.C., Worobey, M., and Rambaut, A. (1999). Phylogenetic evidence for recombination in dengue virus. Molecular biology and evolution *16*, 405–409.

Iritani, N., Seto, Y., Kubo, H., Murakami, T., Haruki, K., Ayata, M., and Ogura, H. (2003). Prevalence of Norwalk-like virus infections in cases of viral gastroenteritis among children in Osaka City, Japan. J. Clin. Microbiol. *41*, 1756–1759.

Ishikawa, M., Janda, M., Krol, M.A., and Ahlquist, P. (1997). In vivo DNA expression of functional brome mosaic virus RNA replicons in *Saccharomyces cerevisiae*. J. Virol. *71*, 7781–7790.

Jia, W., Karaca, K., Parrish, C.R., and Naqi, S.A. (1995). A novel variant of avian infectious bronchitis virus resulting from recombination among three different strains. Arch. Virol. *140*, 259–271.

Jin, M., Xie, H.P., Duan, Z.J., Liu, N., Zhang, Q., Wu, B.S., Li, H.Y., Cheng, W.X., Yang, S.H., Yu, J.M., *et al.* (2008). Emergence of the GII4/2006b variant and recombinant noroviruses in China. J. Med. Virol. *80*, 1997–2004.

Joubert, P., Pautigny, C., Madelaine, M.F., and Rasschaert, D. (2000). Identification of a new cleavage site of the 3C-like protease of rabbit haemorrhagic disease virus. J. Gen. Virol. *81*, 481–488.

Karakasiliotis, I., Paximadi, E., and Markoulatos, P. (2005). Evolution of a rare vaccine-derived multirecombinant poliovirus. J. Gen. Virol. *86*, 3137–3142.

Katayama, K., Miyoshi, T., Uchino, K., Oka, T., Tanaka, T., Takeda, N., and Hansman, G.S. (2004). Novel recombinant sapovirus. Emerg. Infect. Dis. *10*, 1874–1876.

Katayama, K., Shirato-Horikoshi, H., Kojima, S., Kageyama, T., Oka, T., Hoshino, F., Fukushi, S., Shinohara, M., Uchida, K., Suzuki, Y., *et al.* (2002). Phylogenetic analysis of the complete genome of 18 Norwalk-like viruses. Virology *299*, 225–239.

Kew, O., Morris-Glasgow, V., Landaverde, M., Burns, C., Shaw, J., Garib, Z., Andre, J., Blackman, E., Freeman, C.J., Jorba, J., *et al.* (2002). Outbreak of poliomyelitis in Hispaniola associated with circulating type 1 vaccine-derived poliovirus. Science *296*, 356–359.

Kusters, J.G., Jager, E.J., Niesters, H.G., and van der Zeijst, B.A. (1990). Sequence evidence for RNA recombination in field isolates of avian coronavirus infectious bronchitis virus. Vaccine *8*, 605–608.

L'Homme, Y., Sansregret, R., Plante-Fortier, E., Lamontagne, A.M., Ouardani, M., Lacroix, G., and Simard, C. (2009). Genomic characterization of swine caliciviruses representing a new genus of Caliciviridae. Virus Genes *39*, 66–75.

Lai, M.M. (1992). RNA recombination in animal and plant viruses. Microbiol. Rev. *56*, 61–79.

Liu, B., Clarke, I.N., and Lambden, P.R. (1996). Polyprotein processing in Southampton virus: identification of 3C-like protease cleavage sites by in vitro mutagenesis. J. Virol. *70*, 2605–2610.

Liu, B.L., Viljoen, G.J., Clarke, I.N., and Lambden, P.R. (1999). Identification of further proteolytic cleavage sites in the Southampton calicivirus polyprotein by expression of the viral protease in *E. coli*. J. Gen. Virol. *80*, 291–296.

Lole, K.S., Bollinger, R.C., Paranjape, R.S., Gadkari, D., Kulkarni, S.S., Novak, N.G., Ingersoll, R., Sheppard, H.W., and Ray, S.C. (1999). Full-length human immunodeficiency virus type 1 genomes from subtype C-infected seroconverters in India, with evidence of intersubtype recombination. J. Virol. *73*, 152–160.

Martella, V., Lorusso, E., Decaro, N., Elia, G., Radogna, A., D'Abramo, M., Desario, C., Cavalli, A., Corrente, M., Camero, M., *et al.* (2008). Detection and molecular characterization of a canine norovirus. Emerg. Infect. Dis. *14*, 1306–1308.

Martinez, N., Espul, C., Cuello, H., Zhong, W., Jiang, X., Matson, D.O., and Berke, T. (2002). Sequence diversity of human caliciviruses recovered from children with diarrhea in Mendoza, Argentina, 1995–1998. J. Med. Virol. *67*, 289–298.

Maydt, J., and Lengauer, T. (2006). Recco: recombination analysis using cost optimization. Bioinformatics *22*, 1064–1071.

Meyers, G., Wirblich, C., Thiel, H.J., and Thumfart, J.O. (2000). Rabbit hemorrhagic disease virus: genome organization and polyprotein processing of a calicivirus studied after transient expression of cDNA constructs. Virology *276*, 349–363.

Miller, W.A., Dreher, T.W., and Hall, T.C. (1985). Synthesis of brome mosaic virus subgenomic RNA in vitro by internal initiation on (–)(–)sense genomic RNA. Nature *313*, 68–70.

Miller, W.A., and Koev, G. (2000). Synthesis of subgenomic RNAs by positive-strand RNA viruses. Virology *273*, 1–8.

Morse, S.S. (1994). Hantaviruses and the hantavirus outbreak in the United States. A case study in disease emergence. Ann. N Y Acad. Sci. *740*, 199–207.

Muller, B., Klemm, U., Mas Marques, A., and Schreier, E. (2007). Genetic diversity and recombination of murine noroviruses in immunocompromised mice. Arch. Virol. *152*, 1709–1719.

Muller, H.J. (1964). The relation of recombination to mutational advance. Mutat. Res. *106*, 2–9.

Nayak, M.K., Balasubramanian, G., Sahoo, G.C., Bhattacharya, R., Vinje, J., Kobayashi, N., Sarkar,

M.C., Bhattacharya, M.K., and Krishnan, T. (2008). Detection of a novel intergenogroup recombinant Norovirus from Kolkata, India. Virology 377, 117–123.

Neill, J.D. (1990). Nucleotide sequence of a region of the feline calicivirus genome which encodes picornavirus-like RNA-dependent RNA polymerase, cysteine protease and 2C polypeptides. Virus Res. 17, 145–160.

Ng, K.K., Cherney, M.M., Vazquez, A.L., Machin, A., Alonso, J.M., Parra, F., and James, M.N. (2002). Crystal structures of active and inactive conformations of a caliciviral RNA-dependent RNA polymerase. J. Biol. Chem. 277, 1381–1387.

Ng, K.K., Pendas-Franco, N., Rojo, J., Boga, J.A., Machin, A., Alonso, J.M., and Parra, F. (2004). Crystal structure of norwalk virus polymerase reveals the carboxyl terminus in the active site cleft. J. Biol. Chem. 279, 16638–16645.

Nikolenko, G.N., Svarovskaia, E.S., Delviks, K.A., and Pathak, V.K. (2004). Antiretroviral drug resistance mutations in human immunodeficiency virus type 1 reverse transcriptase increase template-switching frequency. J. Virol. 78, 8761–8770.

Noel, J.S., Fankhauser, R.L., Ando, T., Monroe, S.S., and Glass, R.I. (1999). Identification of a distinct common strain of 'Norwalk-like viruses' having a global distribution. J. Infect. Dis. 179, 1334–1344.

Oka, T., Katayama, K., Ogawa, S., Hansman, G.S., Kageyama, T., Miyamura, T., and Takeda, N. (2005a). Cleavage activity of the sapovirus 3C-like protease in *Escherichia coli*. Arch. Virol. 150, 2539–2548.

Oka, T., Katayama, K., Ogawa, S., Hansman, G.S., Kageyama, T., Ushijima, H., Miyamura, T., and Takeda, N. (2005b). Proteolytic processing of sapovirus ORF1 polyprotein. J. Virol. 79, 7283–7290.

Oka, T., Yamamoto, M., Katayama, K., Hansman, G.S., Ogawa, S., Miyamura, T., and Takeda, N. (2006). Identification of the cleavage sites of sapovirus open reading frame 1 polyprotein. J. Gen. Virol. 87, 3329–3338.

Oka, T., Yamamoto, M., Yokoyama, M., Ogawa, S., Hansman, G.S., Katayama, K., Miyashita, K., Takagi, H., Tohya, Y., Sato, H., et al. (2007). Highly conserved configuration of catalytic amino acid residues among calicivirus-encoded proteases. J. Virol. 81, 6798–6806.

Oliver, S.L., Asobayire, E., Dastjerdi, A.M., and Bridger, J.C. (2006a). Genomic characterization of the unclassified bovine enteric virus Newbury agent-1 (Newbury1) endorses a new genus in the family Caliciviridae. Virology 350, 240–250.

Oliver, S.L., Batten, C.A., Deng, Y., Elschner, M., Otto, P., Charpilienne, A., Clarke, I.N., Bridger, J.C., and Lambden, P.R. (2006b). Genotype 1 and genotype 2 bovine noroviruses are antigenically distinct but share a cross-reactive epitope with human noroviruses. J. Clin. Microbiol. 44, 992–998.

Oliver, S.L., Brown, D.W., Green, J., and Bridger, J.C. (2004). A chimeric bovine enteric calicivirus: evidence for genomic recombination in genogroup III of the Norovirus genus of the Caliciviridae. Virology 326, 231–239.

Osmanov, S., Pattou, C., Walker, N., Schwardlander, B., and Esparza, J. (2002). Estimated global distribution and regional spread of HIV-1 genetic subtypes in the year 2000. J. Acquir. Immune Defic. Syndr. 29, 184–190.

Phan, T.G., Kuroiwa, T., Kaneshi, K., Ueda, Y., Nakaya, S., Nishimura, S., Yamamoto, A., Sugita, K., Nishimura, T., Yagyu, F., et al. (2006a). Changing distribution of norovirus genotypes and genetic analysis of recombinant GIIb among infants and children with diarrhea in Japan. J. Med. Virol. 78, 971–978.

Phan, T.G., Takanashi, S., Kaneshi, K., Ueda, Y., Nakaya, S., Nishimura, S., Sugita, K., Nishimura, T., Yamamoto, A., Yagyu, F., et al. (2006b). Detection and genetic characterization of norovirus strains circulating among infants and children with acute gastroenteritis in Japan during 2004–2005. Clin. Lab. 52, 519–525.

Phan, T.G., Yagyu, F., Kozlov, V., Kozlov, A., Okitsu, S., Muller, W.E., and Ushijima, H. (2006c). Viral gastroenteritis and genetic characterization of recombinant norovirus circulating in Eastern Russia. Clin. Lab. 52, 247–253.

Raju, R., Subramaniam, S.V., and Hajjou, M. (1995). Genesis of Sindbis virus by in vivo recombination of nonreplicative RNA precursors. J. Virol. 69, 7391–7401.

Reuter, G., Krisztalovics, K., Vennema, H., Koopmans, M., and Szucs, G. (2005). Evidence of the etiological predominance of norovirus in gastroenteritis outbreaks – emerging new-variant and recombinant noroviruses in Hungary. J. Med. Virol. 76, 598–607.

Reuter, G., Vennema, H., Koopmans, M., and Szucs, G. (2006). Epidemic spread of recombinant noroviruses with four capsid types in Hungary. J. Clin. Virol. 35, 84–88.

Rohayem, J., Munch, J., and Rethwilm, A. (2005). Evidence of recombination in the norovirus capsid gene. J. Virol. 79, 4977–4990.

Roossinck, M.J. (1997). Mechanisms of plant virus evolution. Annu. Rev. Phytopathol. 35, 191–209.

Rubio, T., Borja, M., Scholthof, H.B., Feldstein, P.A., Morris, T.J., and Jackson, A.O. (1999). Broad-spectrum protection against tombusviruses elicited by defective interfering RNAs in transgenic plants. J. Virol. 73, 5070–5078.

Sasaki, Y., Kai, A., Hayashi, Y., Shinkai, T., Noguchi, Y., Hasegawa, M., Sadamasu, K., Mori, K., Tabei, Y., Nagashima, M., et al. (2006). Multiple viral infections and genomic divergence among noroviruses during an outbreak of acute gastroenteritis. J. Clin. Microbiol. 44, 790–797.

Scholtissek, C. (1995). Molecular evolution of influenza viruses. Virus Genes 11, 209–215.

Smith, J.M. (1992). Analyzing the mosaic structure of genes. J. Mol. Evol. 34, 126–129.

Someya, Y., Takeda, N., and Miyamura, T. (2002). Identification of active-site amino acid residues in the Chiba virus 3C-like protease. J. Virol. 76, 5949–5958.

Sosnovtsev, S.V., Belliot, G., Chang, K.O., Prikhodko, V.G., Thackray, L.B., Wobus, C.E., Karst, S.M., Virgin, H.W., and Green, K.Y. (2006). Cleavage map and proteolytic processing of the murine norovirus nonstructural polyprotein in infected cells. J. Virol. *80*, 7816–7831.

Sosnovtsev, S.V., and Green, K.Y. (2000). Identification and genomic mapping of the ORF3 and VPg proteins in feline calicivirus virions. Virology *277*, 193–203.

Thackray, L.B., Wobus, C.E., Chachu, K.A., Liu, B., Alegre, E.R., Henderson, K.S., Kelley, S.T., and Virgin, H.W.t. (2007). Murine noroviruses comprising a single genogroup exhibit biological diversity despite limited sequence divergence. J. Virol. *81*, 10460–10473.

Tsugawa, T., Numata-Kinoshita, K., Honma, S., Nakata, S., Tatsumi, M., Sakai, Y., Natori, K., Takeda, N., Kobayashi, S., and Tsutsumi, H. (2006). Virological, serological, and clinical features of an outbreak of acute gastroenteritis due to recombinant genogroup II norovirus in an infant home. J. Clin. Microbiol. *44*, 177–182.

Tu, E.T., Bull, R.A., Greening, G.E., Hewitt, J., Lyon, M.J., Marshall, J.A., McIver, C.J., Rawlinson, W.D., and White, P.A. (2008). Epidemics of gastroenteritis during 2006 were associated with the spread of norovirus GII.4 variants 2006a and 2006b. Clin. Infect. Dis. *46*, 413–420.

van den Berg, H., Lodder, W., van der Poel, W., Vennema, H., and de Roda Husman, A.M. (2005). Genetic diversity of noroviruses in raw and treated sewage water. Res. Microbiol. *156*, 532–540.

Vinje, J., Hamidjaja, R.A., and Sobsey, M.D. (2004). Development and application of a capsid VP1 (region D) based reverse transcription PCR assay for genotyping of genogroup I and II noroviruses. J. Virol. Methods *116*, 109–117.

Wang, Q.H., Han, M.G., Cheetham, S., Souza, M., Funk, J.A., and Saif, L.J. (2005). Porcine noroviruses related to human noroviruses. Emerg. Infect. Dis. *11*, 1874–1881.

Waters, A., Coughlan, S., and Hall, W.W. (2007). Characterisation of a novel recombination event in the norovirus polymerase gene. Virology *363*, 11–14.

Weiss, B.G., and Schlesinger, S. (1991). Recombination between Sindbis virus RNAs. J. Virol. *65*, 4017–4025.

White, K.A., and Morris, T.J. (1994). Recombination between defective tombusvirus RNAs generates functional hybrid genomes. Proc. Natl. Acad. Sci. U.S.A. *91*, 3642–3646.

Widdowson, M.A., Cramer, E.H., Hadley, L., Bresee, J.S., Beard, R.S., Bulens, S.N., Charles, M., Chege, W., Isakbaeva, E., Wright, J.G., *et al.* (2004). Outbreaks of acute gastroenteritis on cruise ships and on land: identification of a predominant circulating strain of norovirus – United States, 2002. J. Infect. Dis. *190*, 27–36.

Widdowson, M.A., Rockx, B., Schepp, R., van der Poel, W.H., Vinje, J., van Duynhoven, Y.T., and Koopmans, M.P. (2005). Detection of serum antibodies to bovine norovirus in veterinarians and the general population in the Netherlands. J. Med. Virol. *76*, 119–128.

Worobey, M., and Holmes, E.C. (1999). Evolutionary aspects of recombination in RNA viruses. J. Gen. Virol. *80*, 2535–2543.

Zheng, D.P., Ando, T., Fankhauser, R.L., Beard, R.S., Glass, R.I., and Monroe, S.S. (2006). Norovirus classification and proposed strain nomenclature. Virology *346*, 312–323.

Proteolytic Cleavage and Viral Proteins

Stanislav V. Sosnovtsev

4

Abstract

Caliciviruses are icosahedral non-enveloped viruses with a positive-sense single-stranded RNA genome that does not exceed 8.6 kb. Despite its small size, the virus genome encodes a number of non-structural proteins that successfully facilitate and regulate mechanisms required for efficient virus amplification. Although caliciviruses show significant genetic diversity, they share a common protein expression strategy. Recent findings have shown that the non-structural proteins of caliciviruses are produced by autocatalytic cleavage of a polyprotein encoded by ORF1 of the virus genome. A single virus protease structurally similar to a class of viral chymotrypsin-like cysteine proteases mediates these cleavages, and in some caliciviruses, adds to a release of the virus capsid protein. The temporal regulation of viral protein synthesis relies on the specificity of the protease and may be modulated by additional viral and cellular factors. The proteolytic processing results not only in the synthesis of the mature virus proteins, but also their precursors, whose functions have yet to be determined. Almost all calicivirus proteins have been identified as components of the virus replication complexes; however, their roles in replication are not entirely understood and remain an active and crucial target of calicivirus research.

Introduction

The family *Caliciviridae* is part of the picorna-like superfamily of positive-strand RNA viruses. Today, *Caliciviridae* is considered to be comprised of four genera: *Lagovirus*, *Vesivirus*, *Sapovirus* and *Norovirus* (Green *et al.*, 2000), although additional genera probably exist (Farkas *et al.*, 2008; Oliver *et al.*, 2006; Smiley *et al.*, 2002). Caliciviruses replicate in numerous animal hosts as well as in humans. An increase in the recognition of the disease burden associated with calicivirus infection in humans and the recent emergence of new strains has resulted in an intensified interest in understanding the mechanisms of calicivirus replication. The purpose of this chapter is to summarize current knowledge of the synthesis and proteolytic maturation of calicivirus proteins during virus replication.

Virion characteristics

Despite obvious differences in host range and pathogenicity, caliciviruses share several structural and replication features. Electron microscopy (EM) studies have played an important role in early epidemiological investigations, especially for the non-cultivable caliciviruses, and have provided the initial information on the morphological characteristics of calicivirus particles. Electrophoretic analysis of the calicivirus proteins revealed that they contained a single major structural protein with a molecular mass of approximately 60–62 kDa (Bachrach and Hess, 1973). An additional minor protein with a molecular weight of 15 kDa has also been consistently detected in virus preparations (Burroughs and Brown, 1974). By analogy with the picornaviruses, the icosahedral protein shell of the calicivirus particles was predicted to contain 180 copies of a major capsid protein (Bachrach and Hess, 1973). This structural prediction was later confirmed in X-ray crystallographic and cryo-EM studies (Chen *et al.*, 2004; Prasad *et al.*, 1994a; Prasad *et al.*, 1994b; Thouvenin *et al.*, 1997) in which it was shown that the virus particles were assembled from 90 dimers of the major capsid protein, VP1. The X-ray structure of VP1 showed that the protein had a conserved modular organization with two major structural domains, the N-terminal shell (S) and the C-terminal protruding (P) arch (Chen *et al.*, 2006; Prasad *et al.*, 1999). While the P domain formed bridging dimers on the surface of the virion, the S domain was found to be involved in intermolecular interactions of the capsid protein monomers that provide the icosahedral scaffold of the virion. Interestingly, protein–protein interactions of the properly folded capsid protein molecules were shown to be sufficient for the self-assembly of the virus particles. Consistent with that, expression of many calicivirus VP1 genes in plants, insect, yeast or mammalian cell systems resulted in the production of empty virus-like particles (VLPs) morphologically and antigenically indistinguishable from the virion (Boga *et al.*, 1997; Geissler *et al.*, 1999; Jiang *et al.*, 1992; Zhang *et al.*, 2006). Moreover, expression of the S domain alone resulted in the assembly of the icosahedral shell and production of 27-nm particles (Bertolotti-Ciarlet *et al.*, 2002).

The protein shell of the calicivirus virion was found to encapsidate a single-stranded RNA molecule (Adldinger *et al.*, 1969; Wawrzkiewicz *et al.*, 1968). Early sucrose gradient sedimentation studies of the RNA isolated from virions showed that the molecules migrated with velocity coefficients of 35–38S consistent with the RNA being approximately 2.6×10^3 kDa in molecular weight (Ehresmann and Schaffer, 1977; Wawrzkiewicz *et al.*, 1968). In later studies, the length of the calicivirus RNA genome has been reported to vary from 6.7 to 8.5 kb (Farkas *et al.*, 2008; Green, 2007). The shortest calicivirus genome described so far belongs to the Tulane virus isolated from an asymptomatic rhesus macaque (Farkas *et al.*, 2008), and the longest to the canine calicivirus isolated from a dog with diarrhoea (Matsuura *et al.*, 2002).

Affinity chromatography of the purified virion 36S RNA revealed that up to 80% of the molecules could be bound to oligo (dT)-cellulose indicating that calicivirus RNAs were polyadenylated (Ehresmann and Schaffer, 1977). Together with the discovery that purified calicivirus RNA could be infectious; this finding suggested that the 36S RNA might serve as mRNA for the initial rounds of viral protein synthesis (Adldinger *et al.*, 1969; Ehresmann and Schaffer, 1977; Wawrzkiewicz *et al.*, 1968). Most of the eukaryotic mRNA molecules have a cap group ($7m$GpppN) at their 5′ ends that promotes their recognition by the translation machinery of the cell. Analysis of the RNase digestion products of the calicivirus ^{32}P-labelled RNA failed to detect the presence of the methylated cap groups at their 5′ termini (Black *et al.*, 1978). Instead, it was demonstrated that similar to picornaviruses, the 5′ end of calicivirus RNA molecule was covalently linked to a small protein encoded by the virus genome, VPg (Burroughs and Brown, 1978).

Early events of infection

The strategy caliciviruses employ for membrane penetration and delivery of their RNA genomes into the cell cytoplasm is not completely understood. However, data obtained in experiments with inhibitors of endosome acidification such as chloroquine and bafilomycin A1 have indicated that caliciviruses might use pH-dependent mechanisms for the uncoating and release of RNA from the endosomes into the cytoplasm. Consistent with that, an increase of endosomal pH by chloroquine and bafilomycin A1 could be responsible for the observed inhibitory effect of these drugs at the early stages of FCV replication (Kreutz and Seal, 1995; Stuart and Brown, 2006).

The released calicivirus RNA genome might function directly as a template RNA for the initiation of viral protein synthesis. The first evidence that supported this conclusion came from *in vitro* translation experiments of RNA extracted from purified vesicular exanthema of swine virus (VESV) particles, in which several proteins ranging in molecular weight from 20 to 100 kDa were observed (Black *et al.*, 1978). Similar results were obtained in experiments with genomic RNA purified from RHDV virions; when translated *in vitro*, the extracted RNA programmed synthesis of several RHDV proteins including capsid protein (Boniotti *et al.*, 1994; Wirblich *et al.*, 1996).

At present, little is known about the transport of the calicivirus virion RNA to the future sites of virus replication and translation. It is possible that one of the factors that defines the transport of the calicivirus RNA within the infected cell might be linked to the interactions between the small VPg protein present at the 5' end of the calicivirus RNAs and host cell proteins involved in the initiation of the translation. Search for protein–protein interaction partners using the yeast two-hybrid system and pull-down assays revealed that calicivirus VPg proteins could interact with the eIF3, eIF4GI and eIF4E eukaryotic translation initiation factors (Chaudhry *et al.*, 2006; Daughenbaugh *et al.*, 2003, 2006; Goodfellow *et al.*, 2005). Cellular RNA chaperone proteins shown to bind to the 5' and 3' ends of the calicivirus genome are yet another group of factors that might contribute to the virus RNA transport (Gutierrez-Escolano *et al.*, 2000; Gutierrez-Escolano *et al.*, 2003; Karakasiliotis *et al.*, 2006). As a consequence of these interactions, calicivirus RNA might be targeted to the specific sites where formation of translation initiation complexes occurs with a subsequent recruitment of ribosomal subunits.

Strategy of calicivirus gene expression, initiation of protein translation and its control

The RNA genomes of caliciviruses share similar organization. The non-structural proteins are encoded in the large ORF1 located towards the 5' end of the genome, and structural proteins are encoded towards the 3' end. While genomic RNA serves as a template for synthesis of the large non-structural polyprotein encoded by the ORF1, structural proteins are mainly produced from the 3' end co-terminal subgenomic RNA. Despite similarities observed in the localization of structural and non-structural protein genes, caliciviruses can be divided into two major groups according to the organization of the ORF that encodes the major capsid protein, VP1. In the genomes of noroviruses and vesiviruses, the VP1 protein is encoded in a separate ORF, ORF2, which is expressed from the subgenomic template. In the genomes of sapoviruses and lagoviruses, the sequence encoding the VP1

protein is fused to the 3' end of the ORF1, and capsid protein can be expressed from the genomic template. Similar contiguous organization of the non-structural and capsid protein genes has also been reported for the genomes of bovine enteric caliciviruses, NB, Newbury1 and TCG (Oliver *et al.*, 2006; Smiley *et al.*, 2002). Two modes of VP1 expression would lead to the synthesis of the proteins with slightly different N-termini; however, these changes in the VP1 sequence have no effect on protein folding and do not result in loss of capsid self-assembly (Hansman *et al.*, 2008; Sibilia *et al.*, 1995). Interestingly, expression of VP1 from genomic RNA might also take place during replication of norovirus and vesivirus. Two groups of researchers reported synthesis of the ORF2-encoded protein in *in vitro* experiments with templates containing ORF1 and ORF2 of these viruses, indicating that the corresponding ORF2s could be expressed possibly through a termination–reinitiation mechanism (McCormick *et al.*, 2008; Sosnovtsev *et al.*, 1998). It is not clear whether the existence of an additional mechanism of VP1 expression confers any advantage for calicivirus replication. Successful generation of calicivirus replicons where parts of the capsid protein gene were replaced either with GFP or neomycin resistance genes showed that replication of the virus RNA did not require the presence of most of the ORF2 sequence (Chang *et al.*, 2008; Chang *et al.*, 2006). On the other hand, modifications of the sapovirus infectious cDNA clone resulting in the separation of polymerase and capsid genes with a stop codon had a deleterious effect on virus recovery (K.O. Chang, personal communication). It was proposed that some of the functions of the calicivirus capsid protein might be associated with initiation of the virus replication cycle; therefore, its synthesis might be critical for the early steps of the infection (Casais *et al.*, 2008; McCormick *et al.*, 2008).

In eukaryotic cells, the formation of a cap-binding complex around the RNA 5' end cap structure and recruitment of the small ribosomal unit is one of the factors that determine the efficiency of mRNA translation initiation. To outcompete host cell mRNA translation and launch the expression of their own proteins, several positive-sense RNA viruses use a cap-independent mechanism that involves the recognition and binding of the ribosomal initiation complex to the region within 5′ non-coding sequence of the viral RNA. These regions, called internal ribosome entry sites (IRES), are characterized by the presence of several coordinated secondary structure motifs and usually require quite lengthy non-coding sequences. In contrast, almost all caliciviruses have a very short 5′ end non-coding region, whose length argues against the presence of an IRES element. Analysis of the RNA folding performed for the 5′ end sequences of the calicivirus genomic and subgenomic RNA predicted conserved secondary structures near the beginning of each reading frame (Jiang et al., 1993; Pletneva et al., 2001; Simmonds et al., 2008). The presence of conserved sequence elements in the 5′ end of the viral genomes has been associated with their role as RNA replication signals. Consequently, secondary structures found in the 5′ end of the calicivirus genomes have been suggested to be important for virus replication (Simmonds et al., 2008). Their possible role in the translational regulation has been also suggested; furthermore, it was demonstrated that the first 110 nucleotides of the norovirus genome could bind several cellular proteins known to have a stimulating effect on IRES-dependent translation of virus proteins (Belsham and Sonenberg, 2000; Gutierrez-Escolano et al., 2000). However, no experimental data have been reported so far that would support the existence of an IRES-dependent translation initiation mechanism for caliciviruses. Regulation of calicivirus RNA translation initiation is thought to be mediated by the VPg protein covalently linked to the 5′ ends of both genomic and subgenomic RNAs. Interactions of the calicivirus VPg with host-cell translation initiation factors such as eIF3, eIF4GI, and eIF4E have recently been described in vitro and in infected cells (Chaudhry et al., 2006; Daughenbaugh et al., 2003, 2006; Goodfellow et al., 2005). It is likely that the calicivirus VPg might also be a part of a control mechanism that directly regulates the rate of viral RNA 5′ end processing by the ribosome. The short distance between the RNA 5′ end and the first AUG codon of the viral ORFs suggests that the landing ribosomal unit would inevitably be in close contact with the VPg protein attached to the exact 5′ end. As a result, translation initiation might be modulated by the VPg folding and by its interactions with other viral or host cell proteins.

Several studies have produced evidence suggesting that translation of the calicivirus non-structural polyprotein starts at the first in-frame AUG codon of the ORF1. Thus, synthesis of the 16-kDa N-terminal protein predicted to be translated from the first AUG codon of the ORF1 was observed in non-productive infection of rabbit primary hepatocytes with RHDV (Konig et al., 1998). The protein was specifically recognized by serum raised against the N-terminal region (9–112 aa) of the RHDV polyprotein, and the molecular size of the observed protein corresponded to the predicted sequence of the first in-line protein encoded in the ORF1 of RHDV (Konig et al., 1998). Further support for initiation of the ORF1 translation at the first AUG codon was obtained from recent studies of two other cultivable caliciviruses, murine norovirus and FCV. Comparative analysis of the murine norovirus ORF1 N-terminal proteins produced in the infected cells and those synthesized in vitro from an individually cloned gene showed that these proteins had identical gel-electrophoretic mobilities (Sosnovtsev et al., 2006). In addition, indirect evidence for the translational initiation at the first AUG codon of the ORF1 was provided in studies of the FCV replication using this virus's reverse genetics system. Mutagenesis of the corresponding codon residues in the infectious FCV cDNA clone resulted in a complete abrogation of viral replication (Sosnovtsev et al., 2002). Of interest, initiation of the translation of the ORF1 at the first in-frame AUG codon has been suggested in numerous in vitro translational studies of the non-cultivable caliciviruses (Liu et al., 1996; Oka et al., 2005b; Salim et al., 2008). However, it should be noted that the ORF1s of several caliciviruses carry two or three adjacent AUG codons that could be recognized as alternative translation initiation sites. Given that initiation of protein synthesis at alternative AUG codons has been reported in the presence of short 5′ end non-coding sequence and suboptimal (under scanning model) sequence context (Kozak, 1991), mapping of the precise starting point of the ORF1 polyprotein synthesis for these caliciviruses will require further investigation.

Role of calicivirus 3C-like protease in viral protein maturation, and experimental approaches used for its characterization

Initiation at the first AUG codon and successful translation of the entire calicivirus ORF1 would result in synthesis of the large polyprotein that includes domains of the non-structural proteins in the case of vesiviruses and noroviruses, and also a capsid protein in the case of sapoviruses and lagoviruses. Likewise, the expected sizes of the intact ORF1 polyproteins would vary significantly, ranging between 162 and 257 kDa. However, expression of the full-length polyprotein has never been observed in virus-infected cells grown under normal conditions. Instead, during synthesis, the ORF1 polyprotein undergoes efficient autocatalytic processing giving rise to several non-structural proteins (Liu *et al.*, 1996; Martin Alonso *et al.*, 1996; Sosnovtsev *et al.*, 1998; Wirblich *et al.*, 1996). *In vitro* analysis of the ORF1 polyprotein expression showed that proteolytic activity involved in its processing was encoded by the sequence located upstream of the virus RNA polymerase gene in the C-terminal region of the ORF1. An alignment of the corresponding region with sequences of the known viral proteases revealed the presence of the limited sequence similarity with the 3C proteases of the picornaviruses (Boniotti *et al.*, 1994; Farkas *et al.*, 2008; Jiang *et al.*, 1993; Liu *et al.*, 1995; Neill, 1992; Smiley *et al.*, 2002). Interestingly, a sequence database search of the calicivirus ORF1s showed that the rest of the polyprotein sequence had no significant homology to other reported protease sequences, indicating that 3C-like (3CL) protease was the only proteolytic enzyme encoded in the virus ORF1. Consistent with that, the *in vitro* translation of the ORF1 sequences truncated upstream of the putative protease gene resulted in the synthesis of unprocessed proteins (Liu *et al.*, 1996; Martin Alonso *et al.*, 1996; Sosnovtseva *et al.*, 1999).

In addition to analogous genome location and the presence of limited sequence homology between the calicivirus 3CL and picornavirus 3C proteases, the alignment of their sequences suggested that the calicivirus enzymes could have a similar functional organization of the active site (Boniotti *et al.*, 1994; Neill, 1990). Picornavirus 3C proteases have been shown to have overall folding similar to that of the cellular chymotrypsin-like serine proteases. However, in contrast to a serine protease, where active site residues include the serine–histidine–aspartate catalytic triad, the picornavirus 3C enzymes contain a cysteine-histidine-aspartate/glutamate triad and employ a cysteine residue as a catalytic nucleophile (Allaire *et al.*, 1994; Birtley *et al.*, 2005; Matthews *et al.*, 1994; Mosimann *et al.*, 1997). To confirm the involvement of the cysteine residue in the catalysis of the proteolytic reaction by calicivirus protease and therefore verify the enzyme identity, several research groups used site-directed mutagenesis to substitute the predicted active site cysteine residue with a number of amino acids. Replacement of the predicted cysteines with glycine residues at positions 1212 and 1238 (ORF1) in RHDV (lagovirus) and Southampton virus (norovirus), respectively, completely abolished activity of their proteases (Boniotti *et al.*, 1994; Liu *et al.*, 1996). A similar effect was observed in mutagenesis experiments targeting cysteines at positions 1193 and 1171 (ORF1) in FCV (vesivirus) and Mc10 (sapovirus), respectively (Oka *et al.*, 2005b; Sosnovtseva *et al.*, 1999). Furthermore, the expression of full-length calicivirus ORF1 clones carrying a mutated protease gene led to the synthesis of the entire ORF1 polyprotein, usually undetectable when the active protease was present (Blakeney *et al.*, 2003; Liu *et al.*, 1996; Oka *et al.*, 2005b; Sosnovtsev *et al.*, 2002).

Testing calicivirus protease activity in the presence of inhibitors specific to different protease classes further confirmed its similarity to the picornavirus 3C protease. The calicivirus 3CL was found to have either low or no sensitivity to inhibitors of serine proteases (leupeptin, chymostatin, pefabloc SC, and aprotinin). Also, shown to be ineffective in blocking calicivirus protease were inhibitors specific to metalloproteases (EDTA, phosphoramidon and bestatin), aspartate proteases (pepstatin) and papain-like cysteine proteases (cystatin, E-64, and antipain). However, similar with picornavirus 3C proteases, the activity of the calicivirus 3CL enzyme could easily be inhibited with classical cysteine protease inhibitors (ZnCl$_2$, N-ethylmaleimide, *p*-chloromercuribenzoic acid, and methyl methanethiosulphonate), targeting the cysteine thiol group in the active site of the enzyme (Blakeney *et al.*, 2003; Someya *et al.*, 2005; Sosnovtsev *et al.*, 1998). The hypothesized structural and biochemical similarity between the calicivirus and serine proteases found additional support when RHDV protease was observed to retain proteolytic activity after replacement of its active site cysteine with a serine residue (Boniotti *et al.*, 1994).

Progress in research of calicivirus replication has been hampered for years due to the lack of suitable animal and cell culture models. Most of the calicivirus protein maturation studies have been limited to an *in vitro* expression analysis of the virus ORF1-derived proteins. Until recently, the reliability of the *in vitro* generated data could be assessed only for a few animal caliciviruses whose growth could be analysed in cell culture. The recent discovery of the cell lines supporting the replication of the murine norovirus 1 (MNV-1) and RHDV (Liu *et al.*, 2006; Wobus *et al.*, 2004) promises to both verify a wealth of information accumulated in *in vitro* experiments and identify new general protein maturation mechanisms shared by caliciviruses across the entire family.

The starting point in the *in vitro* analysis of the proteolytic maturation of the calicivirus proteins was an expression of the extended ORF1 regions undergoing autocatalytic processing followed by identification of the cleavage products using a variety of biochemical techniques. The latter have included radioactive labelling of the synthesized proteins, their immunological analysis using panels of region-specific sera, and mapping borders of the proteolytic products by their direct N-terminal sequencing. In addition, a number of the cleavage sites were identified and verified using site-directed mutagenesis of the putative scissile bond sequences (Clarke and Lambden, 2000; Liu *et al.*, 1996; Martin Alonso *et al.*, 1996; Meyers *et al.*, 2000; Oka *et al.*, 2006; Sosnovtsev *et al.*, 2002; Wirblich *et al.*, 1996).

Quantitative data related to the rate of processing at different cleavage sites in *in vitro* studies were assessed either by direct measurement of the signals produced by individual protein bands or by measurement of the activity of the marker protein, e.g. luciferase, released by proteolytic cleavage (Belliot *et al.*, 2003; Joubert *et al.*, 2000). Further biochemical and structural characterization of the calicivirus proteases was facilitated by the successful bacterial expression of the active enzymes. Co-expression of the protease fused to different ORF1 polyprotein regions tagged with specific epitopes helped to purify and identify additional proteolytic products as well as to map corresponding cleavage sites including those that defined the borders of the protease gene itself (Liu *et al.*, 1999b; Martin Alonso *et al.*, 1996; Oka *et al.*, 2005a; Sosnovtseva *et al.*, 1999; Wirblich *et al.*, 1995). The proteases of lagovirus and norovirus were identified in bacterial expression studies as 15- and 19-kDa proteins, respectively (Liu *et al.*, 1999b; Wirblich *et al.*, 1995). In contrast, proteases of vesivirus and sapovirus were found to be expressed as 71- and 70-kDa bifunctional precursor proteins, respectively, that also contained polymerase homology domains (Oka *et al.*, 2005a; Sosnovtseva *et al.*, 1999).

A number of *trans* cleavage assays have been developed that have assisted in the elucidation of the enzyme substrate specificities. In general, most of these assays were based on the incubation of a bacterially expressed recombinant protease with radiolabelled *in vitro* synthesized precursor proteins that carried either putative cleavage sites or their modified versions (Belliot *et al.*, 2003; Blakeney *et al.*, 2003; Sosnovtsev *et al.*, 1998; Wirblich *et al.*, 1995). Another approach involved the testing of the protease activity with a set of synthetic peptides that contained cleavage site-specific amino acid sequence followed by mass-spectrometry analysis of the proteolytic products (Robel *et al.*, 2008; Scheffler *et al.*, 2007). An additional modification of the *trans* cleavage assay included the use of fluorogenic peptides. In this case, the increase in fluorescence upon hydrolysis of the scissile bond was directly monitored in real time allowing comparison of the kinetic characteristics of the reaction for the different mutant proteases (Someya *et al.*, 2008). Calicivirus reverse genetics systems have been used also to identify cleavage events essential for virus protein maturation and replication. Of interest, mutagenesis of the majority of the proteolytic cleavage sites sequences in an infectious FCV cDNA clone had a negative effect on virus recovery indicating that corresponding cleavages were critical for virus replication (Sosnovtsev et al., 2002; Sosnovtsev et al., 1998). Similarly, a mutation that eliminated a cleavage site between the protease and polymerase proteins in an infectious cDNA clone of the MNV-1 virus was shown to be lethal for the virus (Ward et al., 2007). At the same time, mutagenesis of the cryptic cleavage sites between functional domains of the FCV protease and polymerase proteins identified in in vitro studies has proved them to be non-essential for productive virus replication (Sosnovtsev et al., 2002).

Proteolytic processing maps

The combination of data produced in in vitro translation and bacterial expression systems led to the generation of proteolytic maps for several caliciviruses (Fig. 4.1). The lagovirus ORF1 processing map established in in vitro translation, bacterial expression and ex vivo studies contained seven cleavage sites (Fig. 4.1) (Meyers et al., 2000; Wirblich et al., 1996). Five cleavage sites corresponding to mature stable proteins that could be detected in infected cells were reported for the ORF1 polyproteins of norovirus and vesivirus (Fig. 4.1) (Belliot et al., 2003; Liu et al., 1999b; Seah et al., 2003; Sosnovtsev et al., 2002, 2006). Recent studies of the Mc10 strain showed that the sapovirus ORF1 polyprotein had at least six cleavage sites (Fig. 4.1) (Oka et al., 2005b, 2006). The number of individual proteins encoded by the calicivirus ORF1s varies slightly among the virus genera. Moreover, for some of the viruses, processing of certain protein borders is inefficient and stable precursors have been reported as mature forms. Owing to variation in sizes, most nomenclature schemes for protein identification used their molecular weights, therefore adding to the complexity of protein description and comparison. We proposed a simplified universal nomenclature that would consider the maximal number of reported cleavage sites shown and predicted to be essential in virus replication. The suggested order and nomenclature of the proteins encoded by the calicivirus ORF1 polyprotein would be as follows NH$_2$-NS1–NS2–NS3–NS4–NS5–NS6–NS7–(VP1)-COOH (Fig. 4.1) (Sosnovtsev et al., 2006).

The NS1 and NS2 proteins are the ORF1 N-terminal proteins that are individually expressed by vesivirus, lagovirus, and sapovirus (Oka et al., 2005b; Sosnovtsev et al., 2002; Wirblich et al., 1996). In contrast, in the replication cycle of the MNV-1 strain, these two proteins are expressed as part of one larger protein, NS1–2 with a molecular mass of 38 kDa (Sosnovtsev et al., 2006). Consistent with that, a lack of processing of the ORF1 N-terminal protein has been observed also in numerous in vitro expression studies of the norovirus non-structural polyprotein (Belliot et al., 2003; Blakeney et al., 2003; Hardy et al., 2002; Liu et al., 1996). Following the NS2 in the polyprotein, the NS3 protein has been identified unambiguously as a 37–40 kDa mature stable product for all caliciviruses (Liu et al., 1996; Oka et al., 2005b; Sosnovtsev et al., 2002; Wirblich et al., 1996). While the release of the NS3 protein was shown to be quite efficient, varying cleavage efficiencies have been observed between the NS4–NS5–NS6–NS7 proteins in the ORF1 C-terminal protein block. Relatively slow processing was found at the NS4–NS5 border leading to the accumulation of the corresponding intermediate, NS4–NS5, in cells infected with vesivirus and norovirus (Sosnovtsev et al., 2006; Sosnovtsev et al., 2002). Unprocessed p46 corresponding to the sapovirus NS4–NS5 protein has been also observed in in vitro translation studies (Oka et al., 2005b). Expression analyses of the lagovirus ORF1 clones have showed the presence of an analogous 41–43 kDa precursor in proteolytic products generated in bacterial cells and in in vitro (Konig et al., 1998; Martin Alonso et al., 1996; Wirblich et al., 1996). The second frequently reported proteolytic product derived from the C-terminal part of the ORF1 polyprotein consisted of the NS6 and NS7 sequences. The NS6–NS7 (NS6–7) protein has been described as a stable complex in vesivirus and sapovirus-infected cells (Casais et al., 2008; Chang et al., 2005; Oehmig et al., 2003; Sosnovtseva et al., 1999). It has also been detected as a stable intermediate in products of the proteolytic processing of the lagovirus and norovirus ORF1 polyproteins (Belliot et al., 2003; Konig et al., 1998; Martin Alonso et al., 1996; Wirblich et al., 1996). However, in vitro expression analysis has revealed that the lagovirus and norovirus NS6–NS7 precursors undergo further processing (Belliot et al., 2003; Liu et al., 1999b;

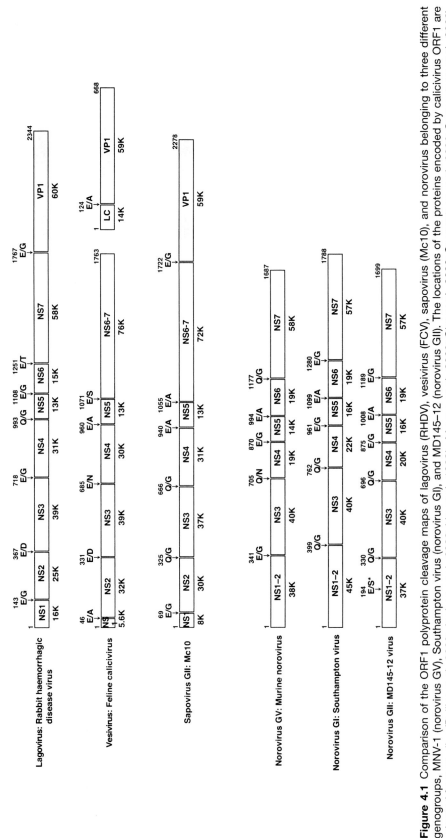

Figure 4.1 Comparison of the ORF1 polyprotein cleavage maps of lagovirus (RHDV), vesivirus (FCV), sapovirus (Mc10), and norovirus belonging to three different genogroups, MNV-1 (norovirus GV), Southampton virus (norovirus GI), and MD145-12 (norovirus GII). The locations of the proteins encoded by calicivirus ORF1 are adapted from studies (Belliot et al., 2003; Joubert et al., 2000; Liu et al., 1999b; Meyers et al., 2000; Oka et al., 2006; Sosnovtsev et al., 2002), and cleavage sites are indicated with arrows. The position of possible cleavage site, E^{194}/S^{195}, recognized by the norovirus NS6 in the NS1-2 protein of Camberwell virus (Seah et al., 2003) is denoted with asterisk. The E^{124}/A^{125} site cleaved by FCV protease during the virus capsid protein maturation is shown also for the ORF2-encoded precursor. Protein coding sequences are drawn as boxes, with the molecular mass of the encoded protein shown below. The names of the proteins are given inside boxes in nomenclature introduced for murine norovirus (Sosnovtsev et al., 2006).

Meyers *et al.*, 2000; Sosnovtsev *et al.*, 2006). More complete processing of the virus NS6–NS7 protein has been observed in mammalian cells transiently expressing corresponding ORF1 polyproteins (Chang *et al.*, 2006; Katayama *et al.*, 2006; Meyers *et al.*, 2000; Seah *et al.*, 1999). Furthermore, immunoanalysis of virus protein synthesis in lagovirus and norovirus-infected cells confirmed the presence of mature forms of the NS6 and NS7 proteins (Konig *et al.*, 1998; Sosnovtsev *et al.*, 2006).

The VP1 is the next and last in the gene order of the lagovirus and sapovirus ORF1 proteolytic maps (Fig. 4.1). The removal of this protein has been described as an efficient event that probably occurs co-translationally (Boniotti *et al.*, 1994; Clarke and Lambden, 2000; Oka *et al.*, 2005b). In *in vitro* experiments with the RHDV ORF1, the VP1 expression could be detected within minutes of the start of translation (Meyers *et al.*, 2000). The efficiency of this cleavage was also confirmed in studies of the lagovirus and sapovirus-infected cells, where most of the VP1 and NS7 or NS6–NS7 proteins was detected as discrete products (Chang *et al.*, 2005; Konig *et al.*, 1998).

The proteolytic cascade of the ORF1 polyprotein processing

Single-stranded positive-sense RNA viruses utilize a number of strategies to maximize the use of genetic information encoded by their small RNA genomes. One of them involves temporal control of the proteolytic processing of their polyproteins that allows sequential release of the virus proteins or their precursors. Using such control, the replicating virus generates not only fully processed, mature forms of the virus proteins, but a number of intermediates, which might have additional functions. It also allows the virus to manage temporal availability of those or other forms of the virus proteins required at certain stages of its replication. Mechanisms employed to control the proteolytic maturation of the virus proteins are based on differences in the rates of processing of individual cleavage sites of the virus polyproteins, which are defined by structural and sequence context factors. They also rely on participation of either more than one virus protease or several forms of the same protease with distinct cleavage site affinities. Some of the cleavages reportedly

occur only in association with RNA replication. In addition, the cleavage efficiency of the virus protease may be modulated by cellular factors.

Sequential processing of the calicivirus ORF1 polyprotein has been extensively studied using cell-free translation systems. *In vitro* analyses of the ORF1 expression demonstrated that calicivirus protease was capable of concurrent proteolytic cleavage of the synthesized polyprotein (Blakeney *et al.*, 2003; Clarke and Lambden, 2000; Martin Alonso *et al.*, 1996; Oka *et al.*, 2005b; Sosnovtseva *et al.*, 1999; Wirblich *et al.*, 1996). Moreover, in time-course experiments, an early appearance and accumulation of N-terminal cleavage products was observed before the detection of stable precursors or mature forms of the virus protease. At early time points, the protease was detected as multiple disperse protein bands corresponding to the high molecular-mass precursors that contained the upstream parts of the ORF1 polyprotein extending into the protease domains (Blakeney *et al.*, 2003; Hardy *et al.*, 2002; Sosnovtseva *et al.*, 1999). These observations suggested that calicivirus protease-mediated cleavage resulting in release of the N-terminal proteins takes place before the translation of the ORF1 reaches the 3' end of the sequence and, apparently, does not require the presence of the NS7 domain. Similar conclusions were drawn from expression studies of the ORF1 templates carrying a truncated NS7 gene where efficient processing of the N-terminal part of the ORF1 polyprotein was observed in the absence of the NS7 domain (Hardy *et al.*, 2002; Sosnovtseva *et al.*, 1999).

The two recognized forms of the calicivirus protease (NS6 and NS6–NS7 precursor) represent proteolytically active enzymes that can mediate processing of the majority of the cleavage sites described in the virus ORF1 polyprotein (Belliot *et al.*, 2003; Robel *et al.*, 2008; Scheffler *et al.*, 2007; Seah *et al.*, 2003; Sosnovtseva *et al.*, 1999). Furthermore, proteolytic activity was also described for the larger precursor molecules supporting the suggestion that processing of the polyprotein cleavage sites does not require the autocatalytic release of the nascent protease. Thus, blocking release of the FCV NS6-7 protein through mutagenesis of the protease N-terminal sequence in the infectious cDNA clone abrogated recovery of the virus; however, it had no effect on the processing of the N-terminal part of the FCV ORF1 polyprotein (Sosnovtsev et al., 2002). Similarly, the norovirus NS6 protein blocked by mutagenesis at its N- and C-termini was capable of the efficient release of the NS1–2 and NS3 proteins (Belliot et al., 2003). It was suggested that intermediate precursor forms of the virus protease protein might play slightly different roles in the calicivirus replication cycle as described for other positive-sense RNA viruses. In addition, the different rates of proteolytic cleavage exhibited by these precursors might be employed by the virus to control the maturation of the non-structural proteins. In support of that idea, variation in the processing rates of certain ORF1 polyprotein cleavage sites has been reported for NS6 and NS6–NS7 protease forms of sapovirus and norovirus (Belliot et al., 2003; Robel et al., 2008; Scheffler et al., 2007).

The kinetics of calicivirus polyprotein processing has not been studied in detail for viruses from all genera. However, a comprehensive in vitro analysis of the ORF1 processing has been performed for the lagovirus and norovirus (Belliot et al., 2003; Blakeney et al., 2003; Hardy et al., 2002; Meyers et al., 2000; Scheffler et al., 2007). In addition, quantitative analyses of the protease substrate specificity using sets of synthetic peptides displaying cleavage site sequences have allowed for certain predictions for the order of the sequential processing mediated by the sapovirus and norovirus enzymes (Robel et al., 2008; Scheffler et al., 2007). These data and those accumulated thus far for other caliciviruses suggest similarities in the overall maturation strategy of their non-structural proteins. Thus, most of the in vitro analyses of the polyprotein processing showed that the first cleavage steps included rapid processing of the ORF1 polyprotein at the NS2–NS3 and NS3–NS4 junctions (Fig. 4.2) (Belliot et al., 2003; Blakeney et al., 2003; Hardy et al., 2002; Liu et al., 1996; Meyers et al., 2000). Processing with similar levels of efficiency has also been observed for the NS1–NS2 bond in lagovirus, vesivirus and sapovirus polyproteins (Fig. 4.2) (Meyers et al., 2000; Oka et al., 2005b; Sosnovtsev et al., 2002). Of interest, in a peptide trans cleavage assay, the sapovirus protease cleaved the NS1-NS2 bond more efficiently than those of NS2–NS3 and NS3–NS4 suggesting

that the NS1 might be among the first proteins released by proteolytic processing of the virus ORF1 (Robel et al., 2008). However, in the case of norovirus NS1–NS2 (NS1–2), in vitro expression of the corresponding ORF1 polyprotein has not led to the identification of the cleavage site recognized by the virus protease, therefore, indicating that this protein might remain unprocessed during infection (Fig. 4.2) (Liu et al., 1996).

Processing of the released C-terminal NS4–NS5-NS6–NS7 precursor has been described to occur at comparatively lower rates, and consistent with that, corresponding cleavages have sometimes been referred to as secondary cleavage events (Belliot et al., 2003; Blakeney et al., 2003). With some variations, further cleavage of this precursor has been reported to result in two stable forms corresponding to NS4–NS5 and NS6–NS7 proteins analogous to the picornavirus 3AB and 3CD proteins (Fig. 4.2). Processing rates observed at the NS5–NS6 junction showed some variation and depended on the expression system employed for the cleavage analysis. High efficiency of the cleavage at this scissile bond has been reported in bacterial and in vitro studies of lagovirus and vesivirus polyproteins (Martin Alonso et al., 1996; Sosnovtseva et al., 1999; Wirblich et al., 1995; Wirblich et al., 1996). Similar observations have been made for the processing of the sapovirus ORF1 polyprotein; though, efficient release and accumulation of the 72 kDa NS6–7 protein were clearly detected only with expression of the N-terminally truncated clones (Oka et al., 2005a; Oka et al., 2005b). Of interest, rapid cleavage at the NS5–NS6 junction was also observed in bacterial expression studies of norovirus ORF1 clones (Belliot et al., 2003; Liu et al., 1999b; Someya et al., 2000, 2002). However, different research groups reported varying rates of processing of this bond in in vitro. Thus, lack of the cleavage at the NS5–NS6 junction was demonstrated in in vitro translation studies of the norovirus genogroup I strains, Southampton and Norwalk viruses (Blakeney et al., 2003; Hardy et al., 2002; Liu et al., 1996). In contrast, this cleavage was readily detected in in vitro studies of norovirus from genogroups II (MD145–12) and V (MNV-1) (Belliot et al., 2003; Sosnovtsev et al., 2006). The observed processing rate variation might

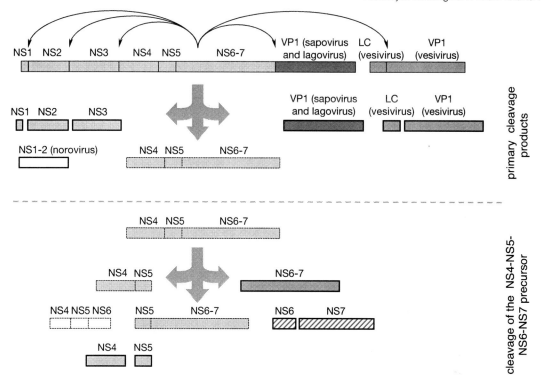

Figure 4.2 Model for the cascade of proteolytic processing of the calicivirus ORF1 polyprotein. Co-translational processing of the calicivirus ORF1 polyproteins, sometimes referred to as primary cleavage events, releases either NS1–2 (norovirus) or NS1 and NS2 (sapovirus, vesivirus, and lagovirus), and NS3 proteins. Simultaneously, the virus protease mediates efficient removal of virus capsid protein fused at the C-terminus of the sapovirus and lagovirus ORF1 polyproteins and processes the vesivirus capsid protein precursor (both shown as dark grey boxes). The NS4–NS5-NS6–NS7 precursor undergoes further autocatalytic processing that reportedly occurs through a number of *cis* or *trans* cleavages. At least two secondary cleavage pathways have been proposed in the processing of the C-terminal protein block. The main proteolytic products of its cleavage include the NS4–NS5 and NS6–NS7 proteins. The reported synthesis of the NS5-NS6–NS7 intermediate indicates that in one pathway the NS4 protein can be removed first. Similarly, the protease reportedly generates the NS4–NS5–NS6 and NS7 proteins when cleaving the NS6–NS7 bond. Sapovirus and vesivirus have the NS6–NS7 (NS6-7) protein as a final form of the virus protease and polymerase. Further processing of the intermediate products generates mature forms for the rest of the virus proteins. White colour boxes indicate norovirus proteins and hatched boxes represent proteins and their precursors common for noroviruses and lagoviruses. Boxes with solid line borders correspond to mature virus proteins and boxes with dashed line borders to their precursors.

relate to considerable divergence of the norovirus ORF1 sequences and might reflect different folding of the virus polyproteins synthesized *in vitro*. It is also possible that optimal presentation of some of the cleavage sites in the norovirus polyprotein requires the presence of the cellular factors or active viral replication. Consistent with this idea, more efficient processing at the NS5–NS6 junction was observed in experiments with transiently transfected mammalian cells, in norovirus replicon-bearing cells and during norovirus infection (Chang *et al.*, 2006; Katayama *et al.*, 2006; Seah *et al.*, 1999; Sosnovtsev *et al.*, 2006).

As mentioned above, processing of the NS6–NS7 protein was documented only for lagovirus and norovirus. In contrast, proteolytic products of the NS4–NS5 protein cleavage were identified for all caliciviruses (Fig. 4.2). The observed *in vitro* processing of the NS4–NS5 scissile bond was relatively slow, and consistent with a post-translational cleavage. It was demonstrated that the NS4–NS5 junction has been cleaved in both bacterial and *in vitro* expressions of the vesivirus ORF1 polyprotein (Sosnovtsev *et al.*, 1998; Sosnovtseva *et al.*, 1999). Furthermore, mature forms of these proteins were detected in vesivirus-infected cells (Casais *et al.*, 2008; Sosnovtsev *et al.*, 2002). Of interest, analysis of the RHDV NS4–NS5 protein processing revealed the existence of two alternative cleavage sites in the sequence of this precursor. In infected hepatocytes and during transient expression of the RHDV ORF1, the protein was shown to undergo proteolytic processing that gave rise to two pairs of the proteolytic products, p29 and p13, and p23 and p18. While the first pair corresponded to the mature RHDV NS4 and NS5 proteins, the latter was represented by the modified forms of these proteins and was produced by unknown proteolytic activity (Konig *et al.*, 1998; Meyers *et al.*, 2000; Thumfart and Meyers, 2002). The observed processing rate of the sapovirus NS4–NS5 site (similar to the NS5–NS6 cleavage) has been reportedly dependent on the experimental setup of the expression analysis. Thus, expression of the Mc10 strain ORF1 polyprotein *in vitro* provided evidence for the synthesis of the virus mature NS4 and NS5 proteins (Oka *et al.*, 2005b), while *trans* cleavage assay showed that the peptide containing the NS4–NS5 cleavage site sequence was resistant to cleavage by the sapovirus protease (Robel *et al.*, 2008). Similar resistance was observed in a peptide *trans* cleavage assay for the peptide corresponding to the norovirus NS4–NS5 cleavage site (Scheffler *et al.*, 2007); however, the corresponding norovirus protein border was shown to be cleaved by the virus protease *in trans* in *in vitro* experiments (Belliot *et al.*, 2003). In addition, mature forms of either protein were identified in norovirus-infected cells (Sosnovtsev *et al.*, 2006). Release of the mature forms of the calicivirus NS4 and NS5 proteins was apparently dependent on the *trans* cleavage of the NS4–NS5 protein. However, one cannot exclude the possibility that certain amounts of the mature proteins could be produced by the consecutive N-terminal cleavages of the larger NS4–NS5–NS6–NS7 and NS4–NS5–NS6 precursors. Several studies have identified these intermediates among the proteolytic products generated by the cleavage of the C-terminal part of the ORF1 polyprotein *in vitro* (Fig. 4.2). The role these intermediates play in the calicivirus replication cycle is not clear and remains to be investigated; however, their presence in infected cells has been confirmed by immunoprecipitation analysis (Belliot *et al.*,

2003; Oka *et al.*, 2005b; Sosnovtsev *et al.*, 2006; Sosnovtsev *et al.*, 2002).

The NS6–NS7 precursor processing in lagovirus and norovirus might also be dependent on post-translational cleavage *in trans*. The lagovirus NS6–NS7 protein was cleaved very inefficiently in bacteria and was found to be resistant to the cleavage in *in vitro* experiments (Wirblich *et al.*, 1995; Wirblich *et al.*, 1996). Yet, synthesis of the NS7 protein was easily detected in BHK21 cells transiently expressing the RHDV ORF1 polyprotein as well as in the virus-infected hepatocytes (Konig *et al.*, 1998; Meyers *et al.*, 2000). Kinetic analysis of the ORF1 transient expression showed synthesis of similar amounts of NS6–NS7 and NS7 proteins during translation and suggested steady-state conversion of the NS6–NS7 precursor (Meyers *et al.*, 2000). Similar conversion was observed in time-course experiments targeting an *in vitro* translation of the norovirus ORF1 (Belliot *et al.*, 2003). The amount of the processed NS7 protein has been observed to increase with time of reaction or when the *trans* cleavage was stimulated by addition of bacterially expressed recombinant enzyme. The ability of the norovirus protease to recognize the NS6–NS7 border *in trans* was demonstrated in another set of experiments where recombinant enzyme successfully processed cleavage sites of the precursor carrying inactivated protease sequence (Belliot *et al.*, 2003). Nevertheless, it should be noted that some of the bacterial expression studies suggested that the cleavage of the norovirus NS6–NS7 border might occur co-translationally and probably *in cis* (Blakeney *et al.*, 2003).

Co-translational processing of NS2–NS3, NS3–NS4 and, in some cases, NS1–NS2 and NS5–NS6 cleavage sites suggested that recognition of these sites and catalysis of the cleavage reaction might occur through a monomolecular mechanism and, therefore, qualify them as *cis* cleavage events. In infected cells, where replication of virus is compartmentalized, and sites of virus replication and virus protein synthesis may carry increased amount of protease molecules, most cleavages are likely to occur *in trans*. However, *cis* cleavages might be important for the virus replication at its early stages, when the concentration of non-structural proteins is still low, and the speed of the virus protein maturation might be affected by the proportion of *trans* versus *cis* cleavages (monomolecular vs. bimolecular cleavages). It is conceivable that the ratio between *cis* and *trans* cleavages coupled to dissimilar specificities of NS6 and NS6–NS7 protease forms might influence the order of processing and release of the virus proteins encoded by ORF1 (Belliot *et al.*, 2003; Robel *et al.*, 2008; Scheffler *et al.*, 2007).

There is no direct experimental data showing that the calicivirus protease is capable of true *cis* cleavage of scissile bonds. Nevertheless, differential recognition of cleavage sites by selected norovirus protease mutants suggests that processing of the virus NS5–NS6 scissile bond is mediated by an intramolecular mechanism analogous to that proposed for the autocleavage of the picornavirus 3C protease (Khan *et al.*, 1999; Matthews *et al.*, 1994; Someya *et al.*, 2002, 2008). Accordingly, an introduction of amino acid substitutions into the N-terminal sequence of the NS6 gene in the position 54 (E54I, E54L or E54P) resulted in the enzyme's loss of its processing activity towards the NS4–NS5 site, which is presumably cleaved *in trans*. However, it had no effect on the cleavage of the NS5–NS6 bond and therefore, on the release of the protease N-terminus. One of the possible explanations suggested by the researchers was that *cis* cleavage and release of the NS6 were associated with structural changes in the protease molecule, and introduced amino acid mutations blocked further conformational transition required for the *trans* cleavage activity (Someya *et al.*, 2008).

The ORF1 polyprotein processing: involvement of cellular factors and cellular proteases

For a number of viruses, proteolytic processing of their polyproteins is regulated by cellular proteins and virus proteins other than the viral protease (Lackner *et al.*, 2005; Liljestrom and Garoff, 1991; Love *et al.*, 1998; Moehring *et al.*, 1993; Morgenstern *et al.*, 1997; Rinck *et al.*, 2001). For caliciviruses, it been demonstrated that the virus NS6 protease was capable of mediating the ORF1 polyprotein processing in the absence of any host-cell factor. The observation was supported by a comparative analysis of the ORF1

proteolytic processing products derived in *in vitro* translation reactions with virus proteins observed in infected cells. Nevertheless, possible involvement of cellular factors has been proposed in the transient expression studies of the ORF1 of the Camberwell virus. Proteolytic processing of the corresponding ORF1 polyprotein in transfected COS cells generated fully processed NS6 and NS7 proteins (Seah *et al.*, 1999). Moreover, in a similar set of experiments, the same group of researchers described an additional cleavage event in the maturation of the virus NS1–2 protein. The protein was probably processed at Glu194/Ser195, and the cleavage was mediated by virus protease expressed *in trans* (Seah *et al.*, 2003). These findings suggested the presence in transfected cells of an 'environment' that was responsible either for enhancing or modulating of the virus protease cleavage efficiency.

A number of studies have shown that most of the calicivirus polyprotein cleavages are mediated by the virus-encoded protease. However, there have been several reports indicating that proteolytic activities other than that of the NS6 might be involved in a processing of the virus non-structural proteins. A non-canonical cleavage site has been mapped for the NS4–NS5 precursor protein of the RHDV (Thumfart and Meyers, 2002). The NS4-specific serum recognized an additional truncated version of this protein in the RHDV-infected hepatocytes and transfected BHK21 cells transiently expressing the virus ORF1 (Konig *et al.*, 1998; Meyers *et al.*, 2000; Thumfart and Meyers, 2002). However, the corresponding cleavage has not been observed in *in vitro* studies and the origin of this proteolytic activity has not been identified (Thumfart and Meyers, 2002). Of interest, transient expression of the vesivirus NS4 protein has been shown to induce caspase activation in transfected CRFK cells (Sosnovtsev *et al.*, 2004). The mapped RHDV cleavage site sequence, VASDNVDRGDQGVD, is consistent with caspase recognition motif DXXD, suggesting that the expression of the RHDV NS4 might also result in caspase activation with further processing of the protein by one of the activated caspases.

Virus-induced apoptosis has been implicated in caspase-mediated cleavage of the norovirus NS1–2 protein. It was demonstrated that MNV-1 replication in RAW264.7 cells results in their significant morphological and biochemical changes at 12–24 hours post infection (Bok *et al.*, 2009; Wobus *et al.*, 2004). The observed changes have been attributed to the induction of apoptosis and shown to be associated with the caspase activation (Bok *et al.*, 2009). Accordingly, an additional internal cleavage of the NS1–2 protein observed at the late stages of the MNV-1 infection has been linked to the activated murine caspase-3 (Sosnovtsev *et al.*, 2006). In *in vitro* assays, the caspase-3 was found to recognize at least two cleavage sites in the corresponding protein sequence (Sosnovtsev *et al.*, 2006). Of interest, caspase-3-mediated cleavage was not the only processing event observed for this protein that was not mediated by the virus protease. At late stages of infection, the MNV-1 38-kDa NS1–2 protein has been converted and replaced with a slightly smaller 36-kDa protein (Sosnovtsev *et al.*, 2006). The nature of this conversion is not clear; however, one of the possible explanations includes an additional cleavage mediated by an unknown protease that might be activated by changes in the intracellular environment inflicted by the progress of virus infection and induction of cell death pathways.

Onset of apoptosis in the virus-infected cells might be responsible for the appearance of the truncated form of the vesivirus capsid protein. Two lines of evidence suggested that activated caspases −2 or −6 could be involved in the processing of the FCV 62-kDa VP1 protein into the smaller 40-kDa form found in virus-infected CRFK cells (Carter *et al.*, 1989). First, the 40-kDa cleavage product could be generated *in vitro*, when the full-length protein was incubated with either of these enzymes, and, second, the corresponding cleavage was successfully blocked by the use of peptide caspase inhibitors (Al-Molawi *et al.*, 2003).

The role, if any, of apoptosis-related processing of the calicivirus proteins in virus replication has not been established, and the caspase-mediated cleavages might simply reflect a by-product of defence mechanism of the cell triggered by virus infection. One cannot exclude the possibility at this point that these cleavage events might have their own functional importance that can be exploited by the virus. Further work is required to characterize processing of virus proteins by host-cell proteases and to establish its presumed role in virus replication.

Involvement of calicivirus protease in maturation of virus capsid protein

Calicivirus proteases are thought to be involved in the maturation of the lagovirus and sapovirus capsid proteins synthesized as part of the ORF1 polyprotein. As mentioned above, the expression of the RHDV ORF1 results in the release of p60 identified as the virus capsid protein, VP1 (Boniotti *et al.*, 1994). The genome organization of sapovirus suggests also that their capsid proteins might be derived by proteolytic cleavage at the polymerase–capsid protein borders (Liu *et al.*, 1995). In support of this hypothesis, expression of the human sapovirus entire ORF1 or its C-terminal part in mammalian and insect cells led to the cleavage of the fused capsid protein from the non-structural polyprotein and assembly of empty capsid particles (Hansman *et al.*, 2008; Oka *et al.*, 2009). The NS6 protease is also responsible for the processing of the precursor of capsid protein in vesivirus (Fig. 4.2). This protein is abundantly expressed in the infected cells from the ORF2 of the virus subgenomic RNA. The maturation process involves rapid proteolytic removal of the N-terminal leader sequence (LC) (Carter, 1989; Matsuura *et al.*, 2000; Neill *et al.*, 1991; Sosnovtsev *et al.*, 1998). In FCV, direct N-terminal sequencing of the VP1 protein derived from virions and followed by mutagenesis studies showed that this processing consisted of an unique cleavage event at the Glu124/Ala125 dipeptide site in the capsid precursor sequence (Carter, 1989; Sosnovtsev *et al.*, 1998). The observed cleavage of the precursor was very efficient, and a full-length product of the ORF2 could not be detected in the virus-infected cells unless they were grown at elevated temperatures or in the presence of a processing inhibitor, such as *p*-fluorophenylalanine (Carter, 1989; Carter *et al.*, 1992; Fretz and Schaffer, 1978; Tohya *et al.*, 1999). In part, the observed efficiency could be explained by the co-translational cleavage of the growing precursor molecule in infected cells. Furthermore, in *in vitro* experiments the NS6–7 protein could process the capsid precursor *in trans* indicating that the cleavage site was still accessible for the protease after synthesis of the entire protein (Sosnovtsev et al., 1998). It is possible that conformational presentation of the cleavage site to the protease is one of the main factors that define this cleavage efficiency. In support of this, the protease did not recognize several similar dipeptide sites located next to the primary cleavage site (Sosnovtsev et al., 1998). It appears that complete cleavage of the FCV capsid precursor was critical for the production of infectious virions. Thus, mutations of the Glu124/Ala125 dipeptide sequence resulting in the incomplete cleavage of the capsid precursor interfered with virion maturation and production of viable virus progeny. The main obstacle was probably associated with the fact that only the cleaved form of the capsid could assemble into infectious virions (Sosnovtsev et al., 1998). Supporting this notion, no precursor molecules could be detected in CsCl-purified FCV virions (Sosnovtsev and Green, 2000; Tohya et al., 1999).

Cellular targets

Many positive-sense single-stranded RNA viruses employ their proteases to control the level of host cell protein expression. For example, picornavirus proteases cleave components of the cap–binding complex, resulting in the inhibition of translation of capped mRNAs and cellular protein synthesis. It is not known whether calicivirus replication in general is associated with a shutoff of host-cell translation. However, studies of vesivirus growth using the FCV cell culture model showed that virus infection was associated with a marked decrease in host-cell protein synthesis over time (Kuyumcu-Martinez et al., 2004; Willcocks et al., 2004). Analysis of the expression of the eIF4GI and eIF4GII factors in FCV-infected cells demonstrated that, similarly to picornavirus infection, both eIF4G proteins were cleaved. However, the observed cleavage occurred at a site different from that of picornaviruses, closer to the N-terminus of the protein. As a result, it produced a larger fragment that retained both eIF4E and eIF3 binding sites (Willcocks et al., 2004). The functional role of such eIF4G cleavage in the mRNA expression shutoff remains unclear since the generated cleavage product would retain an ability to bring capped mRNA molecules and ribosomes together through eIF4E–eIF4G–eIF3

interactions. The eIF4G protein is also cleaved during apoptosis; however, the proteolytic profile of the FCV-induced eIF4G cleavage differed from that observed in apoptotic cells suggesting a virus-induced proteolytic activity specific for eIF4G (Lloyd, 2006; Willcocks et al., 2004). Of interest, in vitro experiments showed that calicivirus protease does not directly cleave eIF4G protein, but rather targets the PABP protein that plays an important role in stimulation of translational initiation (Kuyumcu-Martinez et al., 2004). Recombinant proteases from two caliciviruses, MD145–12 and FCV, were shown to recognize and cleave PABP generating proteolytic products similar to the ones observed for poliovirus 3C (Kuyumcu-Martinez et al., 2004). The recognized scissile bonds mapped to Gln537/Gly538 and Gln413/Gly414 pairs for the NS6 of MD145–12 and to Gln437/Gly438 pair for the NS6–7 of FCV indicating that the corresponding cleavages resulted in removal of the PABP C-terminal domain that binds eIF4B and eRF3 factors involved in recycling translational initiation on capped polyadenylated mRNA. Analysis of the PABP cleavage in FCV-infected cells showed that the appearance of the proteolytic products of this cleavage correlated with the induction of shutoff of the host-cell protein synthesis. Consistent with that, addition of the norovirus NS6 to the HeLa cell extract had an inhibitory effect on translation of both endogenous and exogenous mRNAs. However, the level of translation of polyadenylated mRNAs could be restored by addition of the recombinant PABP. Comparative analysis of the different pools of PABP demonstrated that calicivirus NS6 recognized PABP molecules associated with ribosomes more efficiently suggesting direct targeting of cellular mRNA translation (Kuyumcu-Martinez et al., 2004).

Substrate specificity of the calicivirus protease

Similar to the picornavirus 3C protease, the specificity of the calicivirus enzyme is bound to the context of the cleavage site primary sequence. Naturally occurring scissile bonds recognized by the calicivirus protease are restricted to dipeptides that contain Glu or Gln residues in the P1 position of the cleavage site and a variety of amino acid residues in the P1′ position (Table

Table 4.1 Comparison of cleavage sites identified in the calicivirus ORF1 polyproteins

Cleavage site	Lagovirus P4 P3 P2 P1 – P1'P2'P3'P4'	Norovirus P4 P3 P2 P1 – P1'P2'P3'P4'	Vesivirus P4 P3 P2 P1 – P1'P2'P3'P4'	Sapovirus P4 P3 P2 P1 – P1'P2'P3'P4'
NS1–NS2	P I F **E** – **G** E V D	N A	I R A **E** – **A** C P S	F T E **E** – **G** L L D
NS2–NS3	D T F **E** – **D** S V P	F H L **Q** – **G** P E D	F R S **E** – **D** V A N	F Q S **Q** – **G** P T S
NS3–NS4	A S F **E** – **G** A N K	F Q L **Q** – **G** K M Y	F E A **E** – **N** G H S	F K E **Q** – **G** N E H
NS4–NS5	K A F **Q** – **G** V K G	A T M **E** – **G** K N K	P K S **E** – **A** K G K	R E E **E** – **A** K G K
NS5–NS6	N D Y **E** – **G** L P G	I S F **E** – **A** P P T	F A E **E** – **S** G P G	Y D E **E** – **A** P T P
NS6–NS7	G V Y **E** – **T** S N F	T T L **E** – **G** G D K	N A	N A
NS7–VP1 or LC–VP1	N V M **E** – **G** K A R	N A	F R L **E** – **A** D D G	F E M **E** – **G** L G Q

The P4–P1 and P1'–P4' amino acid residues (nomenclature of Schechter and Berger, 1967) of the ORF1 polyprotein cleavage sites of the lagovirus RHDV FRG, norovirus Southampton, vesivirus FCV Urbana, and sapovirus Mc10 strains as well as those of the FCV capsid precursor cleavage site from publications listed in the legend to Fig. 4.1. The scissile bond amino acids are shown in bold.

4.1) (nomenclature of Schechter and Berger, 1967 (Schechter and Berger, 1967)). Genome sequencing and mapping of the cleavage sites for several caliciviruses from different genera showed some variation in ratio and distribution of the Glu and Gln sites processed by the virus proteases in the corresponding polyproteins. Thus, all of the cleavage sites mapped in the FCV ORF1 polyprotein contained Glu residue in position P1. In addition, Glu was found in the dipeptide mapped as a cleavage site of the virus capsid precursor (Table 4.1) (Sosnovtsev et al., 2002). A similar preference has been reported for the RHDV ORF1 polyprotein. With the exception of the NS4–NS5 cleavage site where the same position was occupied by Gln residue, the P1 positions in the rest of the virus polyprotein cleavage sites carried Glu residues (Table 4.1) (Meyers et al., 2000). Sapovirus and norovirus sequences showed a 'mixed' pattern in relation to the P1 position preference. Most of their established proteolytic maps contained the Gln dipeptides in the N-terminal part of the ORF1 polyprotein at the NS2–NS3 and NS3–NS4 junctions, while Glu residues were found in cleavage sites of the NS4–NS5–NS6–NS7 precursor (Table 4.1) (Liu et al., 1999b; Oka et al., 2006). Nevertheless, certain deviations from this pattern have been reported. Of interest, cleavage site mapped at the border of sapovirus NS1-NS2 proteins contained

Glu–Gly pair. Another variation was observed in proteolytic processing of the MNV-1 proteins where the scissile bond between the NS2 and NS3 proteins contained a Glu residue and the bond between NS6 and NS7 contained a Gln residue in the P1 position. The existence of two types of cleavage sites suggested the regulation of proteolytic processing at the dipeptide cleavage sequence level. However, in vitro expression of the norovirus ORF1 Glu/Gln exchange mutants showed that the virus NS6 recognized both types of the cleavage sites with a similar efficiency (Hardy et al., 2002). In addition to Gln and Glu scissile bonds, in vitro studies demonstrated that calicivirus proteases could recognize, however, with less efficiency, cleavage sites where the P1 position contained Asp (RHDV NS5–NS6 and FCV capsid precursor cleavage sites (Sosnovtsev et al., 1998; Wirblich et al., 1995)) or Asn (norovirus NS1–2–NS3 site; Hardy et al., 2002) residues. However, occurrence of these cleavage sites in vivo has not been documented.

The experiments with calicivirus proteases showed that the enzymes were less selective towards changes in the P1' position of the cleavage sites and tolerated a number of amino acid substitutions. Studies of the RHDV NS6 showed that the enzyme could process cleavage sites containing ten different amino acid residues in the P1' position, although, with varying efficiency levels (Wirblich *et al.*, 1995). The analysis of cleavage efficiency revealed comparable levels of inhibition for the similar substitutions, and indicated a strong protease preference for amino acids with small side chains in this position. The latter observation suggested that the P1'-based substrate specificity was rather related to conformational constraints than to specific recognition of the P1' amino acid by the substrate binding site of the enzyme (Wirblich *et al.*, 1995). Of interest, the experiments with the FCV NS6–7 protein showed that amino acid residues with large side chain groups were well tolerated by the vesivirus enzyme in the P1' position (Sosnovtsev *et al.*, 1998). It is likely that calicivirus proteases have different P1'-position recognition requirements that might be based on their structural differences.

Despite the experimentally established ability of calicivirus proteases to recognize a number of different amino acid residues in the P1' position, their naturally occurring variation is limited to a few amino acids. Thus, in general, proteases of the sapovirus, norovirus and lagovirus have been reported to cleave sites that carry Ala and Gly residues in the P1' position. The corresponding dipeptides have been found conserved among strains that belong to different genogroups of these viruses' genera. In addition to these amino acids, the P1' position of the cleavage sites recognized by the MNV-1 NS6 and FCV NS6–7 has been reported to contain dipeptides with Ser, Thr, Asn and Asp residues (Sosnovtsev *et al.*, 2002).

The P1 and P1' positions are not the only elements of the cleavage site primary sequence that are important for its recognition and processing. Structural studies of a number of virus proteases demonstrated that their substrate recognition specificities rely also on the binding of amino acid residues preceding the scissile bonds. One of the factors involved in defining cleavage efficiency of the related picornavirus 3C proteases has been associated with the presence of small (poliovirus) or large (hepatitis A) hydrophobic residues in the P4 position of the 3C cleavage sites (Jewell *et al.*, 1992; Pallai *et al.*, 1989). The P4 positions are also thought to play an important role in facilitating cleavages in the vesivirus, norovirus and sapovirus (Belliot *et al.*, 2003; Hardy *et al.*, 2002; Oka *et al.*, 2006; Sosnovtsev *et al.*, 2002). Mapping of the cleavage sites recognized by the FCV protease revealed that all of them contained a large hydrophobic residue at the P4 position, and for four of them this position was occupied by Phe residues (Sosnovtsev *et al.*, 2002). Similarly, the P4 position of sapovirus cleavage sites has also been reported to carry either Phe or Tyr residues (Oka *et al.*, 2006). Hydrophobic amino acids appeared to be conserved also in the P4 position in the norovirus cleavage sites, with Phe and Tyr residues found in the corresponding positions of the primary cleavage sites of the human strains. Their critical role in the cleavage efficiency has been suggested in mutagenesis experiments that targeted the NS1–2–NS3 cleavage site of the norovirus polyprotein. The corresponding cleavage was significantly inhibited by the replacement of the Phe with Gly and showed comparable efficiency when the P4 position was occupied by a large hydrophobic residue (Hardy *et al.*, 2002).

The presence of conserved Phe and Tyr residues has been also described for the RHDV; though, they were observed in the P2 position of the cleavage sites. Of interest, the same position (P2) in cleavage sites of several human norovirus contained a conserved Leu residue. However, in both cases, *in vitro* assays showed that this position played a minor role in determining the efficiency of protease recognition of the corresponding cleavage sites (Hardy *et al.*, 2002; Wirblich *et al.*, 1995).

Conformational factors in substrate recognition

It is likely that efficient recognition of the protein borders by the protease is influenced by factors beyond the sequence context of the cleavage sites. Apparently, not all Gln and Glu cleavage dipeptides are processed in the calicivirus polyproteins, even though some of them carry conserved hydrophobic amino acid residues in the putative P4 positions. The protease preference for particular scissile bonds was suggestively associated with the presence of the conformational determinants required for proper folding and exposure of the cleavage site regions. Several observations consistent with that idea have been made in *in vitro* translational studies. Thus, processing of the NS2–NS3 border was found to be sensitive to mutations in the adjacent regions. Accordingly, an efficiency of the cleavage at this site in the sapovirus Mc10 ORF1 polyprotein was decreased by mutations introduced into the predicted helix domain of the NS3 sequence (Oka *et al.*, 2006). Similarly, truncation of the N-terminal sequence in FCV ORF1 polyprotein led to the alternative processing of the virus NS2–NS3 region (Sosnovtseva *et al.*, 1999). Alternative processing of NS2 and NS3 proteins was observed also for norovirus polyproteins where introduced amino acid changes blocked cleavage at the authentic NS2–NS3 border (Hardy *et al.*, 2002; Seah *et al.*, 2003). The introduced mutations were thought to alter conformation of the polyprotein so that different cleavage sites become accessible for the virus protease (Hardy *et al.*, 2002). The proper folding might also be a factor influencing the processing of the C-terminal part of the calicivirus ORF1 polyprotein. It appeared that mutations blocking release of the sapovirus VP1 protein had an inhibitory effect on production of the NS4–NS5 intermediate (Oka *et al.*, 2006). In addition, an atypical cleavage pattern was observed in *in vitro* experiments targeting expression of the truncated C-terminal fragments of the MNV-1 ORF1 polyprotein (Sosnovtsev, unpublished). Indirect evidence that supported the role the adjoining protein sequences might play in the cleavage site presentation has come from peptide *trans* cleavage assays where peptide substrates carrying sequences corresponding to the sapovirus NS4–NS5 and norovirus NS4–NS5 and NS6–NS7 junctions were found to be resistant to *trans* cleavage by NS6 or NS6–7 proteins (Robel *et al.*, 2008; Scheffler *et al.*, 2007). In contrast, cleavage at these junctions occurred more or less efficiently in reactions containing the entire ORF1 (Belliot *et al.*, 2003; Oka *et al.*, 2005b).

Substrate specificity and structural characteristics of the calicivirus protease

The substrate specificity of a protease is defined by the structural characteristics of its active centre. In general, functions of the protease active centre include substrate binding, stabilization and proper positioning of the scissile bond in the proximity of the enzyme catalytic residues and the catalysis of hydrolysis itself. The mechanism of hydrolysis of the peptide bond by viral 3C-like proteases is thought to involve a nucleophilic

attack of the carbonyl carbon atom by the thiol group of the Cys residue, the formation of a tetrahedral intermediate followed by release of the substrate C-terminal part, and the generation of an intermediate acyl-enzyme form. The final deacylation step that includes release of the regenerated enzyme is mediated by nucleophilic attack of the acyl-enzyme intermediate by a water molecule. Catalytic activity of the sulphur nucleophile is assisted by an imidazole ring of the neighbouring His residue working as a general base catalyst. In its turn, the His residue acquiring a positive charge is stabilized by the negatively charged carboxyl group of either Glu or Asp residues of the catalytic site. Three residues, Cys, His and Glu/Asp, constitute a catalytic triad, directly involved in formation of a charge-transfer system and in catalysis of the cleavage. The catalytic triad represents a hallmark feature of the 3C-like proteases that is conserved in most of their sequences. Two more structural features involved in stabilization and orientation of the scissile bond play an important role in proteolytic activity of the 3C-like proteases. The first one is an oxyanion binding site formed by the amino acid residues adjacent to the active site Cys residue, and the second one is a substrate-binding subsite 1 (S1) for the P1 amino acid, which in most cases is represented by the Gln residue. The S1-site conserved His residue is involved in interaction with a side chain carbonyl oxygen of the Gln residue, and therefore, in the recognition of the P1 residue of the scissile bond. The amino acid residues that determine positioning and charge characteristics of the S1 sites vary among the proteases reflecting divergence in their sequences. In addition, differences in protease sequence result in a significant variation in structural characteristics of the rest of the 3C-like protease subsites implicated in substrate binding.

The localization of residues involved in formation of the calicivirus protease substrate-binding site and its catalytic centre was first addressed using site-directed mutagenesis. In these studies, amino acid residues subjected to modification were selected based on homology of the calicivirus NS6 protein sequences to those of picornavirus 3C proteases with resolved crystal structures. In general, the picornavirus and calicivirus proteases exhibit a low degree of sequence similarity; nevertheless, the amino acid residues corresponding to the motifs critical for the activity of picornavirus 3C proteases have been found conserved in all calicivirus NS6 protein sequences (Boniotti et al., 1994; Neill, 1990; Oka et al., 2007).

Based on sequence alignment, residues His27, Asp44 and Cys104 were predicted to be part of the catalytic triad of the RHDV protease. Subsequent expression analysis of the corresponding protease mutants showed these residues to be essential for the activity of the enzyme (Boniotti et al., 1994). Similar conclusions have been drawn in mutagenesis experiments with the sapovirus and vesivirus enzymes. His31, Glu52, Cys116, and His39, Glu60, Cys122 have been mapped as components of a catalytic triad for the sapovirus and vesivirus protease, respectively (Oka et al., 2007; Sosnovtsev et al., 1998).

In the norovirus protease, His30 and Cys139 have been unambiguously identified as essential catalytic residues. In contrast, controversial results have been reported on the functional importance of the third member of the catalytic triad, the acidic residue. Sequence alignment analysis showed that the conserved Glu residue was present in the norovirus protease sequences at position 54 corresponding to that of the acidic amino acid of the picornavirus 3C protease active centre (Hardy et al., 2002). Replacement of this residue with Gly abolished processing of the norovirus polyprotein (Hardy et al., 2002). At the same time, an Ala-mutagenesis analysis of the Chiba virus (another norovirus) protease did not identify this residue as important for the preservation of the proteolytic activity (Someya et al., 2002). Later experiments that employed a site-saturation mutagenesis confirmed that the Glu54 residue in the Chiba virus protease sequence was not essential for the enzymatic activity, but rather important for the efficiency of cleavage (Someya et al., 2008).

Mutagenesis-based predictions were afterward confirmed by the structural studies of the norovirus enzyme. The crystal structure of the protease has been resolved recently for two different human norovirus strains. First, the 2.8-Å-resolution crystal structure protease was determined for the Chiba virus (Nakamura et al., 2005), and later, the structure of the norovirus protease was resolved to 1.5 Å for the protein alone and to 2.2 Å for the protein crystallized together with the protease inhibitor AEBSF [4-(2-aminoethyl)-benzenesulphonyl fluoride] (Zeitler et al., 2006). Further analysis of the protein folding is covered in Chapter 5.

Briefly, X-ray analysis confirmed the structural similarity of the norovirus protease to that of the chymotrypsin-like group proteases with a Cys residue in the active centre. The established three-dimensional structure showed that the protease sequence had a characteristic two-domain folding. The identified domains were mapped to N- and C-terminal halves of the protease sequence and consisted of five and six beta-strands, respectively (Nakamura et al., 2005; Zeitler et al., 2006). Connected by loops, the beta-strands of the N-terminal domain formed an anti-parallel twisted beta-sheet, while a similar beta-sheet of the protein C-terminal part appeared as a beta-barrel (Nakamura et al., 2005; Zeitler et al., 2006). The latter was apparently involved in tail–to-tail interactions between protease monomers in a dimer structure observed in the norovirus protease crystals (Zeitler et al., 2006).

The enzyme substrate binding and catalytic sites were located in a cleft between these domains, and the latter, as predicted, included His30, Glu54, and Cys139 residues as main components of a catalytic triad (Nakamura et al., 2005; Zeitler et al., 2006). The oxyanion hole important for stabilization of the P1 residue carbonyl oxygen was formed by a conserved motif of amino acids adjacent to the catalytic Cys139 residue, Gly137-Asp138-Cys139-Gly140 (Nakamura et al., 2005; Zeitler et al., 2006).

Structural modelling of the protease–substrate interactions revealed that the enzyme specificity was defined by two S1 and S2 sites, accommodating P1 and P2 residues of the substrate cleavage site (Nakamura et al., 2005). Consistent with mutagenesis data, the structure analysis showed that the S1 pocket specificity was dependent on the His157 residue capable of forming a hydrogen bond with the P1 residue. Of interest, substitution of the His157 residues in the norovirus protease sequence with Tyr, Gln, Arg, Ser, and Pro residues drastically reduced efficiency of cleavage, but did not abolish it completely (Someya et al., 2002). Analysis of the S1 site structure revealed that the P1 residue could also form a hydrogen bond with the conserved Thr134 residue suggesting that this amino acid might play role in defining the P1 specificity of the norovirus protease (Nakamura *et al.*, 2005). The modelling also suggested participation of the hydrogen bonds of the protease 158–162 residues in binding P5-P2 residues of the substrate (Nakamura *et al.*, 2005). In agreement with experimental data, the model demonstrated the absence of a size restriction in the binding subsites for the side chains of the substrate P4 and P3 residues. A similar observation has been made for the S2 pocket whose hydrophobic characteristics and relatively large size explained the ability of the norovirus protease to recognize bulky hydrophobic amino acid residues in the P2 position of the cleavage sites (Nakamura *et al.*, 2005). Examination of the protease co-crystallized with AEBSF revealed that enzyme–substrate interactions might induce structural changes in the active site of the protease. Of interest, binding of the AEBSF molecule resulted in movement of the Glu54 and His30 residues, disrupting a hydrogen bond observed in the native protease structure. In addition, it led to a conformational change in Arg112 residue participating in the S2 pocket formation (Nakamura *et al.*, 2005; Zeitler *et al.*, 2006).

Determination of the crystal structure of the norovirus protease made it possible to predict folding of its vesivirus, sapovirus and lagovirus counterparts. Homology modelling, based on structural alignment of the calicivirus NS6 amino acid sequences, revealed a structural conservation of the spatial organization of the calicivirus enzyme active centre. Predicted folding of the sapovirus and vesivirus NS6 molecules showed that amino acid residues identified by mutagenesis as essential for the protease activity (His31, Glu52, Cys116, and His131 in sapovirus and His39, Glu60, Cys122, and His137 in FCV) occupied positions on the inner surface of the cleft of the molecules, in an arrangement similar to that of the active centre residues of the norovirus enzyme (Oka *et al.*, 2007). Analogous predictions have been made for the arrangement of the residues of the lagovirus protease active centre; however, in RHDV, the acidic residue of the catalytic triad was represented by an Asp residue (Oka *et al.*,

2007). Conservation of the catalytic residues and their configuration in the active centre of the calicivirus proteases indicates the crucial role these residues play in the functionality of this protein. It also shows that despite a relatively low level of sequence similarity, the structural elements of the calicivirus NS6 proteins were preserved during evolution. Structural and sequence similarities reflect the existence of a common mechanism of the proteolytic catalysis shared among calicivirus proteases. Future structural and biochemical studies of calicivirus proteases can lead to better understanding of the enzyme functioning and possible development of the new virus growth inhibitors.

Non-structural proteins

Proteolytic processing of the calicivirus ORF1 polyprotein mediated by the virus 3C-like protease represents a multi-faceted mechanism designed to deliver highly regulated control of the synthesis of the virus mature non-structural proteins and their functionally important precursors. Recent progress in generation of the proteolytic cleavage maps has allowed structure-functional and biochemical studies of the individual non-structural proteins. Prediction of the ORF1-encoded proteins functions has been assisted by their comparative sequence analysis. An initial alignment of the calicivirus non-structural polyprotein sequences with those of the related picornaviruses, demonstrated the presence of the functional motifs corresponding to the picornavirus 2C NTPase (NS3), 3C protease (NS6), and 3D polymerase (NS7) (Neill, 1990). Expression and characterization of the corresponding (NS3, NS6, NS7) recombinant calicivirus proteins provided further support for the predicted protein identities (Boniotti *et al.*, 1994; Marin *et al.*, 2000; Pfister and Wimmer, 2001; Vazquez *et al.*, 1998). The functions of the proteins encoded by the N-terminal (NS1, NS2) and middle (NS4, NS5) parts of the calicivirus ORF1 could not be predicted based on their sequence alignments due to lack of significant similarity with the sequences of their picornavirus counterparts. However, in the case of the NS5 protein, its linkage to the virus RNA molecules and localization of the protein gene immediately upstream from the protease sequence suggested that this protein might function similarly to the picornavirus 3B protein (VPg), participating in RNA synthesis and packaging (Dunham *et al.*, 1998). Known features of the calicivirus non-structural proteins are briefly described below and summarized in Fig. 4.3.

NS1–2

As mentioned above, the calicivirus N-terminal proteins may exist either as two distinct products (NS1 and NS2) or as a single protein (NS1–2). The vesivirus, lagovirus and sapovirus proteins encoded by the N-terminal part of the ORF1 undergo efficient proteolytic processing resulting in generation of approximately 28–32 kDa NS2 proteins and small N-terminal NS1 products, which vary in size from 5.6 to 16 kDa between genera. The cleavage is mediated by the virus NS6 protease and constitutes an event essential for virus replication (Oka *et al.*, 2006; Sosnovtsev *et al.*, 2002; Wirblich *et al.*, 1996). The function of the NS1 protein is unknown and cannot be predicted based on the protein sequence since the protein sequence shows no significant similarity with any established functional sequence motifs. Moreover, the level of similarity of NS1 amino acid sequences between different genera does not exceed that of random sequences making it difficult to identify conserved sequence motifs.

The norovirus counterpart of the NS1 and NS2 proteins, NS1–2 protein (known as Nterm, or p48 for norovirus) also shows considerable variation in calculated mass and sequence among strains of different genogroups, I (45 kDa), II, III and V (35–37 kDa) (Belliot *et al.*, 2003; Dingle *et al.*, 1995; Liu *et al.*, 1996; Liu *et al.*, 1999a; Oliver *et al.*, 2007; Pletneva *et al.*, 2001; Seah *et al.*, 1999). Based on sequence alignment, the norovirus NS1–2 can be divided into three domains: the highly variable N-terminus, the middle part that contains H-box and NC motifs found in tumour suppressor proteins such as H-rev107, and a C-terminal part with a hydrophobic region predicted to mediate binding of this protein to lipid membranes (Ettayebi and Hardy, 2003; Fernandez-Vega *et al.*, 2004). It is likely that the norovirus NS1–2 protein, similar to the analogous picornavirus 2B protein, participates in intracellular membrane changes that occur during virus replication. An association of the NS1–2

protein with membranes has been observed in transient expression experiments in COS7 cells (Ettayebi and Hardy, 2003). Of interest, analysis of the expression of the norovirus NS1–2 protein in transfected cells showed co-localization of this protein with markers of the Golgi complex. Accumulation of the norovirus NS1–2 induced an apparent disassembly of the Golgi complex into a number of discrete aggregates (Fernandez-Vega *et al.*, 2004). Expression of the norovirus NS1–2 deletion mutants suggested that the protein targeting to the Golgi complex relies on a signal located in the C-terminal hydrophobic region; whereas, disassembly of this organelle might require the presence of the protein N-terminal sequences (Fernandez-Vega *et al.*, 2004). The mechanism of this disruption remains unclear, however, norovirus NS1–2 inhibition of the cell surface expression of the vesicular stomatitis virus glycoprotein G indicated that the virus protein might interfere with cellular protein trafficking (Ettayebi and Hardy, 2003). In addition, it was shown that the norovirus NS1–2 membrane binding and Golgi targeting might also be assisted by the protein–protein interactions with the SNARE regulator vesicle-associated membrane protein-associated protein A (VAP-A). The latter was identified as an NS1–2 binding partner using a yeast two-hybrid system screening approach (Ettayebi and Hardy, 2003). The ability of the VAP-A to bind proteins implicated in membrane vesicle fusion and ER/Golgi transport including SNAREs has been reported (Weir *et al.*, 2001).

The role of the NS1–2–membrane interactions in the virus cycle remains unclear. However, the finding of this protein and its vesivirus counterpart, NS2, in the RNA-synthesis-competent membranous fraction of the norovirus and vesivirus-infected cells may suggest that the hydrophobic C-terminal part of these proteins is involved in recruitment of the membrane vesicles during the assembly of the virus replication complexes (Green *et al.*, 2002; Wobus *et al.*, 2004). Moreover, NS2 protein–protein interactions with NS3, NS4 and NS6-7 proteins reported in a yeast two-hybrid analysis of the FCV proteins suggests that NS2 (NS1–2) might function as a scaffolding protein during replication (Ettayebi and Hardy, 2003; Kaiser *et al.*, 2006).

NS1	NS2	NS3	NS4	NS5	NS6	NS7
Membrane association Replication complexes Golgi localization Golgi disassembly Interaction with virus NS3, NS4 and NS6-7 proteins Binding of cellular VAP-A protein Cleaved by caspase-3 NS1, NS2, NS1-2 forms Dimerization		NTPase activity Membrane association Replication complexes Interaction with virus NS2 protein RNA binding	Predicted membrane association Replication complexes Interaction with virus NS2 protein NS4, NS4-NS5, NS4-NS5-NS6, NS4-NS5-NS6-NS7 forms	VPg (virus protein linked to genome) Virion assembly Initiation of translation Replication complexes Priming of RNA synthesis Interaction with viral NS6-NS7 and ORF2 proteins Interaction with cellular eIF3, eIF4GI, eIF4E,eIF2a and S6 proteins NS5, NS4-NS5, NS4-NS5-NS6, NS5-NS6-NS7, NS4-NS5-NS6-NS7, NS5-NS6 forms	Protease activity (3C-like) involved in maturation of virus non-structural and structural proteins Replication complexes Interaction of NS6-NS7 form with virus NS2, NS5, ORF2 and VP2 proteins Cleavage of cellular PABP protein NS6, NS4-NS5-NS6, NS6-NS7 forms Dimerization	Polymerase activity (3D-like RNA-dependent RNA polymerase) Replication complexes Interaction of NS6-NS7 form with virus NS2, NS5, ORF2, and VP2 proteins NS7, NS6-NS7, NS5-NS6-NS7, NS4-NS5-NS6-NS7 forms Dimerization

Figure 4.3 Properties and functions of the calicivirus non-structural proteins.

NS3

The NS3 protein maps to one of the less divergent regions of the calicivirus ORF1 sequence. The sequence similarity with picornavirus 2C proteins and conserved sequence motifs associated with NTPase activity place this protein within a family of putative viral helicases, which in turn belongs to a large group of ubiquitous enzymes that mediate unwinding of double stranded nucleic acid molecules (Gorbalenya *et al.*, 1990; Neill, 1990). So far, no experimental evidence has been obtained that the calicivirus NS3 can function as a helicase and can successfully separate an RNA duplex (Pfister and Wimmer, 2001). However, two groups of researchers reported the presence of the NTP binding and NTPase activity for this protein, functional characteristics thought to be coupled to helicase unwinding activity (Marin *et al.*, 2000; Pfister and Wimmer, 2001).

The role of the calicivirus NTPase during replication is not known; however, an analysis of murine norovirus NS3 protein transient expression in RAW264.7 cells showed that the NS3 synthesis induced changes in intracellular membrane structure and promoted the formation of membranous vesicles in transfected cells. Consistent with that, in both infected and transfected cells, the protein was found in fractions of membrane-associated proteins. The deletion and point-mutation analysis of the NS3 sequence identified the N-terminal part of the protein as a region that mediates its membrane association (Sosnovtsev *et al.*, 2007). The presence of the N-terminal cluster of hydrophobic amino acids is a conserved structural feature shared by all calicivirus NS3 proteins. Considering its location in the protein sequence, it is possible that this protein is anchored to intracellular membranes via its N-terminus. Of interest, the N-terminal domain appears to play an important role in oligomerization of the related SF3 helicase of the simian virus 40 (Gai *et al.*, 2004). Whether the calicivirus NS3 in infected cells undergoes oligomerization is yet unknown; however, the N-terminal 80 amino acid sequence of the FCV NS3 has been implicated in interactions with another virus protein, NS2 (Kaiser *et al.*, 2006). Given that NS2 interacts with a number of nonstructural proteins and might also associate with host cellular membranes, the NS3–NS2 complex might serve as a membrane-anchored platform for calicivirus RNA synthesis.

NS4

The NS4 protein is one of the least characterized calicivirus proteins. Ranging in size from 18 to 30 kDa, the protein is encoded by the second most divergent region of the calicivirus ORF1 and shares very little, if any, sequence similarity with its picornavirus 3A protein counterpart. Despite the lack of sequence relatedness, the NS4 protein might play a role similar to that of 3A in virus replication. The latter has been shown to interact with intracellular membranes and has been implicated in inhibition of a cellular protein transport. Computer analysis of the NS4 sequence predicted the presence of hydrophobic domain, suggesting membrane localization of the protein. Consistent with that and similar to the picornavirus 3AB, the NS4–NS5(VPg) precursor might serve as a membrane anchor for the VPg protein. The NS4 interactions with cellular membranes and other virus nonstructural proteins have been proposed to assist VPg-mediated virus RNA synthesis. Interestingly, the NS4 protein has been found in both cleaved and precursor forms in calicivirus-infected cells (Sosnovtsev *et al.*, 2002, 2006). Furthermore, both forms of protein were identified as membrane-associated components of the VeV replication complexes (Green *et al.*, 2002). Given that yeast two-hybrid system analysis of the NS2 interactions identified NS4 protein as its binding partner, the NS4-NS5 precursor might also function as the source of VPg delivered to the calicivirus replication complexes (Kaiser *et al.*, 2006).

NS5

The calicivirus NS5 is a small, approximately 13–15 kDa in size, protein that can be found covalently linked to the 5'-end of the viral genomic and subgenomic RNAs. The sequence of the calicivirus VPg does not show significant similarity to those of the picornaviruses and other RNA viruses. Nevertheless, similar to the picornavirus VPg, the calicivirus protein is predicted to participate in initiation of the viral RNA synthesis. Several recent studies reported efficient nucleotidylylation of the calicivirus VPg in *in vitro* assays that employed bacterially expressed viral polymerase or its precursor (Belliot *et al.*, 2008; Machin *et al.*, 2001; Rohayem *et al.*, 2006b). The nucleotidylylated form of the protein is thought to serve as a primer for template-dependent synthesis of the viral RNA. Consistent with that, modification of the conserved Tyr residue involved in covalent linkage of the VPg molecule to the virus RNA has been shown to have a deleterious effect on virus replication (Mitra *et al.*, 2004).

In contrast to picornaviruses, calicivirus protein translation during infection is thought to be dependent on the presence of VPg at the 5′ end of the viral RNA. In experiments with protease K treatment of the calicivirus RNA isolated from purified virions, proteolytic removal of the protein led to drastic reductions in translation efficiency *in vitro* and as a result, to a loss of infectivity in RNA transfection assays (Burroughs and Brown, 1978; Dunham *et al.*, 1998; Herbert *et al.*, 1997). As described above, search for protein–protein interaction partners using yeast two-hybrid system and pull-down assays revealed that calicivirus VPg proteins could interact with the eIF3, eIF4GI, eIF4E, eIF2a eukaryotic translation initiation factors and the ribosomal protein S6 (Chaudhry *et al.*, 2006; Daughenbaugh *et al.*, 2003, 2006; Goodfellow *et al.*, 2005). Detailed description of the current state of the VPg research is provided in the 'Murine norovirus translation, replication and reverse genetics' chapter (Chapter 11) of this book.

NS6

The calicivirus NS6 is the viral protease. The structure and functional role of this enzyme are reviewed in the text above.

NS7

The calicivirus NS7 is the last in the gene order of the calicivirus non-structural proteins encoded by the virus ORF1. The significant sequence similarity between this protein and the 3D RNA-dependent RNA polymerase (RdRp) of picornaviruses suggested that NS7 could have a similar role in virus replication and be responsible for the synthesis of the virus both plus- and minus-strand RNAs (Neill, 1990; Vazquez *et al.*, 1998). Based on the mature form of the NS7 protein, caliciviruses have been shown to form two distinctive groups. The NS6–NS7 (NS6–7)

precursor synthesized in vesivirus-infected cells represents a mature and stable bifunctional enzyme that shows no sign of additional processing during an infection (Martin-Alonso *et al.*, 2005; Oehmig *et al.*, 2003; Sosnovtsev *et al.*, 2002; Sosnovtseva *et al.*, 1999; Wei *et al.*, 2001). The 76-kDa NS6–NS7 protein has been reported also as a major form of the virus polymerase in porcine enteric calicivirus (sapovirus)-infected cells (Chang *et al.*, 2005). Because vesivirus NS6–7 protein has been shown to exhibit activities characteristic of an RdRp, it is likely that further processing of the vesivirus and, perhaps, sapovirus NS6–7 proteins is not required for productive virus replication (Wei *et al.*, 2001). The evolutionary advantage, if any, for the synthesis of a stable bifunctional protease-polymerase precursor in some caliciviruses is not known. The picornavirus analogue of the NS6–7 protein, 3CD is required for the successful uridylylation of the picornavirus VPg (Marcotte *et al.*, 2007; Paul *et al.*, 2000), and it is possible that the NS6–7 precursor may have a similar critical role for all the caliciviruses.

The NS6–NS7 precursor synthesized in lagovirus and norovirus-infected cells clearly undergoes further cleavage to produce a fully processed NS7 protein. Nevertheless, these viruses have been reported to generate small amounts of the NS6–NS7 protein during their replication (Konig *et al.*, 1998; Sosnovtsev *et al.*, 2006). Comparative biochemical studies of the recombinant enzymes showed that both NS7 and NS6–7 forms of the norovirus polymerase had similar characteristics (Belliot *et al.*, 2005).

The calicivirus polymerase has been expressed in an enzymatically active form in both bacterial and baculovirus expression systems (Belliot *et al.*, 2005; Fukushi *et al.*, 2004; Morales *et al.*, 2004; Vazquez *et al.*, 1998; Wei *et al.*, 2001). As expected, recombinant enzymes exhibited template-dependent RNA polymerase activity sensitive to EDTA (Belliot *et al.*, 2005; Fukushi *et al.*, 2004; Vazquez *et al.*, 2000; Vazquez *et al.*, 1998; Wei *et al.*, 2001). They could initiate synthesis from the RNA templates in the presence or absence of an oligonucleotide primer (Belliot *et al.*, 2005; Fukushi *et al.*, 2004; Fullerton *et al.*, 2007; Rohayem *et al.*, 2006b; Vazquez *et al.*, 1998; Wei *et al.*, 2001). Copy-back elongation of RNA templates by the NS7 in the absence of added primer has been reported to result in synthesis of a self-complementary double-stranded product (Belliot *et al.*, 2005; Vazquez *et al.*, 1998; Wei *et al.*, 2001). The polymerase has been also shown to direct synthesis of the plus-strand subgenomic RNA using an internal initiation mechanism on the minus-strand RNA template (Morales *et al.*, 2004). It was shown that initiation of the subgenomic RNA synthesis relied on recognition by the polymerase of the signal encoded in the 50-nt sequence located upstream of the transcription start (Morales *et al.*, 2004). Similar to the picornavirus 3D, the calicivirus polymerase has been shown to nucleotidylylate the NS5 protein, VPg, (Belliot *et al.*, 2008; Machin *et al.*, 2001; Rohayem *et al.*, 2006b). In addition, a recent study provided evidence that the norovirus polymerase could employ modified VPg protein to prime RNA synthesis (Rohayem *et al.*, 2006b). The efficiency of the norovirus VPg modification has been shown to be dependent on the presence of poly(A)-tailed RNA molecules and could be significantly enhanced by the sequences present in the 3′ end of the virus genome (Belliot *et al.*, 2008; Rohayem *et al.*, 2006b).

Another activity described for the calicivirus NS7 includes non-template transferase-like addition of nucleotides at the 3′ terminus of single-stranded RNA (Fullerton *et al.*, 2007; Rohayem *et al.*, 2006a, Rohayem *et al.*, 2006b). The virus polymerases are suggested to employ this activity for the initiation of RNA synthesis using a copy-back mechanism (Fullerton *et al.*, 2007). Furthermore, addition of the poly (C) stretch to the 3′ end of the RNA template by the norovirus polymerase was proposed to assist the enzyme in *de novo* initiation of the RNA synthesis (Rohayem *et al.*, 2006b).

The structure of the polymerase has been resolved to the atomic level by X-ray crystallography for lagovirus and norovirus proteins (Ng *et al.*, 2002; Ng *et al.*, 2004). Also, a similar analysis was performed for the truncated version of the sapovirus NS6–7 protein that contained predicted polymerase domains (Fullerton *et al.*, 2007). Established structures of the calicivirus NS7 proteins revealed the presence of the classic 'right hand' domain organization described for RdRp, which included a finger, palm, and thumb domains. The structural characteristics of the calicivirus NS7 protein are discussed in more detail in Chapter 5.

Conclusion

The maturation of the calicivirus non-structural proteins is largely dependent on the activity of the only viral protease. It is likely that the proteolytic cleavage of the ORF1 polyprotein resulting in generation of the mature virus proteins and their precursors is a finely tuned process designed to provide the virus replication machinery with its required components in an efficient and timely manner. The efficiency of this process is tightly linked to the protease-substrate recognition and is probably influenced by interactions of the enzyme with virus and cellular factors. Therefore, further studies of coordination and regulation of the proteolytic processing steps will extend our knowledge of the general mechanisms of calicivirus replication. Discovery and characterization of the possible cellular targets recognized by the calicivirus protease might lead to identification of the factors regulating the rate of virus amplification and, perhaps, the cytopathy induced by the virus infection. The absence of efficient vaccines for most of the caliciviruses makes the virus protease an attractive target in the search for potent virus growth inhibitors. The latter can be used as an alternative strategy for the control of infection and spread of the disease during outbreak. In addition, an established reverse genetics systems and the increasing acquisition of structural information for calicivirus proteins will be valuable in the development of attenuated virus strains suitable for future vaccine development.

References

Adldinger, H.K., Lee, K.M., and Gillespie, J.H. (1969). Extraction of infectious ribonucleic acid from a feline picornavirus. Arch. Ges. Virusforsch. *28*, 245–247.

Al-Molawi, N., Beardmore, V.A., Carter, M.J., Kass, G.E., and Roberts, L.O. (2003). Caspase-mediated cleavage of the feline calicivirus capsid protein. J. Gen. Virol. *84*, 1237–1244.

Allaire, M., Chernaia, M.M., Malcolm, B.A., and James, M.N. (1994). Picornaviral 3C cysteine proteinases have a fold similar to chymotrypsin-like serine proteinases. Nature *369*, 72–76.

Bachrach, H.L., and Hess, W.R. (1973). Animal picornaviruses with a single major species of capsid protein. Biochem. Biophys. Res. Commun. *55*, 141–149.

Belliot, G., Sosnovtsev, S.V., Chang, K.O., Babu, V., Uche, U., Arnold, J.J., Cameron, C.E., and Green, K.Y. (2005). Norovirus proteinase-polymerase and polymerase are both active forms of RNA-dependent RNA polymerase. J. Virol. 79, 2393–2403.

Belliot, G., Sosnovtsev, S.V., Chang, K.O., McPhie, P., and Green, K.Y. (2008). Nucleotidylylation of the VPg protein of a human norovirus by its proteinase-polymerase precursor protein. Virology 374, 33–49.

Belliot, G., Sosnovtsev, S.V., Mitra, T., Hammer, C., Garfield, M., and Green, K.Y. (2003). In vitro proteolytic processing of the MD145 norovirus ORF1 nonstructural polyprotein yields stable precursors and products similar to those detected in calicivirus-infected cells. J. Virol. 77, 10957–10974.

Belsham, G.J., and Sonenberg, N. (2000). Picornavirus RNA translation: roles for cellular proteins. Trends Microbiol. 8, 330–335.

Bertolotti-Ciarlet, A., White, L.J., Chen, R., Prasad, B.V., and Estes, M.K. (2002). Structural requirements for the assembly of Norwalk virus-like particles. J. Virol. 76, 4044–4055.

Birtley, J.R., Knox, S.R., Jaulent, A.M., Brick, P., Leatherbarrow, R.J., and Curry, S. (2005). Crystal structure of foot-and-mouth disease virus 3C protease. New insights into catalytic mechanism and cleavage specificity. J. Biol. Chem. 280, 11520–11527.

Black, D.N., Burroughs, J.N., Harris, T.J., and Brown, F. (1978). The structure and replication of calicivirus RNA. Nature 274, 614–615.

Blakeney, S.J., Cahill, A., and Reilly, P.A. (2003). Processing of Norwalk virus nonstructural proteins by a 3C-like cysteine proteinase. Virology 308, 216–224.

Boga, J.A., Martin Alonso, J.M., Casais, R., and Parra, F. (1997). A single dose immunization with rabbit haemorrhagic disease virus major capsid protein produced in Saccharomyces cerevisiae induces protection. J. Gen. Virol. 78, 2315–2318.

Bok, K., Prikhodko, V.G., Green, K.Y., and Sosnovtsev, S.V. (2009). Apoptosis in murine norovirus-infected RAW264.7 cells is associated with downregulation of survivin. J. Virol. 83, 3647–3656.

Boniotti, B., Wirblich, C., Sibilia, M., Meyers, G., Thiel, H.J., and Rossi, C. (1994). Identification and characterization of a 3C-like protease from rabbit hemorrhagic disease virus, a calicivirus. J. Virol. 68, 6487–6495.

Burroughs, J.N., and Brown, F. (1974). Physico-chemical evidence for the re-classification of the caliciviruses. J. Gen. Virol. 22, 281–286.

Burroughs, J.N., and Brown, F. (1978). Presence of a covalently linked protein on calicivirus RNA. J. Gen. Virol. 41, 443–446.

Carter, M.J. (1989). Feline calicivirus protein synthesis investigated by western blotting. Arch. Virol. 108, 69–79.

Carter, M.J., Milton, I.D., Turner, P.C., Meanger, J., Bennett, M., and Gaskell, R.M. (1992). Identification and sequence determination of the capsid protein gene of feline calicivirus. Arch. Virol. 122, 223–235.

Carter, M.J., Routledge, E.G., and Toms, G.L. (1989). Monoclonal antibodies to feline calicivirus. J. Gen. Virol. 70, 2197–2200.

Casais, R., Molleda, L.G., Machin, A., del Barrio, G., Manso, A.G., Dalton, K.P., Coto, A., Alonso, J.M., Prieto, M., and Parra, F. (2008). Structural and functional analysis of virus factories purified from Rabbit vesivirus-infected Vero cells. Virus Res. 137, 112–121.

Chang, K.O., George, D.W., Patton, J.B., Green, K.Y., and Sosnovtsev, S.V. (2008). Leader of the capsid protein in feline calicivirus promotes replication of Norwalk virus in cell culture. J. Virol. 82, 9306–9317.

Chang, K.O., Sosnovtsev, S.S., Belliot, G., Wang, Q., Saif, L.J., and Green, K.Y. (2005). Reverse genetics system for porcine enteric calicivirus, a prototype sapovirus in the Caliciviridae. J. Virol. 79, 1409–1416.

Chang, K.O., Sosnovtsev, S.V., Belliot, G., King, A.D., and Green, K.Y. (2006). Stable expression of a Norwalk virus RNA replicon in a human hepatoma cell line. Virology 353, 463–473.

Chaudhry, Y., Nayak, A., Bordeleau, M.E., Tanaka, J., Pelletier, J., Belsham, G.J., Roberts, L.O., and Goodfellow, I.G. (2006). Caliciviruses differ in their functional requirements for eIF4F components. J. Biol. Chem. 281, 25315–25325.

Chen, R., Neill, J.D., Estes, M.K., and Prasad, B.V. (2006). X-ray structure of a native calicivirus: structural insights into antigenic diversity and host specificity. Proc. Natl. Acad. Sci. U.S.A. 103, 8048–8053.

Chen, R., Neill, J.D., Noel, J.S., Hutson, A.M., Glass, R.I., Estes, M.K., and Prasad, B.V. (2004). Inter- and intra-genus structural variations in caliciviruses and their functional implications. J. Virol. 78, 6469–6479.

Clarke, I.N., and Lambden, P.R. (2000). Organization and expression of calicivirus genes. J. Infect. Dis. 181, S309–S316.

Daughenbaugh, K.F., Fraser, C.S., Hershey, J.W., and Hardy, M.E. (2003). The genome-linked protein VPg of the Norwalk virus binds eIF3, suggesting its role in translation initiation complex recruitment. EMBO J. 22, 2852–2859.

Daughenbaugh, K.F., Wobus, C.E., and Hardy, M.E. (2006). VPg of murine norovirus binds translation initiation factors in infected cells. Virol. J. 3, 33.

Dingle, K.E., Lambden, P.R., Caul, E.O., and Clarke, I.N. (1995). Human enteric Caliciviridae: the complete genome sequence and expression of virus-like particles from a genetic group II small round structured virus. J. Gen. Virol. 76, 2349–2355.

Dunham, D.M., Jiang, X., Berke, T., Smith, A.W., and Matson, D.O. (1998). Genomic mapping of a calicivirus VPg. Arch. Virol. 143, 2421–2430.

Ehresmann, D.W., and Schaffer, F.L. (1977). RNA synthesized in calicivirus-infected cells is atypical of picornaviruses. J. Virol. 22, 572–576.

Ettayebi, K., and Hardy, M.E. (2003). Norwalk virus nonstructural protein p48 forms a complex with the SNARE regulator VAP-A and prevents cell surface expression of vesicular stomatitis virus G protein. J. Virol. 77, 11790–11797.

Farkas, T., Sestak, K., Wei, C., and Jiang, X. (2008). Characterization of a rhesus monkey calicivirus representing a new genus of Caliciviridae. J. Virol. 82, 5408–5416.

Fernandez-Vega, V., Sosnovtsev, S.V., Belliot, G., King, A.D., Mitra, T., Gorbalenya, A., and Green, K.Y.

(2004). Norwalk virus N-terminal nonstructural protein is associated with disassembly of the Golgi complex in transfected cells. J. Virol. *78*, 4827–4837.

Fretz, M., and Schaffer, F.L. (1978). Calicivirus proteins in infected cells: evidence for a capsid polypeptide precursor. Virology *89*, 318–321.

Fukushi, S., Kojima, S., Takai, R., Hoshino, F.B., Oka, T., Takeda, N., Katayama, K., and Kageyama, T. (2004). Poly(A)- and primer-independent RNA polymerase of Norovirus. J. Virol. *78*, 3889–3896.

Fullerton, S.W., Blaschke, M., Coutard, B., Gebhardt, J., Gorbalenya, A., Canard, B., Tucker, P.A., and Rohayem, J. (2007). Structural and functional characterization of sapovirus RNA-dependent RNA polymerase. J. Virol. *81*, 1858–1871.

Gai, D., Zhao, R., Li, D., Finkielstein, C. V., and Chen, X. S. (2004). Mechanisms of conformational change for a replicative hexameric helicase of SV40 large tumour antipen. Cell *119*, 47–60.

Geissler, K., Schneider, K., Fleuchaus, A., Parrish, C.R., Sutter, G., and Truyen, U. (1999). Feline calicivirus capsid protein expression and capsid assembly in cultured feline cells. J. Virol. *73*, 834–838.

Goodfellow, I., Chaudhry, Y., Gioldasi, I., Gerondopoulos, A., Natoni, A., Labrie, L., Laliberte, J.F., and Roberts, L. (2005). Calicivirus translation initiation requires an interaction between VPg and eIF 4 E. EMBO Rep. *6*, 968–972.

Gorbalenya, A.E., Koonin, E.V., and Wolf, Y.I. (1990). A new superfamily of putative NTP-binding domains encoded by genomes of small DNA and RNA viruses. FEBS Lett. *262*, 145–148.

Green, K.Y. (2007). Caliciviridae: The Noroviruses. In Fields Virology, D.M. Knipe, and P.M. Howley, eds. (Philadelphia: Lippincott Williams & Wilkins), pp. 949–979.

Green, K.Y., Ando, T., Balayan, M.S., Berke, T., Clarke, I.N., Estes, M.K., Matson, D.O., Nakata, S., Neill, J.D., Studdert, M.J., *et al.* (2000). Taxonomy of the caliciviruses. J. Infect. Dis. *181*, S322–S330.

Green, K.Y., Mory, A., Fogg, M.H., Weisberg, A., Belliot, G., Wagner, M., Mitra, T., Ehrenfeld, E., Cameron, C.E., and Sosnovtsev, S.V. (2002). Isolation of enzymatically active replication complexes from feline calicivirus-infected cells. J. Virol. *76*, 8582–8595.

Gutierrez-Escolano, A.L., Brito, Z.U., del Angel, R.M., and Jiang, X. (2000). Interaction of cellular proteins with the 5′ end of Norwalk virus genomic RNA. J. Virol. *74*, 8558–8562.

Gutierrez-Escolano, A.L., Vazquez-Ochoa, M., Escobar-Herrera, J., and Hernandez-Acosta, J. (2003). La, PTB, and PAB proteins bind to the 3(′) untranslated region of Norwalk virus genomic RNA. Biochem. Biophys. Res. Commun. *311*, 759–766.

Hansman, G.S., Oka, T., and Takeda, N. (2008). Sapovirus-like particles derived from polyprotein. Virus Res. *137*, 261–265.

Hardy, M.E., Crone, T.J., Brower, J.E., and Ettayebi, K. (2002). Substrate specificity of the Norwalk virus 3C-like proteinase. Virus Res. *89*, 29–39.

Herbert, T.P., Brierley, I., and Brown, T.D. (1997). Identification of a protein linked to the genomic and subgenomic mRNAs of feline calicivirus and its role in translation. J. Gen. Virol. *78*, 1033–1040.

Jewell, D.A., Swietnicki, W., Dunn, B.M., and Malcolm, B.A. (1992). Hepatitis A virus 3C proteinase substrate specificity. Biochemistry *31*, 7862–7869.

Jiang, X., Wang, M., Graham, D.Y., and Estes, M.K. (1992). Expression, self-assembly, and antigenicity of the Norwalk virus capsid protein. J. Virol. *66*, 6527–6532.

Jiang, X., Wang, M., Wang, K., and Estes, M.K. (1993). Sequence and genomic organization of Norwalk virus. Virology *195*, 51–61.

Joubert, P., Pautigny, C., Madelaine, M.F., and Rasschaert, D. (2000). Identification of a new cleavage site of the 3C-like protease of rabbit haemorrhagic disease virus. J. Gen. Virol. *81 Pt 2*, 481–488.

Kaiser, W.J., Chaudhry, Y., Sosnovtsev, S.V., and Goodfellow, I.G. (2006). Analysis of protein–protein interactions in the feline calicivirus replication complex. J. Gen. Virol. *87*, 363–368.

Karakasiliotis, I., Chaudhry, Y., Roberts, L.O., and Goodfellow, I.G. (2006). Feline calicivirus replication: requirement for polypyrimidine tract-binding protein is temperature-dependent. J. Gen. Virol. *87*, 3339–3347.

Katayama, K., Hansman, G.S., Oka, T., Ogawa, S., and Takeda, N. (2006). Investigation of norovirus replication in a human cell line. Arch. Virol. *151*, 1291–1308.

Khan, A.R., Khazanovich-Bernstein, N., Bergmann, E.M., and James, M.N. (1999). Structural aspects of activation pathways of aspartic protease zymogens and viral 3C protease precursors. Proc. Natl. Acad. Sci. U.S.A. *96*, 10968–10975.

Konig, M., Thiel, H.J., and Meyers, G. (1998). Detection of viral proteins after infection of cultured hepatocytes with rabbit hemorrhagic disease virus. J. Virol. *72*, 4492–4497.

Kozak, M. (1991). Structural features in eukaryotic mRNAs that modulate the initiation of translation. J. Biol. Chem. *266*, 19867–19870.

Kreutz, L.C., and Seal, B.S. (1995). The pathway of feline calicivirus entry. Virus Res. *35*, 63–70.

Kuyumcu-Martinez, M., Belliot, G., Sosnovtsev, S.V., Chang, K.O., Green, K.Y., and Lloyd, R.E. (2004). Calicivirus 3C-like proteinase inhibits cellular translation by cleavage of poly(A)-binding protein. J. Virol. *78*, 8172–8182.

Lackner, T., Muller, A., Konig, M., Thiel, H.J., and Tautz, N. (2005). Persistence of bovine viral diarrhea virus is determined by a cellular cofactor of a viral autoprotease. J. Virol. *79*, 9746–9755.

Liljestrom, P., and Garoff, H. (1991). Internally located cleavable signal sequences direct the formation of Semliki Forest virus membrane proteins from a polyprotein precursor. J. Virol. *65*, 147–154.

Liu, B., Clarke, I.N., and Lambden, P.R. (1996). Polyprotein processing in Southampton virus: identification of 3C-like protease cleavage sites by in vitro mutagenesis. J. Virol. *70*, 2605–2610.

Liu, B.L., Clarke, I.N., Caul, E.O., and Lambden, P.R. (1995). Human enteric caliciviruses have a unique genome structure and are distinct from the Norwalk-like viruses. Arch. Virol. *140*, 1345–1356.

Liu, B.L., Lambden, P.R., Gunther, H., Otto, P., Elschner, M., and Clarke, I.N. (1999a). Molecular characterization of a bovine enteric calicivirus: relationship to the Norwalk-like viruses. J. Virol. 73, 819–825.

Liu, B.L., Viljoen, G.J., Clarke, I.N., and Lambden, P.R. (1999b). Identification of further proteolytic cleavage sites in the Southampton calicivirus polyprotein by expression of the viral protease in E. coli. J. Gen. Virol. 80, 291–296.

Liu, G., Ni, Z., Yun, T., Zhang, Y., Du, Q., Sheng, Z., Liang, H., Hua, J., Li, S., and Chen, J. (2006). Rescued virus from infectious cDNA clone of rabbit hemorrhagic disease Virus is adapted to RK13 cells line. Chinese Sci. Bull. 51, 1698–1702.

Lloyd, R.E. (2006). Translational control by viral proteinases. Virus Res. 119, 76–88.

Love, R.A., Parge, H.E., Wickersham, J.A., Hostomsky, Z., Habuka, N., Moomaw, E.W., Adachi, T., Margosiak, S., Dagostino, E., and Hostomska, Z. (1998). The conformation of hepatitis C virus NS3 proteinase with and without NS4A: a structural basis for the activation of the enzyme by its cofactor. Clin. Diagn. Virol. 10, 151–156.

Machin, A., Martin Alonso, J.M., and Parra, F. (2001). Identification of the amino acid residue involved in rabbit hemorrhagic disease virus VPg uridylylation. J. Biol. Chem. 276, 27787–27792.

Marcotte, L.L., Wass, A.B., Gohara, D.W., Pathak, H.B., Arnold, J.J., Filman, D.J., Cameron, C.E., and Hogle, J.M. (2007). Crystal structure of poliovirus 3CD protein: virally encoded protease and precursor to the RNA-dependent RNA polymerase. J. Virol. 81, 3583–3596.

Marin, M.S., Casais, R., Alonso Martin, J.M., and Parra, F. (2000). ATP binding and ATPase activities associated with recombinant rabbit hemorrhagic disease virus 2C-like polypeptide. J. Virol. 74, 10846–10851.

Martin Alonso, J.M., Casais, R., Boga, J.A., and Parra, F. (1996). Processing of rabbit hemorrhagic disease virus polyprotein. J. Virol. 70, 1261–1265.

Martin-Alonso, J.M., Skilling, D.E., Gonzalez-Molleda, L., del Barrio, G., Machin, A., Keefer, N.K., Matson, D.O., Iversen, P.L., Smith, A.W., and Parra, F. (2005). Isolation and characterization of a new Vesivirus from rabbits. Virology 337, 373–383.

Matsuura, Y., Tohya, Y., Nakamura, K., Shimojima, M., Roerink, F., Mochizuki, M., Takase, K., Akashi, H., and Sugimura, T. (2002). Complete nucleotide sequence, genome organization and phylogenic analysis of the canine calicivirus. Virus Genes 25, 67–73.

Matsuura, Y., Tohya, Y., Onuma, M., Roerink, F., Mochizuki, M., and Sugimura, T. (2000). Expression and processing of the canine calicivirus capsid precursor. J. Gen. Virol. 81, 195–199.

Matthews, D.A., Smith, W.W., Ferre, R.A., Condon, B., Budahazi, G., Sisson, W., Villafranca, J.E., Janson, C.A., McElroy, H.E., Gribskov, C.L., et al. (1994). Structure of human rhinovirus 3C protease reveals a trypsin-like polypeptide fold, RNA-binding site, and means for cleaving precursor polyprotein. Cell 77, 761–771.

McCormick, C.J., Salim, O., Lambden, P.R., and Clarke, I.N. (2008). Translation termination reinitiation

between open reading frame 1 (ORF1) and ORF2 enables capsid expression in a bovine norovirus without the need for production of viral subgenomic RNA. J. Virol. 82, 8917–8921.

Meyers, G., Wirblich, C., Thiel, H.J., and Thumfart, J.O. (2000). Rabbit hemorrhagic disease virus: genome organization and polyprotein processing of a calicivirus studied after transient expression of cDNA constructs. Virology 276, 349–363.

Mitra, T., Sosnovtsev, S.V., and Green, K.Y. (2004). Mutagenesis of Tyrosine 24 in the VPg Protein Is Lethal for Feline Calicivirus. J. Virol. 78, 4931–4935.

Moehring, J.M., Inocencio, N.M., Robertson, B.J., and Moehring, T.J. (1993). Expression of mouse furin in a Chinese hamster cell resistant to Pseudomonas exotoxin A and viruses complements the genetic lesion. J. Biol. Chem. 268, 2590–2594.

Morales, M., Barcena, J., Ramirez, M.A., Boga, J.A., Parra, F., and Torres, J.M. (2004). Synthesis in vitro of rabbit hemorrhagic disease virus subgenomic RNA by internal initiation on (–)sense genomic RNA: mapping of a subgenomic promoter. J. Biol. Chem. 279, 17013–17018.

Morgenstern, K.A., Landro, J.A., Hsiao, K., Lin, C., Gu, Y., Su, M.S., and Thomson, J.A. (1997). Polynucleotide modulation of the protease, nucleoside triphosphatase, and helicase activities of a hepatitis C virus NS3–NS4A complex isolated from transfected COS cells. J. Virol. 71, 3767–3775.

Mosimann, S.C., Cherney, M.M., Sia, S., Plotch, S., and James, M.N. (1997). Refined X-ray crystallographic structure of the poliovirus 3C gene product. J. Mol. Biol. 273, 1032–1047.

Nakamura, K., Someya, Y., Kumasaka, T., Ueno, G., Yamamoto, M., Sato, T., Takeda, N., Miyamura, T., and Tanaka, N. (2005). A norovirus protease structure provides insights into active and substrate binding site integrity. J. Virol. 79, 13685–13693.

Neill, J.D. (1990). Nucleotide sequence of a region of the feline calicivirus genome which encodes picornavirus-like RNA-dependent RNA polymerase, cysteine protease and 2C polypeptides. Virus Res. 17, 145–160.

Neill, J.D. (1992). Nucleotide sequence of the capsid protein gene of two serotypes of San Miguel sea lion virus: identification of conserved and non-conserved amino acid sequences among calicivirus capsid proteins. Virus Res. 24, 211–222.

Neill, J.D., Reardon, I.M., and Heinrikson, R.L. (1991). Nucleotide sequence and expression of the capsid protein gene of feline calicivirus. J. Virol. 65, 5440–5447.

Ng, K.K., Cherney, M.M., Vazquez, A.L., Machin, A., Alonso, J.M., Parra, F., and James, M.N. (2002). Crystal structures of active and inactive conformations of a caliciviral RNA-dependent RNA polymerase. J. Biol. Chem. 277, 1381–1387.

Ng, K.K., Pendas-Franco, N., Rojo, J., Boga, J.A., Machin, A., Alonso, J.M., and Parra, F. (2004). Crystal structure of norwalk virus polymerase reveals the carboxyl terminus in the active site cleft. J. Biol. Chem. 279, 16638–16645.

Oehmig, A., Buttner, M., Weiland, F., Werz, W., Bergemann, K., and Pfaff, E. (2003). Identification of a calicivirus isolate of unknown origin. J. Gen. Virol. *84*, 2837–2845.

Oka, T., Katayama, K., Ogawa, S., Hansman, G.S., Kageyama, T., Miyamura, T., and Takeda, N. (2005a). Cleavage activity of the sapovirus 3C-like protease in *Escherichia coli*. Arch. Virol. *150*, 2539–2548.

Oka, T., Katayama, K., Ogawa, S., Hansman, G.S., Kageyama, T., Ushijima, H., Miyamura, T., and Takeda, N. (2005b). Proteolytic processing of sapovirus ORF1 polyprotein. J. Virol. *79*, 7283–7290.

Oka, T., Yamamoto, M., Katayama, K., Hansman, G.S., Ogawa, S., Miyamura, T., and Takeda, N. (2006). Identification of the cleavage sites of sapovirus open reading frame 1 polyprotein. J. Gen. Virol. *87*, 3329–3338.

Oka, T., Yamamoto, M., Miyashita, K., Ogawa, S., Katayama, K., Wakita, T., and Takeda, N. (2009). Self-assembly of sapovirus recombinant virus-like particles from polyprotein in mammalian cells. Microbiol. Immunol. *53*, 49–52.

Oka, T., Yamamoto, M., Yokoyama, M., Ogawa, S., Hansman, G.S., Katayama, K., Miyashita, K., Takagi, H., Tohya, Y., Sato, H., *et al.* (2007). Highly conserved configuration of catalytic amino acid residues among calicivirus-encoded proteases. J. Virol. *81*, 6798–6806.

Oliver, S.L., Asobayire, E., Charpilienne, A., Cohen, J., and Bridger, J.C. (2007). Complete genomic characterization and antigenic relatedness of genogroup III, genotype 2 bovine noroviruses. Arch. Virol. *152*, 257–272.

Oliver, S.L., Asobayire, E., Dastjerdi, A.M., and Bridger, J.C. (2006). Genomic characterization of the unclassified bovine enteric virus Newbury agent-1 (Newbury1) endorses a new genus in the family Caliciviridae. Virology *350*, 240–250.

Pallai, P.V., Burkhardt, F., Skoog, M., Schreiner, K., Bax, P., Cohen, K.A., Hansen, G., Palladino, D.E., Harris, K.S., Nicklin, M.J., *et al.* (1989). Cleavage of synthetic peptides by purified poliovirus 3C proteinase. J. Biol. Chem. *264*, 9738–9741.

Paul, A.V., Rieder, E., Kim, D.W., van Boom, J.H., and Wimmer, E. (2000). Identification of an RNA hairpin in poliovirus RNA that serves as the primary template in the in vitro uridylylation of VPg. J. Virol. *74*, 10359–10370.

Pfister, T., and Wimmer, E. (2001). Polypeptide p41 of a Norwalk-like virus is a nucleic acid-independent nucleoside triphosphatase. J. Virol. *75*, 1611–1619.

Pletneva, M.A., Sosnovtsev, S.V., and Green, K.Y. (2001). The genome of Hawaii virus and its relationship with other members of the Caliciviridae. Virus Genes *23*, 5–16.

Prasad, B.V., Hardy, M.E., Dokland, T., Bella, J., Rossmann, M.G., and Estes, M.K. (1999). X-ray crystallographic structure of the Norwalk virus capsid. Science *286*, 287–290.

Prasad, B.V., Matson, D.O., and Smith, A.W. (1994a). Three-dimensional structure of calicivirus. J. Mol. Biol. *240*, 256–264.

Prasad, B.V., Rothnagel, R., Jiang, X., and Estes, M.K. (1994b). Three-dimensional structure of baculovirus-expressed Norwalk virus capsids. J. Virol. *68*, 5117–5125.

Rinck, G., Birghan, C., Harada, T., Meyers, G., Thiel, H.J., and Tautz, N. (2001). A cellular J-domain protein modulates polyprotein processing and cytopathogenicity of a pestivirus. J. Virol. *75*, 9470–9482.

Robel, I., Gebhardt, J., Mesters, J.R., Gorbalenya, A., Coutard, B., Canard, B., Hilgenfeld, R., and Rohayem, J. (2008). Functional characterization of the cleavage specificity of the sapovirus chymotrypsin-like protease. J. Virol. *82*, 8085–8093.

Rohayem, J., Jager, K., Robel, I., Scheffler, U., Temme, A., and Rudolph, W. (2006a). Characterization of norovirus 3Dpol RNA-dependent RNA polymerase activity and initiation of RNA synthesis. J. Gen. Virol. *87*, 2621–2630.

Rohayem, J., Robel, I., Jager, K., Scheffler, U., and Rudolph, W. (2006b). Protein-primed and de novo initiation of RNA synthesis by norovirus 3Dpol. J. Virol. *80*, 7060–7069.

Salim, O., Clarke, I.N., and Lambden, P.R. (2008). Functional analysis of the 5′ genomic sequence of a bovine norovirus. PLoS ONE 3, e2169.

Schechter, I., and Berger, A. (1967). On the size of the active site in proteases. I. Papain. Biochem. Biophys. Res. Commun. *27*, 157–162.

Scheffler, U., Rudolph, W., Gebhardt, J., and Rohayem, J. (2007). Differential cleavage of the norovirus polyprotein precursor by two active forms of the viral protease. J. Gen. Virol. *88*, 2013–2018.

Seah, E.L., Marshall, J.A., and Wright, P.J. (1999). Open reading frame 1 of the Norwalk-like virus Camberwell: Completion of sequence and expression in mammalian cells. J. Virol. *73*, 10531–10535.

Seah, E.L., Marshall, J.A., and Wright, P.J. (2003). Trans activity of the norovirus Camberwell proteinase and cleavage of the N-terminal protein encoded by ORF1. J. Virol. *77*, 7150–7155.

Sibilia, M., Boniotti, M.B., Angoscini, P., Capucci, L., and Rossi, C. (1995). Two independent pathways of expression lead to self-assembly of the rabbit hemorrhagic disease virus capsid protein. J. Virol. *69*, 5812–5815.

Simmonds, P., Karakasiliotis, I., Bailey, D., Chaudhry, Y., Evans, D.J., and Goodfellow, I.G. (2008). Bioinformatic and functional analysis of RNA secondary structure elements among different genera of human and animal caliciviruses. Nucleic Acids Res. *36*, 2530–2546.

Smiley, J.R., Chang, K.O., Hayes, J., Vinje, J., and Saif, L.J. (2002). Characterization of an enteropathogenic bovine calicivirus representing a potentially new calicivirus genus. J. Virol. *76*, 10089–10098.

Someya, Y., Takeda, N., and Miyamura, T. (2000). Complete nucleotide sequence of the chiba virus genome and functional expression of the 3C-like protease in *Escherichia coli*. Virology *278*, 490–500.

Someya, Y., Takeda, N., and Miyamura, T. (2002). Identification of active-site amino acid residues in the Chiba virus 3C-like protease. J. Virol. *76*, 5949–5958.

Someya, Y., Takeda, N., and Miyamura, T. (2005). Characterization of the norovirus 3C-like protease. Virus Res. *110*, 91–97.

Someya, Y., Takeda, N., and Wakita, T. (2008). Saturation mutagenesis reveals that GLU54 of norovirus 3C-like protease is not essential for the proteolytic activity. J. Biochem. *144*, 771–780.

Sosnovtsev, S.V., Belliot, G., Chang, K.O., and Green, K.Y. (2004). Induction of apoptosis in feline calicivirus-infected cells: implication of the p30 ('3A-like') protein in caspase activation. Paper presented at: VII International Symposium on Positive Strand RNA Viruses.San Francisco, CA.

Sosnovtsev, S.V., Belliot, G., Chang, K.O., Prikhodko, V.G., Thackray, L.B., Wobus, C.E., Karst, S.M., Virgin, H.W., and Green, K.Y. (2006). Cleavage map and proteolytic processing of the murine norovirus nonstructural polyprotein in infected cells. J. Virol. *80*, 7816–7831.

Sosnovtsev, S.V., Garfield, M., and Green, K.Y. (2002). Processing map and essential cleavage sites of the nonstructural polyprotein encoded by ORF1 of the feline calicivirus genome. J. Virol. *76*, 7060–7072.

Sosnovtsev, S.V., and Green, K.Y. (2000). Identification and genomic mapping of the ORF3 and VPg proteins in feline calicivirus virions. Virology *277*, 193–203.

Sosnovtsev, S.V., Prikhodko, V.G., Kabat, J., Hyde, J., Mackenzie, J., and Green, K.Y. (2007). Murine norovirus NS3 protein is a membrane-associated component of virus replication complexes. Paper presented at: III International Calicivirus Conference.Cancun, Mexico.

Sosnovtsev, S.V., Sosnovtseva, S.A., and Green, K.Y. (1998). Cleavage of the feline calicivirus capsid precursor is mediated by a virus-encoded proteinase. J. Virol. *72*, 3051–3059.

Sosnovtseva, S.A., Sosnovtsev, S.V., and Green, K.Y. (1999). Mapping of the feline calicivirus proteinase responsible for autocatalytic processing of the nonstructural polyprotein and identification of a stable proteinase-polymerase precursor protein. J. Virol. *73*, 6626–6633.

Stuart, A.D., and Brown, T.D. (2006). Entry of feline calicivirus is dependent on clathrin-mediated endocytosis and acidification in endosomes. J. Virol. *80*, 7500–7509.

Thouvenin, E., Laurent, S., Madelaine, M.F., Rasschaert, D., Vautherot, J.F., and Hewat, E.A. (1997). Bivalent binding of a neutralising antibody to a calicivirus involves the torsional flexibility of the antibody hinge. J. Mol. Biol. *270*, 238–246.

Thumfart, J.O., and Meyers, G. (2002). Rabbit hemorrhagic disease virus: identification of a cleavage site in the viral polyprotein that is not processed by the known calicivirus protease. Virology *304*, 352–363.

Tohya, Y., Shinchi, H., Matsuura, Y., Maeda, K., Ishiguro, S., Mochizuki, M., and Sugimura, T. (1999). Analysis of the N-terminal polypeptide of the capsid precursor protein and the ORF3 product of feline calicivirus. J. Vet. Med. Sci. *61*, 1043–1047.

Vazquez, A.L., Alonso, J.M., and Parra, F. (2000). Mutation analysis of the GDD sequence motif of a calicivirus RNA-dependent RNA polymerase. J. Virol. *74*, 3888–3891.

Vazquez, A.L., Martin Alonso, J.M., Casais, R., Boga, J.A., and Parra, F. (1998). Expression of enzymatically active rabbit hemorrhagic disease virus RNA-dependent RNA polymerase in *Escherichia coli*. J. Virol. *72*, 2999–3004.

Ward, V.K., McCormick, C.J., Clarke, I.N., Salim, O., Wobus, C.E., Thackray, L.B., Virgin, H.W.t., and Lambden, P.R. (2007). Recovery of infectious murine norovirus using pol II-driven expression of full-length cDNA. Proc. Natl. Acad. Sci. U.S.A. *104*, 11050–11055.

Wawrzkiewicz, J., Smale, C.J., and Brown, F. (1968). Biochemical and biophysical characteristics of vesicular exanthema virus and the viral ribonucleic acid. Arch. Ges. Viruforsc.h *25*, 337–351.

Wei, L., Huhn, J.S., Mory, A., Pathak, H.B., Sosnovtsev, S.V., Green, K.Y., and Cameron, C.E. (2001). Proteinase-polymerase precursor as the active form of feline calicivirus RNA-dependent RNA polymerase. J. Virol. *75*, 1211–1219.

Weir, M.L., Xie, H., Klip, A., and Trimble, W.S. (2001). VAP-A binds promiscuously to both v- and tSNAREs. Biochem. Biophys. Res. Commun. *286*, 616–621.

Willcocks, M.M., Carter, M.J., and Roberts, L.O. (2004). Cleavage of eukaryotic initiation factor eIF4G and inhibition of host-cell protein synthesis during feline calicivirus infection. J. Gen. Virol. *85*, 1125–1130.

Wirblich, C., Sibilia, M., Boniotti, M.B., Rossi, C., Thiel, H.J., and Meyers, G. (1995). 3C-like protease of rabbit hemorrhagic disease virus: identification of cleavage sites in the ORF1 polyprotein and analysis of cleavage specificity. J. Virol. *69*, 7159–7168.

Wirblich, C., Thiel, H.J., and Meyers, G. (1996). Genetic map of the calicivirus rabbit hemorrhagic disease virus as deduced from in vitro translation studies. J. Virol. *70*, 7974–7983.

Wobus, C.E., Karst, S.M., Thackray, L.B., Chang, K.O., Sosnovtsev, S.V., Belliot, G., Krug, A., Mackenzie, J.M., Green, K.Y., and Virgin, H.W. (2004). Replication of Norovirus in cell culture reveals a tropism for dendritic cells and macrophages. PLoS Biol. *2*, e432.

Zeitler, C.E., Estes, M.K., and Venkataram Prasad, B.V. (2006). X-ray crystallographic structure of the Norwalk virus protease at 1.5-A resolution. J. Virol. *80*, 5050–5058.

Zhang, X., Buehner, N.A., Hutson, A.M., Estes, M.K., and Mason, H.S. (2006). Tomato is a highly effective vehicle for expression and oral immunization with Norwalk virus capsid protein. Plant Biotechnol. J. *4*, 419–432.

Calicivirus Protein Structures

Kenneth K.S. Ng and Francisco Parra

5

Abstract

Sequence analysis and experimentally determined three-dimensional structures of structural and non-structural proteins from a range of caliciviruses help to provide a molecular framework for understanding many aspects of their replication strategies. Structures of intact virions, virus-like particles and capsid fragments, as well as capsid–receptor complexes help to explain basic mechanisms of capsid assembly and receptor recognition. Structural studies of the recombinant viral proteinase and polymerase in complex with substrates and inhibitors provide a basis for understanding substrate recognition and enzymatic mechanisms, thus setting the stage for the design of new antiviral compounds.

Introduction

Sequence analysis and three-dimensional structural information has been critical for elucidating many of the details underlying calicivirus protein function. Cryo-transmission electron microscopy (cryo-TEM) and X-ray crystallography have been the primary techniques used to determine structure-function relationships in the viral capsid and several non-structural proteins. These structural studies provide the fundamental framework necessary for understanding such important aspects of the viral life cycle as viral capsid assembly, the binding of virions to host cell receptors, replication and transcription of viral genomic RNA, and proteolytic processing of the viral polyprotein. Although many proteins and RNA structures important to caliciviruses are still poorly characterized, it is valuable to review the current state of knowledge in calicivirus structural biology to highlight how much has been learned (Tables 5.1–5.4) and to realize how much more needs to be discovered.

Viral capsids

The structures of calicivirus capsids have been investigated in at least a representative member of the four calicivirus genera *Norovirus*, *Vesivirus*, *Lagovirus* and *Sapovirus*. Caliciviruses possess a fairly simple structure in which a single strand of genomic RNA (7.3 to 8.3 kb) of positive polarity is surrounded by a protein capsid of between 27 and 40 nm in diameter. The capsid is composed primarily of a major polypeptide of approximately 60 kDa formerly called VP60 and more recently designated as VP1 (Bertolotti-Ciarlet et al., 2002). Pioneering work in the caliciviruses has shown that 180 copies of VP1 can spontaneously self-assemble into non-infectious virus-like particles (VLPs) (Jiang et al., 1992) with T=3 icosahedral symmetry (Prasad et al., 1999). VLPs are believed to be similar in structure to native virions, with differences probably being modulated by the presence in native virions of the minor structural protein VP2, as well as the genomic RNA which is covalently attached by its 5′ end to the viral genome-linked protein (VPg).

Cryo-TEM reconstructions of VLPs from genogroup I (Norwalk virus; NV) and genogroup II (Grimsby virus) human noroviruses, the Parkville sapovirus, rabbit haemorrhagic disease virus (RHDV), San Miguel sea lion virus (SMSV), murine norovirus (MNV) and feline calicivirus (FCV) as well as native virions from the Pan-1 primate calicivirus, reveal a common overall structure (Barcena *et al.*, 2004; Bhella *et al.*, 2008; Chen *et al.*, 2003a, 2004; Katpally *et al.*, 2008; Prasad *et al.*, 1994, 1996; Thouvenin *et al.*, 1997). A unique feature of the calicivirus capsid is the formation of 'cup-like' depressions, which inspired the name of 'calici' from the Greek word for cup, for the entire family.

X-ray crystal structures determined at 3.4 Å resolution for the NV VLP (Prasad *et al.*, 1999) and 3.2 Å for SMSV native virion (Chen *et al.*, 2006) are consistent with the EM reconstructions and also reveal details about the structure of VP1 protomers and their assembly into the T=3 icosahedral capsid (Fig. 5.1 and Table 5.1). NV and SMSV VP1 both contain an N-terminal arm (NTA) (residues 10–49 in NV and residues 160–200 in SMSV, the latter numbered from the

N-terminus of the capsid precursor), and shell or S-domain (residues 50–225 in NV and 201–361 in SMSV) that contains a β-sandwich fold that is commonly found in viral capsid proteins and is responsible for forming most of the inter-subunit interactions needed for assembling the icosahedral capsid. The remainder of the structure of VP1 forms a protrusion or P-domain that appears to be unique to caliciviruses and helps to generate the 'cup-like' depressions on the viral surface. It is notable that a smaller protrusion domain with an unrelated folding topology has also been seen in the capsid proteins from some tombusviruses, such as tomato bushy stunt virus (Olson *et al.*, 1983), and some nodaviruses (Tang *et al.*, 2002).

The P-domain is composed of two subdomains called P1 and P2 (Fig. 5.1 and Table 5.2). P1 is connected to the S-domain by a single

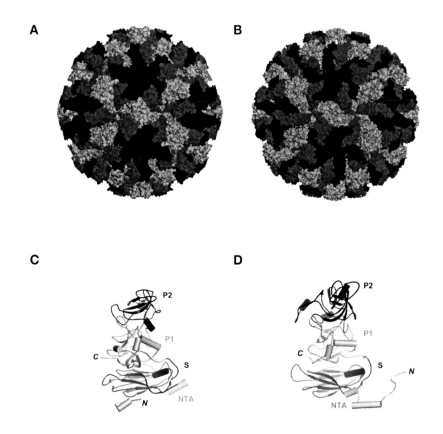

Figure 5.1 Crystal structures of (A) a VLP and (C) a VP1 capsid protomer from human norovirus GI.1 (PDB 1IHM) (Prasad *et al.*, 1999). Crystal structures of (B) a native virion and (D) a VP1 capsid protomer from SMSV (PDB 2GH8) (Chen *et al.*, 2006). In panels (A) and (B), the capsid protomers occupying quasi-equivalent positions A, B and C are coloured black, grey and white, respectively. All figures were prepared using PyMOL (DeLano, 2002).

Table 5.1 Crystal structures of viral capsids

Virus	PDB	Resolution (Å)	Details	Reference
NV GI.1	1IHM	3.4	VLPs	Prasad et al. (1999)
SMSV	2GH8	3.2	Native virion	Chen et al. (2006)

Table 5.2 P domain/P polypeptide structures

Virus	PDB	Resolution (Å)	Details	Reference
VA387 GII.4	2OBR	2.20	P domain	Cao et al. (2007)
	2OBS	2.00	P domain + A-type HBGA	
	2OBT	2.00	P domain + B-type HBGA	
VA387 GII.4	3BQJ	2.70	P polypeptide	Bu et al. (2008)
NV GI.1	3BY1	2.69	P domain	Bu et al. (2008)
	3BY2	2.60	P polypeptide	
	3D26	2.30	P domain + A-type HBGA	
NV GI.1	2ZL5	1.47	Acetate, Ca^{2+}, Mg^{2+}	Choi et al. (2008)
	2ZL6	1.43	H-type HBGA, acetate, Mg^{2+}	
	2ZL7	1.35	A-type HBGA, acetate, Ca^{2+}, Mg^{2+}	

linker segment that can act as a flexible hinge. The orientation of the P-domain to the S-domain in the three different copies in the asymmetric unit of the crystal structure of recombinant NV VLPs differ substantially from each other, as well as from the orientations seen in the three copies of VP1 from crystals of SMSV virions (Chen et al., 2006; Prasad et al., 1999). In terms of linear sequence, the P2 subdomain (residues 279–405 in NV and 414–589 in SMSV) is an insertion that interrupts the N-terminal and C-terminal portions of the P1 subdomain (residues 226–278 and 406–520 in NV, and residues 362–413 and 590–703 in SMSV). The six-stranded β-barrel fold of the P2 subdomain resembles the fold adopted by one of the RNA-binding domains of EF-Tu, but the functional relevance of this similarity is unclear (Chen et al., 2006; Prasad et al., 1999).

The amount of sequence and presumably structural variation is lowest in the S domain, intermediate in the P1 subdomain and highest in the P2 subdomain (Chakravarty et al., 2005; Chen et al., 2003b). This pattern of sequence conservation reflects the different functions of these parts of VP1. The S domain provides most of the intersubunit contacts lying at the threefold, fivefold and sixfold rotational symmetry axes and forms the inner shell of the capsid. In contrast, the P1 and P2 subdomains of each protomer interact with only a single adjacent protomer related by either a local 2-fold rotational symmetry axis or a crystallographic twofold rotational symmetry axis. As required by the T=3 icosahedral symmetry, the three capsid proteins arranged about each 3-fold rotational symmetry axis occupy quasi-equivalent positions and adopt different conformations to help achieve a high degree of sphericity in the overall capsid structure. P1 and especially P2 are also more exposed to the solvent and thus present immunogenic surfaces, as well as binding sites for cellular receptors.

Cellular receptors have been identified for RHDVs, human noroviruses and FCVs. Native RHDV virions and VLPs were first shown to bind to blood group antigens (Rademacher et al., 2008; Ruvoen-Clouet et al., 2000). In human norovirus, a series of studies examining the covariation of capsid sequences in different strains and binding specificities for blood group antigens (Huang *et al.*, 2003; Hutson *et al.*, 2002; Lindesmith *et al.*, 2003; Tan *et al.*, 2009), site-directed mutagenesis studies (Tan *et al.*, 2003) and crystal structures of P domain fragments bound to oligosaccharides have identified the locations of specific carbohydrate-receptor binding pockets in the P2 subdomain (Bu *et al.*, 2008; Cao *et al.*, 2007; Choi *et al.*, 2008). Additional details of these binding interactions are described in the following chapter. In FCV, α2,6-linked sialic acid (Stuart and Brown, 2007) and the feline junctional adhesion molecule-1 (fJAM-1) (Makino *et al.*, 2006) have been identified as receptors, the latter of which also appears to interact with the P2 subdomain of VP1 (Bhella *et al.*, 2008).

Comparisons of capsid structures from a range of caliciviruses using either VLPs or native virions reveal some interesting differences that may be attributed to strain or genus peculiarities, the existence of VP1 precursor sequences in some caliciviruses or the presence of the RNA genome and minor polypeptides such as VP2 and VPg in native virions but not VLPs. Although the specific reasons underlying some of these differences and their potential importance remain poorly defined, these structural variations are intriguing and may suggest important aspects of calicivirus biology related to capsid structure. A difference observed in the SMSV virion structure, which is distinct from other calicivirus capsids, resides in the intersubunit interactions involving the NTAs, a region which has been implicated in providing a switch to facilitate bent and flat conformations of the subunit dimers during capsid assembly. In SMSV, the three NTAs are equally ordered (Chen *et al.*, 2006), whereas in NV only one of the three NTAs is ordered (Prasad *et al.*, 1999). In the case of SMSV, instead of an order-to-disorder transition, a distinct conformational change involving a Pro residue in the protomer occupying quasi-equivalent position B leads to the formation of a ring-like structure around the fivefold axis that may be involved in a switch. Whether these NTA interactions in SMSV are influenced by the presence of the RNA genome or the proteolytic processing of the capsid protein precursor should be further investigated. A second distinct structural variation is seen in the relative orientation of the VP1 P domain with respect to the S domain. In SMSV the orientation of the S domain relative to the P1 subdomain prevents P1 from participating in the dimeric interactions between adjacent capsid protomers, that are seen in NV (Chen *et al.*, 2006; Prasad *et al.*, 1999). However, the additional flexibility between P1 and P2 allows the opposing P2 subdomains in adjacent protomers to interact closely. A tight interaction between the dimer-related P2 subdomains is consistently observed in all calicivirus capsid structures thus far characterized by cryo-EM, suggesting that the dimeric association of the P2 subdomain is a functional requirement.

Recent work on the native murine norovirus (MNV) virion structure has also shown that the P domains of this animal norovirus project further away from the sphere formed by the S domain, thus forming a second shell around the virion (Katpally *et al.*, 2008). This structure appears to be significantly different from the capsid structures of all other caliciviruses, in which the P domains lie directly upon the S domains, connected by a long flexible loop. This striking difference does not seem to be attributable to a difference between the structure of VLPs and native virions, because crystal structures and cryo-TEM reconstructions of SMSV native virions are consistent with each other and do not reveal 'floating P domains' projecting away from the inner shell formed by the S domains. It is possible that the unusual arrangement of P domains found in MNV is norovirus strain-specific, but more structural studies on VLPs and native virions from MNV and other caliciviruses are needed to understand the exact mechanisms controlling interactions between P and S domains, and the resulting effects on capsid conformation and virus biology.

Non-structural proteins

As outlined in the previous chapter, the primary translation product of the calicivirus RNA genome is a polyprotein that is proteolytically processed into at least seven non-structural proteins (NS1–7), as well as VP1 in the lagoviruses and sapoviruses. Although the functions of NS1, NS2 and NS4 are not well defined at present, there is a rapidly growing body of information on both the structure and function of the other non-structural proteins. The NS3 NTPase shares sequence similarity with enzymes in other positive-strand RNA viruses and has been enzymatically characterized in RHDV (Marin *et al.*, 2000) and NV (Pfister and Wimmer, 2001), but structural information is currently lacking. The NS5 VPg protein is attached to the 5'-end of the genomic RNA strand (Burroughs and Brown, 1978; Herbert *et al.*, 1997; Meyers *et al.*, 1991) involved with the initiation of genome replication and is a substrate for nucleotidylation reaction that is catalysed by the NS7 polymerase (Machin *et al.*, 2001) or NS6–7 proteinase–polymerase (Belliot *et al.*, 2008; Machin *et al.*, 2009). VPg has also been involved in protein synthesis initiation acting as a 5'-cap analogue able to recruit translation factors such as eIF3, eIF4F and eIF4E (Chaudhry *et al.*, 2006; Daughenbaugh *et al.*, 2003; Goodfellow *et al.*, 2005) close to the genome 5' region. Structural information is available at present only for the viral proteinase (NS6) and polymerase (NS7), which are located at the penultimate and terminal positions of the polyprotein respectively of the noroviruses and vesiviruses or preceding the region coding for VP1 in the lagoviruses and sapoviruses.

Proteinase

A single chymotrypsin-like proteinase (CLP) domain is encoded by the genomes of all caliciviruses and is responsible for cleaving the polyprotein into the seven major non-structural proteins NS1–7. Although found as a free polypeptide chain, crystal structures of the viral proteinase have been determined (Table 5.3) from three genogroup I noroviruses, Norwalk virus (GI.1) (Zeitler *et al.*, 2006), Southampton virus (GI.2)

and Chiba virus (GI.4) (Nakamura *et al.*, 2005). All three structures are very similar and resemble the CLPs from a wide range of viruses (Allaire *et al.*, 1994; Seipelt *et al.*, 1999) and other organisms. Although the sequence identity between the proteinases from different caliciviruses can be quite low, key residues in the active site, such as the active site cysteine–histidine dyad, and residues forming the S1 specificity pocket, are conserved in all calicivirus proteinases (Fig. 5.2). Homology models can thus be constructed for all calicivirus proteinases based on the NV proteinase crystal structures (Oka *et al.*, 2007).

The norovirus proteinase crystal structures reveal a modified CLP fold in which the N-terminal domain contains seven antiparallel β-strands that form a twisted β-sheet that appears to be intermediate between the abbreviated four-stranded β-sheet found in picornaviral 2A-type proteinases (Petersen *et al.*, 1999) and the more conventional eight-stranded β-barrel found in most viral CLPs (Fig. 5.3). The complete β-barrel formed by the six antiparallel β-strands in the C-terminal domain resembles more closely the structure of other viral CLPs. The N- and C-terminal domains pack together as seen in other CLPs to form an active-site cleft with a series of binding pockets complementary to the side chains of a peptide substrate bound in an extended conformation. At the centre of the active site cleft, the N-terminal domain contributes His30 and the C-terminal domain contributes Cys139 to form a catalytic dyad with an activated thiolate anion nucleophile (Fig. 5.2). Glu54 from the N-terminal domain is on the opposite side of the His30 side chain from Cys139, and site-directed mutagenesis experiments indicate that this residue may assist in catalysis and play a role in substrate specificity, but is not essential

Table 5.3 Proteinase structures

Virus	PDB	Resolution (Å)	Details	Reference
NV GI.1	2FYQ	1.5	Cl-, PO$_4$$^{2-}$	Zeitler *et al.* (2006)
	2FYR	2.2	AEBSF, Cl-, Mg^{2+}	
Norovirus GI.2	2IPH	1.75	EFVQVQ-	Unpublished
Norovirus GI.4	1WQS	2.8	Hg^{2+}, tartrate	Nakamura *et al.* (2005)

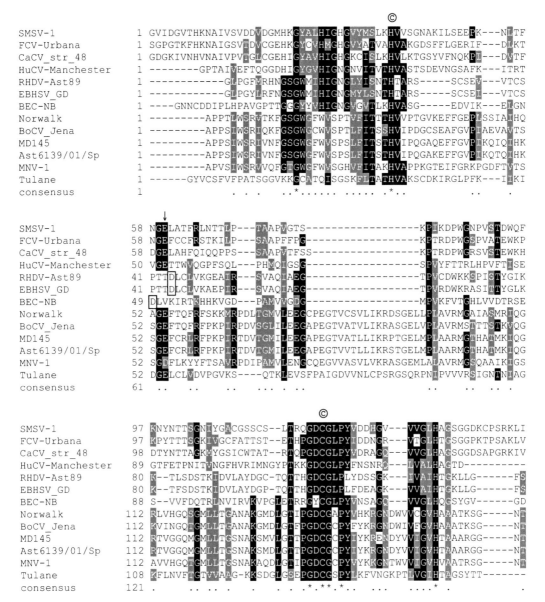

Figure 5.2 Amino acid sequence alignments of NS6 proteinase sequences from representative members of the admitted (*Norovirus*, *Sapovirus*, *Lagovirus* and *Sapovirus*) and two unclassified genera (BEC-NB and Tulane) within the family *Caliciviridae*). Accession numbers are indicated in brackets: SMSV-1 (AF181081), FCV-Urbana (L40021), CaCV_str_48 (AB070225), HuCV-Manchester (X86560), RHDV-Ast89 (Z49271), EBHSV_GD (Z69620), BEC-NB (Q8JN60), Norwalk (M87661), BoCV_Jena (AJ011099), MD145 (AY032605), Ast6139/01/Sp (AJ583672), MNV-1 (AY228235), and Tulane (EU391643). © Residues involved in the catalytic dyad. (↓) Glu-54 residue, putatively assisting in catalysis. Boxed Asp residues indicate putative functional analogues of Glu 54 in *Lagovirus* and BEC-NB.

for proteolysis (Someya *et al.*, 2008). In contrast, other authors indicate that mutation of this residue to Ala abolished the protease activity for certain protein and peptide substrates and is essential for a catalytic triad (Hardy *et al.*, 2002; Zeitler *et al.*, 2006). It is notable that Glu54 does not appear to be conserved in the lagovirus proteinase, which supports the notion that Glu54 may not be essential for catalysis in all calicivirus proteinases (Fig. 5.2).

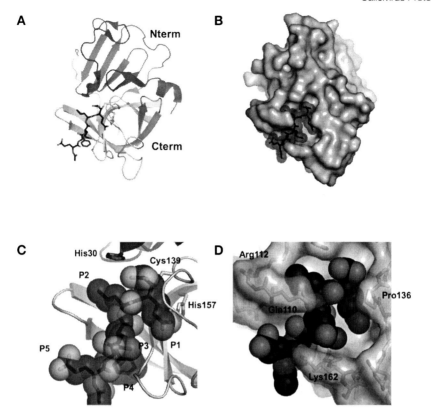

Figure 5.3 Crystal structure of the proteinase from Southampton virus (GI.2) bound to a covalent peptide-based inhibitor (PDB 1IPH) (DOI: 10.2210/pdb2iph/pdb, unpublished). (A) Ribbon diagram representation of the proteinase, with inhibitor and several key active-site residues drawn in stick representation. (B) Solvent-accessible representation of the proteinase, with inhibitor drawn in stick representation. (C) and (D) detailed views of the bound inhibitor and key active-site residues. Carbon atoms in the peptide inhibitor is coloured dark grey, nitrogen atoms are coloured medium grey and oxygen atoms are coloured light grey.

The structure of an unpublished covalent enzyme-inhibitor complex of the GI.2 Southampton virus (2IPH) mimics the naturally occurring acyl-enzyme intermediate and reveals how substrates interact with the S1–S4 specificity pockets (nomenclature according to Schechter and Berger (Schechter and Berger, 1967)) formed by the C-terminal β-barrel (Fig. 5.3). As in many positive-strand RNA viruses, the P1 position of most polyprotein cleavage sites appears to be either glutamic acid or glutamine (Belliot *et al.*, 2003; Liu *et al.*, 1996, 1999; Martin Alonso *et al.*, 1996; Oka *et al.*, 2005a,b, 2006; Sosnovtsev *et al.*, 1998, 1999; Wirblich *et al.*, 1996). The presence of a highly conserved His157 residue at the base of the S1 pocket is an important determinant of substrate specificity and can donate a hydrogen bond to either glutamic acid or glutamine side

chains in the P1 position N-terminal to the cleavage site. Pro136, which is also conserved in most caliciviruses forms an important hydrophobic wall of the S1 pocket, which may differ somewhat in caliciviruses where this residue is replaced by Gln, Lys, His or Arg. The preference for a small residue such as Gly, Ala or Ser at the P1' position following the cleavage site also seems to be defined by the presence of a small S1' specificity pocket in the norovirus proteinase crystal structures, as well as the predicted structures of related calicivirus proteinases (Oka *et al.*, 2007).

The Southampton virus inhibitor complex crystal structure and models of other norovirus proteinases bound to peptide substrates, as inferred by comparison with other CLP–peptide complexes, indicate that the P2, P3 and P4 positions may also be specifically recognized by the

S2, S3 and S4 pockets of calicivirus proteinases (Fig. 5.3). The aliphatic portions of the Gln110 and Arg112 side chains and positively charged guanidinium group of the latter form a well-defined S2 pocket that appears to favour the binding of side chains with some hydrophobic character and a negatively charged group at the P2 position (Robel *et al.*, 2008). Notably, this pocket is not seen in the structures of norovirus proteinases in the absence of bound substrates or inhibitors (Nakamura *et al.*, 2005; Zeitler *et al.*, 2006) and only appears to form when a peptide substrate is bound. The S3 and S4 pockets appear to be more open and less well defined, but these pockets probably exert additional constraints on substrate specificity by favouring long polar side chains at P3 that may form hydrogen bonds with the P1 side chain as well as polar groups lining the S3 pocket. In contrast, large, hydrophobic side chains appear to be favoured at P4, due to the presence of an exposed hydrophobic patch in the S4 pocket.

Polymerase

All caliciviruses encode a RNA-dependent RNA polymerase (RdRp) that is responsible for catalysing the nucleotidyl transfer reaction essential for genome replication and transcription. Crystal structures of the RdRp (Table 5.4) have been determined for RHDV (Ng *et al.*, 2002), human norovirus GII.4 (Ng *et al.*, 2004; Zamyatkin *et al.*, 2008), NV GI (Hogbom *et al.*, 2009) and sapovirus GI (Fullerton *et al.*, 2007). These structures reveal that calicivirus polymerases adopt an overall structure that resembles a cupped right hand, in which portions of the structure correspond to the fingers, palm and thumb. The arrangement of these structural elements is seen in most nucleic acid polymerases and was first described for the Klenow fragment of *E. coli* DNA polymerase I (Ollis *et al.*, 1985; Steitz, 2006) (Fig. 5.4). A distinctive feature of the viral RdRp is the presence of an N-terminal region that bridges the thumb and fingers domains, which are unlinked in most other nucleic acid polymerases (Bruenn, 2003; Ferrer-Orta *et al.*, 2006; Ng *et al.*, 2008). The oligomerization of polymerase protomers, which has been proposed to be important in various RdRp from picornaviruses (Boerner *et al.*, 2005; Hansen *et al.*, 1997; Hobson *et al.*, 2001; Pata *et al.*, 1995) and flaviviruses, has also been suggested to be important in norovirus polymerases (Hogbom *et al.*, 2009; Zamyatkin *et al.*, 2008).

Six sequence and structural motifs (motifs A-F) have been identified in the viral RdRp (Figs 5.4 and 5.5) that contain most of the key residues found at the viral RdRp active site (Bruenn, 2003; Hansen *et al.*, 1997; Kamer and Argos,

Table 5.4 Polymerase structures

Virus	PDB	Resolution (Å)	Details	Reference
NV GI.1	2B43	2.3	Apoenzyme	Hogbom *et al.* (2009)
Norovirus GII.4	1SH0	2.17	Triclinic, apoenzyme	Ng *et al.* (2004)
	1SH2	2.30	Orthorhombic, apoenzyme	
	1SH3	2.95	$MgSO_4$, apoenzyme	
	3BSN	1.80	Mn^{2+}, 5-NO_2-CTP, RNA primer-template	Zamyatkin *et al.* (2008)
	3BSO	1.74	Mn^{2+}, CTP, RNA primer-template	
	3H5X	1.77	6 mM Mn^{2+}, 2′-amino-2′-deoxy-CTP, RNA	Zamyatkin *et al.* (2009)
	3H5Y	1.77	6 mM Mn^{2+}, CTP, RNA	
Sapovirus GI	2CKW	2.3	Apoenzyme	Fullerton *et al.* (2007)
RHDV	1KHV	2.5	Mg^{2+}	Ng *et al.* (2002)
	1KHW	2.7	UTP, Mn^{2+}	

1984; O'Reilly and Kao, 1998). High resolution structures of the norovirus GII.4 polymerase in complex with divalent metal cations, nucleoside triphosphates and a primer-template RNA duplex (Zamyatkin et al., 2009; Zamyatkin et al., 2008) provide a structural framework for understanding the roles of most of these residues. The 1.8-Å-resolution crystal structures determined for these norovirus complexes provide detailed three-dimensional information on the closed complexes with natural substrate nucleotides (CTP) as well as inhibitor nucleotides (5-nitro-CTP and 2′-amino-CTP) that reveal many similarities to the crystal structures of polymerase-RNA-NTP complexes from other RNA viruses (Butcher et al., 2001; Ferrer-Orta et al., 2006; Ferrer-Orta et al., 2007; Ng et al., 2008; Tao et al., 2002). The functions of most of these key residues as predicted by the structures are consistent with

site-directed mutagenesis studies in polymerases from caliciviruses (Fullerton et al., 2007; Lopez Vazquez et al., 2000) and other viruses (Cameron et al., 2002; O'Reilly and Kao, 1998).

Motifs A and C contain highly conserved Asp residues that participate in the 'two-metal ion' mechanism of catalysis found in most nucleic acid polymerases (Steitz, 1998). The side chain carboxylate groups from these Asp residues, as well as main chain carbonyl oxygen atoms coordinate to two divalent metal ions involved in positioning the triphosphate moiety of NTP substrates and the 3′-OH of the primer at the active site, and in polarizing the electrons from these groups to assist in catalysis (Ng et al., 2008; Zamyatkin et al., 2008) (Fig. 5.4). Mutation of these Asp residues leads to the loss of activity or in some cases interesting changes in metal ion binding specificity (Lopez Vazquez et al., 2000). Motif B

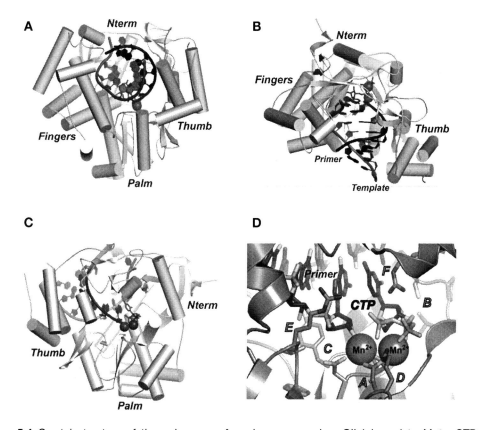

Figure 5.4 Crystal structure of the polymerase from human norovirus GII.4 bound to Mn^{2+}, CTP and primer:template RNA duplex (PDB 3BSO) (Zamyatkin et al., 2008). Ribbon diagrams of the complex viewed from (A) front, (B) top and (C) side views. (D) Detailed view of the polymerase active site, with selected residues from motifs A–F, CTP and RNA primer:template duplex drawn in stick representation, and the two Mn^{2+} ions drawn as spheres.

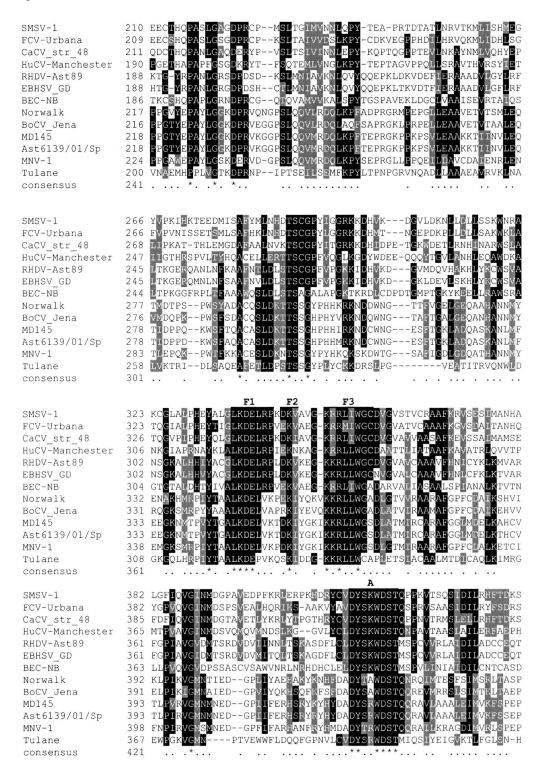

Figure 5.5 Amino acid sequence alignments of NS7 polymerase (RdRp) sequences from representative members of the admitted (*Norovirus*, *Sapovirus*, *Lagovirus* and *Sapovirus*) and two unclassified genera (BEC-NB and Tulane) family *Caliciviridae*. Accession numbers are indicated in brackets: SMSV-1 (AF181081), FCV-Urbana (L40021), CaCV_str_48 (AB070225), HuCV-Manchester (X86560), RHDV-Ast89 (Z49271), EBHSV_GD (Z69620), BEC-NB (Q8JN60), Norwalk (M87661), BoCV_Jena (AJ011099),

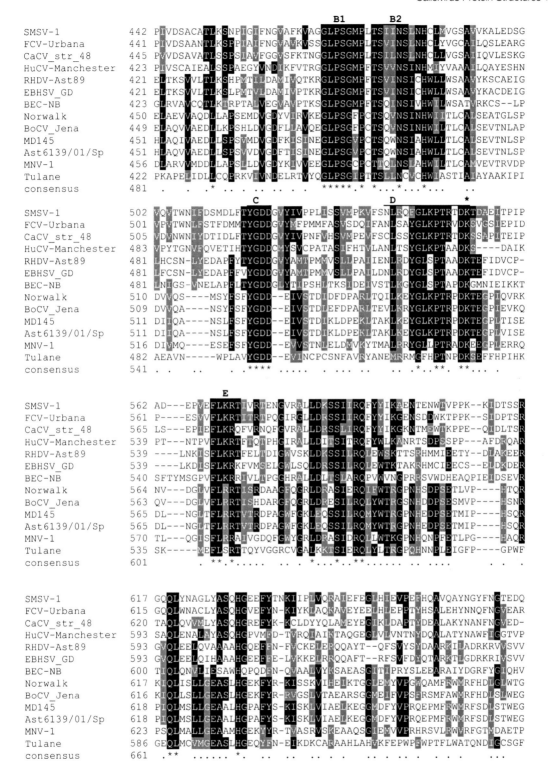

MD145 (AY032605), Ast6139/01/Sp (AJ583672), MNV-1 (AY228235), and Tulane (EU391643). For drawing convenience the N and C-terminal sequences not including relevant conserved sequences have been deleted. Numbers indicated amino acid residues starting from the N-terminal of the NS6-7 precursor, corresponding to residue 1 in Fig. 5.2. The conserved sequences corresponding to RdRp structural motifs (A–F) are indicated.

contains two highly conserved peptide segments that form a binding pocket to recognize the ribose moiety of the NTP. A perfectly conserved Ser residue (Ser300 in human norovirus) forms a hydrogen bond with the 2'-OH group to distinguish NTPs from 2'-deoxy-NTPs (Zamyatkin et al., 2009; Zamyatkin et al., 2008) (Fig. 5.4). Motif D is located in a flexible loop that is usually poorly ordered in crystal structures of the RdRp, but contains a highly conserved residue (Lys374 in human norovirus) that has been implicated as a general acid catalyst found in all nucleic acid polymerases (Castro et al., 2009; Zamyatkin et al., 2008) (Fig. 5.4). It is likely that the flexibility of this loop and the highly conserved Lys residue play important roles in catalysis that are not fully understood at present. Motif E contains several hydrophobic and basic residues at the tip of the β-hairpin located between the palm and thumb domain. These residues seem to form a platform to position the 3'-end of the primer in the active site, and perhaps assist in a conformational change important to the translocation of the RNA duplex during RNA synthesis (Fig. 5.4). The positively charged side chains from basic amino acids in Motif F contribute electrostatic interactions to position the triphosphate moiety of the bound NTP (Fig. 5.4).

One of the major discrepancies observed comparing the genetic organization and the functional studies performed in various members of the family *Caliciviridae* is related to the origin and nature of the physiologically relevant RdRp. The first study characterizing the recombinant form of a calicivirus RdRp showed that in RHDV, the polymerase domain corresponding to the mature 3D polymerase polypeptide previously characterized in picornaviruses was enzymatically active (Lopez Vazquez et al., 1998). In contrast, the active RdRp in FCV was shown to be a proteinase-polymerase fusion protein corresponding to the 3CD polypeptide also seen in picornaviruses (Wei et al., 2001), which was consistent with an analysis of polyprotein processing that indicated the lack of proteolytic processing between the 3C-like proteinase and 3D-like polymerase domains (Green et al., 2002; Sosnovtseva et al., 1999). More recently, it has been shown in NV isolate MD145-12 that both the precursor 3CD and the mature 3D were functional and have

similar RdRp activities (Belliot et al., 2005), including the capacity of the Pro-Pol precursor to nucleotidylylate VPg (Belliot et al., 2008). In RHDV, the 3CD precursor and 3D mature polypeptides have been shown to co-exist within the cell using either RHDV-infected cells (Konig et al., 1998) or transient expression experiments (Meyers et al., 2000). A recent report on the RHDV 3CD precursor (Machin et al., 2009) indicates a tendency towards functional specialization of the two polymerase forms. 3CD is more active than 3D in catalysing VPg uridylylation, whereas the latter enzyme form shows higher activity than its precursor in RNA polymerization. These data suggest that the precursor 3CD seems to be more efficient than its 3D derivative during the initial steps of genome replication, which, in light of our current knowledge, should involve the addition of NTP residues to a VPg protein primer and, once the 3CD precursor is processed into 3D, this latter form of the enzyme develops an increased capacity to act on RNA rather than on protein primers. The conversion of 3CD into 3D would then contribute to an increase in the efficiency of late replication events, which require the extension of the growing RNA chains covalently linked to VPg. Several authors (Lopez Vazquez et al., 1998, 2001; Wei et al., 2001) have provided evidence that the calicivirus polymerases show poor template specificity and can act equally well on non-viral RNAs *in vitro*. In this view, the increased VPg nucleotidylylation activity of 3CD could also be envisaged as part of the virus specificity mechanisms involved in ensuring that the viral RdRp acts primarily on viral genome replication and not on other cellular RNAs. In the future, it will be interesting to further investigate the structural and functional effects of processing 3CD into 3C and 3D that is observed in some caliciviruses (e.g. *Lagovirus* and *Norovirus*) and to compare this with other caliciviruses (*e.g. Vesivirus* and *Sapovirus*) in which the mature RdRp is the bifunctional protein 3CD. It should be also mentioned that, despite the evident similarities of many molecular mechanisms of the calicivirus with respect to related viral systems, such as the picornaviruses, the biological role of the 3CD precursor appears to be very different. Most importantly, the calicivirus 3CD is catalytically active as a polymerase, whereas the picornavirus 3CD is a true precursor that is devoid of polymerase activity and appears to play an essential role as a regulatory protein in genome replication (Marcotte *et al.*, 2007; Oh *et al.*, 2009).

Future work

Although structural studies on proteins from caliciviruses have already yielded many important insights into the molecular basis of calicivirus biology, many important proteins and protein-RNA complexes remain poorly understood. The assembly of capsids in the presence of RNA, VP2 and VPg remains to be explored, and variations on the structures of VP1 and receptor-bound complexes in caliciviruses other than the human noroviruses are also not known. Since little is known about NS1, NS2 and NS4, structural information may help shed light on the functions of these proteins. Although more is known about the functions of the NS3 NTPase and NS5 VPg proteins, structural information on these proteins and their complexes with other proteins or RNA will help reveal important details about the roles of these proteins in the viral life cycle. In addition to the extensive functional and structural studies already performed on NS6 proteinase and NS7 polymerase, many details about how these two proteins interact with each other and with other proteins and RNA structures have yet to be determined. Moreover, the existence of NS6–7 protease-polymerase bifunctional enzymes that are not further processed into mature NS6 and NS7 products in some calicivirus genera (*Vesivirus* and *Sapovirus*) is a striking differential feature and points to the need of determining the structures of these proteins as well as the corresponding NS6–7 precursors in the *Lagovirus* and *Norovirus* members in order to understand the molecular basis of the different molecular strategies used by these viruses within the family *Caliciviridae*.

Conclusion

The past decade has seen great advances in our understanding of the structure and function of calicivirus proteins at the molecular level. Recent advances in electron microscopy, X-ray crystallography and nuclear magnetic resonance are allowing detailed studies on larger and more challenging proteins and protein complexes. Because of these recent developments and most

likely further technical improvements in the coming decade, even greater advances in understanding the molecular structural basis of calicivirus biology can be expected in the near future.

References

Allaire, M., Chernaia, M.M., Malcolm, B.A., and James, M.N. (1994). Picornaviral 3C cysteine proteinases have a fold similar to chymotrypsin-like serine proteinases. Nature *369*, 72–76.

Barcena, J., Verdaguer, N., Roca, R., Morales, M., Angulo, I., Risco, C., Carrascosa, J.L., Torres, J.M., and Caston, J.R. (2004). The coat protein of Rabbit hemorrhagic disease virus contains a molecular switch at the N-terminal region facing the inner surface of the capsid. Virology *322*, 118–134.

Belliot, G., Sosnovtsev, S.V., Chang, K.O., Babu, V., Uche, U., Arnold, J.J., Cameron, C.E., and Green, K.Y. (2005). Norovirus proteinase-polymerase and polymerase are both active forms of RNA-dependent RNA polymerase. J. Virol. *79*, 2393–2403.

Belliot, G., Sosnovtsev, S.V., Chang, K.O., McPhie, P., and Green, K.Y. (2008). Nucleotidylylation of the VPg protein of a human norovirus by its proteinase-polymerase precursor protein. Virology *374*, 33–49.

Belliot, G., Sosnovtsev, S.V., Mitra, T., Hammer, C., Garfield, M., and Green, K.Y. (2003). In vitro proteolytic processing of the MD145 norovirus ORF1 nonstructural polyprotein yields stable precursors and products similar to those detected in calicivirus-infected cells. J. Virol. *77*, 10957–10974.

Bertolotti-Ciarlet, A., White, L.J., Chen, R., Prasad, B.V., and Estes, M.K. (2002). Structural requirements for the assembly of Norwalk virus-like particles. J. Virol. *76*, 4044–4055.

Bhella, D., Gatherer, D., Chaudhry, Y., Pink, R., and Goodfellow, I.G. (2008). Structural insights into calicivirus attachment and uncoating. J. Virol. *82*, 8051–8058.

Boerner, J.E., Lyle, J.M., Daijogo, S., Semler, B.L., Schultz, S.C., Kirkegaard, K., and Richards, O.C. (2005). Allosteric effects of ligands and mutations on poliovirus RNA-dependent RNA polymerase. J. Virol. *79*, 7803–7811.

Bruenn, J.A. (2003). A structural and primary sequence comparison of the viral RNA-dependent RNA polymerases. Nucleic Acids Res. *31*, 1821–1829.

Bu, W., Mamedova, A., Tan, M., Xia, M., Jiang, X., and Hegde, R.S. (2008). Structural basis for the receptor binding specificity of Norwalk virus. J. Virol. *82*, 5340–5347.

Burroughs, J.N., and Brown, F. (1978). Presence of a covalently linked protein on calicivirus RNA. J. Gen. Virol. *41*, 443–446.

Butcher, S.J., Grimes, J.M., Makeyev, E.V., Bamford, D.H., and Stuart, D.I. (2001). A mechanism for initiating RNA-dependent RNA polymerization. Nature *410*, 235–240.

Cameron, C.E., Gohara, D.W., and Arnold, J.J. (2002). Poliovirus RNA-dependent RNA polymerase (3Dpol): Structure, function and mechanism. In

Molecular Biology of Picornaviruses, B. Semler, and E. Wimmer, eds. (Washington, D.C., ASM Press), pp. 255–267.

Cao, S., Lou, Z., Tan, M., Chen, Y., Liu, Y., Zhang, Z., Zhang, X.C., Jiang, X., Li, X., and Rao, Z. (2007). Structural basis for the recognition of blood group trisaccharides by norovirus. J. Virol. *81*, 5949–5957.

Castro, C., Smidansky, E.D., Arnold, J.J., Maksimchuk, K.R., Moustafa, I., Uchida, A., Gotte, M., Konigsberg, W., and Cameron, C.E. (2009). Nucleic acid polymerases use a general acid for nucleotidyl transfer. Nat. Struct. Mol. Biol. *16*, 212–218.

Chakravarty, S., Hutson, A.M., Estes, M.K., and Prasad, B.V. (2005). Evolutionary trace residues in noroviruses: importance in receptor binding, antigenicity, virion assembly, and strain diversity. J. Virol. *79*, 554–568.

Chaudhry, Y., Nayak, A., Bordeleau, M.E., Tanaka, J., Pelletier, J., Belsham, G.J., Roberts, L.O., and Goodfellow, I.G. (2006). Caliciviruses differ in their functional requirements for eIF4F components. J. Biol. Chem. *281*, 25315–25325.

Chen, R., Neill, J.D., Estes, M.K., and Prasad, B.V. (2006). X-ray structure of a native calicivirus: structural insights into antigenic diversity and host specificity. Proc. Natl. Acad. Sci. U.S.A. *103*, 8048–8053.

Chen, R., Neill, J.D., Noel, J.S., Hutson, A.M., Glass, R.I., Estes, M.K., and Prasad, B.V. (2004). Inter- and intragenus structural variations in caliciviruses and their functional implications. J. Virol. *78*, 6469–6479.

Chen, R., Neill, J.D., and Prasad, B.V. (2003a). Crystallization and preliminary crystallographic analysis of San Miguel sea lion virus: an animal calicivirus. J. Struct. Biol. *141*, 143–148.

Chen, R., Neill, J.D., and Venkataram Prasad, B.V. (2003b). Crystallization and preliminary crystallographic analysis of San Miguel sea lion virus: An animal calicivirus. J. Struct. Biol. *141*, 143–148.

Choi, J.M., Hutson, A.M., Estes, M.K., and Prasad, B.V. (2008). Atomic resolution structural characterization of recognition of histo-blood group antigens by Norwalk virus. Proc. Natl. Acad. Sci. U.S.A. *105*, 9175–9180.

Daughenbaugh, K.F., Fraser, C.S., Hershey, J.W., and Hardy, M.E. (2003). The genome-linked protein VPg of the Norwalk virus binds eIF3, suggesting its role in translation initiation complex recruitment. EMBO J. *22*, 2852–2859.

DeLano, W.L. (2002). The PyMOL Molecular Graphics System (San Carlos, CA: DeLano Scientific).

Ferrer-Orta, C., Arias, A., Escarmis, C., and Verdaguer, N. (2006). A comparison of viral RNA-dependent RNA polymerases. Curr. Opin. Struct. Biol. *16*, 27–34.

Ferrer-Orta, C., Arias, A., Perez-Luque, R., Escarmis, C., Domingo, E., and Verdaguer, N. (2007). Sequential structures provide insights into the fidelity of RNA replication. Proc. Natl. Acad. Sci. U.S.A. *104*, 9463–9468.

Fullerton, S.W., Blaschke, M., Coutard, B., Gebhardt, J., Gorbalenya, A., Canard, B., Tucker, P.A., and Rohayem, J. (2007). Structural and functional characterization of sapovirus RNA-dependent RNA polymerase. J. Virol. *81*, 1858–1871.

Goodfellow, I., Chaudhry, Y., Gioldasi, I., Gerondopoulos, A., Natoni, A., Labrie, L., Laliberte, J.F., and Roberts, L. (2005). Calicivirus translation initiation requires an interaction between VPg and eIF 4 E. EMBO Rep. *6*, 968–972.

Green, K.Y., Mory, A., Fogg, M.H., Weisberg, A., Belliot, G., Wagner, M., Mitra, T., Ehrenfeld, E., Cameron, C.E., and Sosnovtsev, S.V. (2002). Isolation of enzymatically active replication complexes from feline calicivirus-infected cells. J. Virol. *76*, 8582–8595.

Hansen, J.L., Long, A.M., and Schultz, S.C. (1997). Structure of the RNA-dependent RNA polymerase of poliovirus. Structure *5*, 1109–1122.

Hardy, M.E., Crone, T.J., Brower, J.E., and Ettayebi, K. (2002). Substrate specificity of the Norwalk virus 3C-like proteinase. Virus Res. *89*, 29–39.

Herbert, T.P., Brierley, I., and Brown, T.D. (1997). Identification of a protein linked to the genomic and subgenomic mRNAs of feline calicivirus and its role in translation. J. Gen. Virol. *78*, 1033–1040.

Hobson, S.D., Rosenblum, E.S., Richards, O.C., Richmond, K., Kirkegaard, K., and Schultz, S.C. (2001). Oligomeric structures of poliovirus polymerase are important for function. EMBO J. *20*, 1153–1163.

Hogbom, M., Jager, K., Robel, I., Unge, T., and Rohayem, J. (2009). The active form of the norovirus RNA-dependent RNA polymerase is a homodimer with cooperative activity. J. Gen. Virol. *90*, 281–291.

Huang, P., Farkas, T., Marionneau, S., Zhong, W., Ruvoen-Clouet, N., Morrow, A.L., Altaye, M., Pickering, L.K., Newburg, D.S., LePendu, J., *et al.* (2003). Noroviruses bind to human ABO, Lewis, and secretor histo-blood group antigens: identification of 4 distinct strain-specific patterns. J. Infect. Dis. *188*, 19–31.

Hutson, A.M., Atmar, R.L., Graham, D.Y., and Estes, M.K. (2002). Norwalk virus infection and disease is associated with ABO histo-blood group type. J. Infect. Dis. *185*, 1335–1337.

Jiang, X., Wang, M., Graham, D.Y., and Estes, M.K. (1992). Expression, self-assembly, and antigenicity of the Norwalk virus capsid protein. J. Virol. *66*, 6527–6532.

Kamer, G., and Argos, P. (1984). Primary structural comparison of RNA-dependent polymerases from plant, animal and bacterial viruses. Nucleic Acids Res. *12*, 7269–7282.

Katpally, U., Wobus, C.E., Dryden, K., Virgin, H.W.t., and Smith, T.J. (2008). Structure of antibody-neutralized murine norovirus and unexpected differences from viruslike particles. J. Virol. *82*, 2079–2088.

Konig, M., Thiel, H.J., and Meyers, G. (1998). Detection of viral proteins after infection of cultured hepatocytes with rabbit hemorrhagic disease virus. J. Virol. *72*, 4492–4497.

Lindesmith, L., Moe, C., Marionneau, S., Ruvoen, N., Jiang, X., Lindblad, L., Stewart, P., LePendu, J., and Baric, R. (2003). Human susceptibility and resistance to Norwalk virus infection. Nat. Med. *9*, 548–553.

Liu, B., Clarke, I.N., and Lambden, P.R. (1996). Polyprotein processing in Southampton virus: identification of 3C-like protease cleavage sites by in vitro mutagenesis. J. Virol. *70*, 2605–2610.

Liu, B.L., Viljoen, G.J., Clarke, I.N., and Lambden, P.R. (1999). Identification of further proteolytic cleavage sites in the Southampton calicivirus polyprotein by expression of the viral protease in *E. coli.* J. Gen. Virol. *80 (Pt 2)*, 291–296.

Lopez Vazquez, A., Alonso, J.M., and Parra, F. (2000). Mutation analysis of the GDD sequence motif of a calicivirus RNA-dependent RNA polymerase. J. Virol. *74*, 3888–3891.

Lopez Vazquez, A., Martin Alonso, J.M., Casais, R., Boga, J.A., and Parra, F. (1998). Expression of enzymatically active rabbit hemorrhagic disease virus RNA-dependent RNA polymerase in *Escherichia coli.* J. Virol. *72*, 2999–3004.

Lopez Vazquez, A.L., Martin Alonso, J.M., and Parra, F. (2001). Characterisation of the RNA-dependent RNA polymerase from Rabbit hemorrhagic disease virus produced in *Escherichia coli.* Arch. Virol. *146*, 59–69.

Machin, A., Martin Alonso, J.M., Dalton, K.P., and Parra, F. (2009). Functional differences between precursor and mature forms of the RNA-dependent RNA polymerase from rabbit hemorrhagic disease virus. J. Gen. Virol. *90*, 2114–2118.

Machin, A., Martin Alonso, J.M., and Parra, F. (2001). Identification of the amino acid residue involved in rabbit hemorrhagic disease virus VPg uridylylation. J. Biol. Chem. *276*, 27787–27792.

Makino, A., Shimojima, M., Miyazawa, T., Kato, K., Tohya, Y., and Akashi, H. (2006). Junctional adhesion molecule 1 is a functional receptor for feline calicivirus. J. Virol. *80*, 4482–4490.

Marcotte, L.L., Wass, A.B., Gohara, D.W., Pathak, H.B., Arnold, J.J., Filman, D.J., Cameron, C.E., and Hogle, J.M. (2007). Crystal structure of poliovirus 3CD protein: virally encoded protease and precursor to the RNA-dependent RNA polymerase. J. Virol. *81*, 3583–3596.

Marin, M.S., Casais, R., Alonso, J.M., and Parra, F. (2000). ATP binding and ATPase activities associated with recombinant rabbit hemorrhagic disease virus 2C-like polypeptide. J. Virol. *74*, 10846–10851.

Martin Alonso, J.M., Casais, R., Boga, J.A., and Parra, F. (1996). Processing of rabbit hemorrhagic disease virus polyprotein. J. Virol. *70*, 1261–1265.

Meyers, G., Wirblich, C., and Thiel, H.J. (1991). Genomic and subgenomic RNAs of rabbit hemorrhagic disease virus are both protein-linked and packaged into particles. Virology *184*, 677–686.

Meyers, G., Wirblich, C., Thiel, H.J., and Thumfart, J.O. (2000). Rabbit hemorrhagic disease virus: genome organization and polyprotein processing of a calicivirus studied after transient expression of cDNA constructs. Virology *276*, 349–363.

Nakamura, K., Someya, Y., Kumasaka, T., Ueno, G., Yamamoto, M., Sato, T., Takeda, N., Miyamura, T., and Tanaka, N. (2005). A norovirus protease structure provides insights into active and substrate binding site integrity. J. Virol. *79*, 13685–13693.

Ng, K.K., Cherney, M.M., Vazquez, A.L., Machin, A., Alonso, J.M., Parra, F., and James, M.N. (2002).

Crystal structures of active and inactive conformations of a caliciviral RNA-dependent RNA polymerase. J. Biol. Chem. *277*, 1381–1387.

Ng, K.K., Pendas-Franco, N., Rojo, J., Boga, J.A., Machin, A., Alonso, J.M., and Parra, F. (2004). Crystal structure of norwalk virus polymerase reveals the carboxyl terminus in the active site cleft. J. Biol. Chem. *279*, 16638–16645.

Ng, K.K.S., Arnold, J.J., and Cameron, C.E. (2008). Structure and mechanism of RNA-dependent RNA polymerases. Curr. Top. Microbiol. Immunol. *320*, 137–156.

O'Reilly, E.K., and Kao, C.C. (1998). Analysis of RNA-dependent RNA polymerase structure and function as guided by known polymerase structures and computer predictions of secondary structure. Virology *252*, 287–303.

Oh, H.S., Pathak, H.B., Goodfellow, I.G., Arnold, J.J., and Cameron, C.E. (2009). Insight into poliovirus genome replication and encapsidation obtained from studies of 3B-3C cleavage site mutants. J. Virol. *83*, 9370–9387.

Oka, T., Katayama, K., Ogawa, S., Hansman, G.S., Kageyama, T., Miyamura, T., and Takeda, N. (2005a). Cleavage activity of the sapovirus 3C-like protease in *Escherichia coli.* Arch. Virol. *150*, 2539–2548.

Oka, T., Katayama, K., Ogawa, S., Hansman, G.S., Kageyama, T., Ushijima, H., Miyamura, T., and Takeda, N. (2005b). Proteolytic processing of sapovirus ORF1 polyprotein. J. Virol. *79*, 7283–7290.

Oka, T., Yamamoto, M., Katayama, K., Hansman, G.S., Ogawa, S., Miyamura, T., and Takeda, N. (2006). Identification of the cleavage sites of sapovirus open reading frame 1 polyprotein. J. Gen. Virol. *87*, 3329–3338.

Oka, T., Yamamoto, M., Yokoyama, M., Ogawa, S., Hansman, G.S., Katayama, K., Miyashita, K., Takagi, H., Tohya, Y., Sato, H., *et al.* (2007). Highly conserved configuration of catalytic amino acid residues among calicivirus-encoded proteases. J. Virol. *81*, 6798–6806.

Ollis, D.L., Brick, P., Hamlin, R., Xuong, N.G., and Steitz, T.A. (1985). Structure of large fragment of *Escherichia coli* DNA polymerase I complexed with dTMP. Nature *313*, 762–766.

Olson, A.J., Bricogne, G., and Harrison, S.C. (1983). Structure of tomato busy stunt virus IV. The virus particle at 2.9 A resolution. J. Mol. Biol. *171*, 61–93.

Pata, J.D., Schultz, S.C., and Kirkegaard, K. (1995). Functional oligomerization of poliovirus RNA-dependent RNA polymerase. Rna *1*, 466–477.

Petersen, J.F., Cherney, M.M., Liebig, H.D., Skern, T., Kuechler, E., and James, M.N. (1999). The structure of the 2A proteinase from a common cold virus: a proteinase responsible for the shut-off of host-cell protein synthesis. EMBO J. *18*, 5463–5475.

Pfister, T., and Wimmer, E. (2001). Polypeptide p41 of a Norwalk-like virus is a nucleic acid-independent nucleoside triphosphatase. J. Virol. *75*, 1611–1619.

Prasad, B.V., Hardy, M.E., Dokland, T., Bella, J., Rossmann, M.G., and Estes, M.K. (1999). X-ray crystallographic structure of the Norwalk virus capsid. Science *286*, 287–290.

Prasad, B.V., Hardy, M.E., Jiang, X., and Estes, M.K. (1996). Structure of Norwalk virus. Arch. Virol. Suppl. *12*, 237–242.

Prasad, B.V., Rothnagel, R., Jiang, X., and Estes, M.K. (1994). Three-dimensional structure of baculovirus-expressed Norwalk virus capsids. J. Virol. *68*, 5117–5125.

Rademacher, C., Krishna, N.R., Palcic, M., Parra, F., and Peters, T. (2008). NMR experiments reveal the molecular basis of receptor recognition by a calicivirus. J. Am. Chem. Soc. *130*, 3669–3675.

Robel, I., Gebhardt, J., Mesters, J.R., Gorbalenya, A., Coutard, B., Canard, B., Hilgenfeld, R., and Rohayem, J. (2008). Functional characterization of the cleavage specificity of the sapovirus chymotrypsin-like protease. J. Virol. *82*, 8085–8093.

Ruvoen-Clouet, N., Ganiere, J.P., Andre-Fontaine, G., Blanchard, D., and Le Pendu, J. (2000). Binding of rabbit hemorrhagic disease virus to antigens of the ABH histo-blood group family. J. Virol. *74*, 11950–11954.

Schechter, I., and Berger, A. (1967). On the size of the active site in proteases. I. Papain. Biochem. Biophys. Res. Commun. *27*, 157–162.

Seipelt, J., Guarne, A., Bergmann, E., James, M., Sommergruber, W., Fita, I., and Skern, T. (1999). The structures of picornaviral proteinases. Virus Res. *62*, 159–168.

Someya, Y., Takeda, N., and Wakita, T. (2008). Saturation mutagenesis reveals that GLU54 of norovirus 3C-like protease is not essential for the proteolytic activity. J. Biochem. *144*, 771–780.

Sosnovtsev, S.V., Sosnovtseva, S.A., and Green, K.Y. (1998). Cleavage of the feline calicivirus capsid precursor is mediated by a virus-encoded proteinase. J. Virol. *72*, 3051–3059.

Sosnovtseva, S.A., Sosnovtsev, S.V., and Green, K.Y. (1999). Mapping of the feline calicivirus proteinase responsible for autocatalytic processing of the nonstructural polyprotein and identification of a stable proteinase-polymerase precursor protein. J. Virol. *73*, 6626–6633.

Steitz, T.A. (1998). A mechanism for all polymerases. Nature *391*, 231–232.

Steitz, T.A. (2006). Visualizing polynucleotide polymerase machines at work. EMBO J. *25*, 3458–3468.

Stuart, A.D., and Brown, T.D. (2007). Alpha2, 6-linked sialic acid acts as a receptor for Feline calicivirus. J. Gen. Virol. *88*, 177–186.

Tan, M., Huang, P., Meller, J., Zhong, W., Farkas, T., and Jiang, X. (2003). Mutations within the P2 domain of norovirus capsid affect binding to human histo-blood group antigens: evidence for a binding pocket. J. Virol. *77*, 12562–12571.

Tan, M., Xia, M., Chen, Y., Bu, W., Hegde, R.S., Meller, J., Li, X., and Jiang, X. (2009). Conservation of carbohydrate binding interfaces: evidence of human HBGA selection in norovirus evolution. PLoS ONE *4*, e5058.

Tang, L., Lin, C.S., Krishna, N.K., Yeager, M., Schneemann, A., and Johnson, J.E. (2002). Virus-like particles of a fish nodavirus display a capsid subunit domain organization different from that of insect nodaviruses. J. Virol. *76*, 6370–6375.

Tao, Y., Farsetta, D.L., Nibert, M.L., and Harrison, S.C. (2002). RNA synthesis in a cage – structural studies of reovirus polymerase lambda3. Cell *111*, 733–745.

Thouvenin, E., Laurent, S., Madelaine, M.F., Rasschaert, D., Vautherot, J.F., and Hewat, E.A. (1997). Bivalent binding of a neutralising antibody to a calicivirus involves the torsional flexibility of the antibody hinge. J. Mol. Biol. *270*, 238–246.

Wei, L., Huhn, J.S., Mory, A., Pathak, H.B., Sosnovtsev, S.V., Green, K.Y., and Cameron, C.E. (2001). Proteinase-polymerase precursor as the active form of feline calicivirus RNA-dependent RNA polymerase. J. Virol. *75*, 1211–1219.

Wirblich, C., Sibilia, M., Boniotti, M.B., Rossi, C., Thiel, H.J., and Meyers, G. (1995). 3C-like protease of rabbit hemorrhagic disease virus: identification of cleavage sites in the ORF1 polyprotein and analysis of cleavage specificity. J. Virol. *69*, 7159–7168.

Wirblich, C., Thiel, H.J., and Meyers, G. (1996). Genetic map of the calicivirus rabbit hemorrhagic disease virus as deduced from in vitro translation studies. J. Virol. *70*, 7974–7983.

Zamyatkin, D.F., Parra, F., Machin, A., Grochulski, P., and Ng, K.K. (2009). Binding of 2′-amino-2′-deoxycytidine-5′-triphosphate to norovirus polymerase induces rearrangement of the active site. J. Mol. Biol. *390*, 10–16.

Zamyatkin, D.F., Parra, F., Martin Alonso, J.M., Harki, D.A., Peterson, B.R., Grochulski, P., and Ng, K.K. (2008). Structural insights into mechanisms of catalysis and inhibition in norwalk virus polymerase. J. Biol. Chem. *283*, 7705–7712.

Zeitler, C.E., Estes, M.K., and Venkataram Prasad, B.V. (2006). X-ray crystallographic structure of the Norwalk virus protease at 1.5-A resolution. J. Virol. *80*, 5050–5058.

Virus–Host Interaction and Cellular Receptors of Caliciviruses

6

Ming Tan and Xi Jiang

Abstract

Caliciviruses are a diverse virus family with a wide range of host and tissue tropisms. Most calicivirus genera recognize a carbohydrate ligand for attachment, including the A, B, H and Lewis histo-blood group antigens (HBGAs) and heparan sulphate for the human noroviruses, the H type 2 antigen for the rabbit haemorrhagic disease virus (genus *Lagovirus*), the type B antigen for the Tulane virus (a potential new genus), and sialic acid for feline calicivirus (FCV; genus *Vesivirus*) and murine norovirus (MNV; genus *Norovirus*). Following attachment, FCV recognizes also a cell surface protein, the junctional adhesion molecule 1 (JAM-1), as a functional receptor or co-receptor potentially for penetration or entry into host cells. Some human noroviruses interact also with a 105-kDa membrane protein, but its role in viral penetration/entry into host cells remains unknown. The genetic and structural analyses of selected strains of norovirus and FCV have generated new insights into virus–host interactions that chart the course for innovative research in the development of effective strategies to control and prevent calicivirus infection and illness.

Introduction

The initial steps of virus infection characteristically involve the attachment of virions to host cell receptor(s), followed by entry of virions through the cell membrane and release of the genetic material into the cytoplasm for replication. The attachment of virions to host cells relies on specific recognition of cellular receptor(s), while the penetration of virions is often mediated by a membrane protein that specifically interacts with virions as a receptor or co-receptor. The virus–host interactions of many caliciviruses have been analysed, with most data generated from studies of the human noroviruses, the feline calicivirus (FCV) and the rabbit haemorrhagic disease virus (RHDV). All caliciviruses analysed thus far exhibit variable interactions with polymorphic carbohydrates on host cell surfaces. Human noroviruses in the genus *Norovirus* recognize human histo-blood group antigens (HBGAs) in the ABO, Lewis and secretor families, certain sialyl-modified HBGAs, and heparan sulphate (reviewed in Estes *et al.*, 2006; Hutson *et al.*, 2004a; Tan and Jiang, 2005a, 2007). RHDV in the genus *Lagovirus* binds to H type 2 antigen (Ruvoen-Clouet *et al.*, 2000). The recently discovered Tulane virus recognizes the type B antigen (Farkas *et al.*, 2008a). Murine norovirus (MNV) in the genus *Norovirus* (Taube *et al.*, 2009) and FCV in the genus *Vesivirus* (Stuart and Brown, 2007) bind to sialic acid on permissive cells. In addition, FCV recognizes a cellular membrane protein, the junctional adhesion molecule 1 (JAM-1) as the host receptor for infection (Bhella *et al.*, 2008; Makino *et al.*, 2006; Ossiboff and Parker, 2007). A 105-kDa membrane protein was shown to interact with human norovirus virus-like particles (VLPs) (Tamura *et al.*, 2000), although its detailed function in virus infection remains to be defined.

Many studies in defining calicivirus–host interactions, especially those of the non-cultivable noroviruses, have utilized *in vitro* binding and blocking assays with recombinant viral capsid proteins as probes. However, compelling evidence of the association of HBGA binding with norovirus infection and clinical illness was obtained in human volunteer challenge studies (Hutson *et al.*, 2005; Hutson *et al.*, 2002; Lindesmith *et al.*, 2003) and in outbreak investigations of acute gastroenteritis (Rockx *et al.*, 2005; Tan *et al.*, 2008b; Thorven *et al.*, 2005). A field study of RHDV in wild rabbit populations found evidence for a genetic link between expression of H type 2 HBGA and susceptibility to disease (Guillon *et al.*, 2009). Direct evidence of JAM-1 as a functional receptor for FCV was discovered when expression of JAM-1 in non-permissive cells rendered them permissive for FCV (Makino *et al.*, 2006; Ossiboff and Parker, 2007).

Following the initial studies that linked calicivirus binding to various ligands and receptors, rapid advances were made in solving the structures of the binding interface with X-ray crystallography, cryo-electron microscopy (cryo-EM), NMR and genetic mutagenesis (Bhella *et al.*, 2008; Bu *et al.*, 2008; Cao *et al.*, 2007; Choi *et al.*, 2008; Ossiboff and Parker, 2007; Rademacher *et al.*, 2008; Tan *et al.*, 2003, 2008c, 2009a). These studies have broadened the understanding of the virus/host interaction, evolution and immunology, and may lead to innovative strategies to control and prevent calicivirus-associated disease. This chapter will review the studies that led to the recognition of carbohydrates as important binding ligands for the caliciviruses, with an emphasis on the recognition of HBGA molecules by the human noroviruses, a major calicivirus pathogen.

Carbohydrates as cellular receptors or co-receptors of the caliciviruses

Carbohydrates are abundant in living organisms. Depending on their compositions and complexities, individual carbohydrates are named by individual sugar residues in specific orders and linkages appearing in the molecules or simply as mono-, di-, tri-, oligo- and polysaccharides, respectively. One major group of carbohydrates is the HBGAs that are the determinants of human and animal blood types and are abundant in bodily fluids such as blood, saliva and milk. Similar HBGAs are found also on variable cell surfaces and mucosal epithelia, which earned their name 'histo-blood group antigens' from the original

name 'blood group antigens'. The most well studied HBGAs are those of the ABO, Lewis and secretor families. Additional chemical modifications of these antigens through such as sialyl and/or acetyl groups further define the structures and functions of these molecules. Carbohydrates are involved in many important biological functions such as storage of energy, cell-to-cell signalling and cell-to-extracellular matrix interactions. Increasingly, data show that carbohydrates linked to glycoproteins and glycolipids on the mucosal surfaces serve also as receptors for a number of bacterial and viral pathogens (Le Pendu, 2004). Furthermore, recent studies indicate that soluble carbohydrates and glycans found in bodily fluids such as human milk may serve as decoy receptors in the prevention of infection with numerous bacterial and viral pathogens in breast fed infants (Jiang *et al.*, 2004; Morrow *et al.*, 2004a,b, 2005; Newburg, 1999, 2005; Newburg *et al.*, 1998; Newburg and Walker, 2007; Saravanan *et al.*, 2008). In this chapter, the major classes of the human HBGAs will be discussed owing to their heavy involvement in the interactions of caliciviruses with their hosts.

The polymorphic human HBGAs

HBGAs are complex carbohydrates present at the outermost ends of N- or O-linked glycans or glycolipids (Le Pendu, 2004; Ravn and Dabelsteen, 2000). In humans the biosynthesis of HBGAs begins with a disaccharide precursor (Fig. 6.1) that can be sorted into four major types: type 1 (Galβ1–3GlcNAcβ), type 2 (Galβ1–4GlcNAcβ), type 3 (Galβ1–3GalNAcα) and type 4 (Galβ1–3GalNAcβ). Individual monosaccharides are sequentially added to the disaccharide precursor in defined linkages catalysed by a specific glycosyltransferase. For example, the α-1,3- or α-1,4-fucosyltransferases (FUT3) are responsible for the addition of a fucose residue to the type 1 or type 2 disaccharide precursors at an α-1,3 or α-1,4 linkage, resulting in trisaccharides called Lewis a (Le^a) or Lewis x (Le^x) antigens, respectively.

Another fucosyltransferase (FUT2) adds a fucose to the same disaccharide precursors at an α-1,2 linkage, resulting in trisaccharides of H types 1 and 2, also called secretor antigens. Further addition of another fucose residue to the H type 1 or 2 antigens catalysed by the

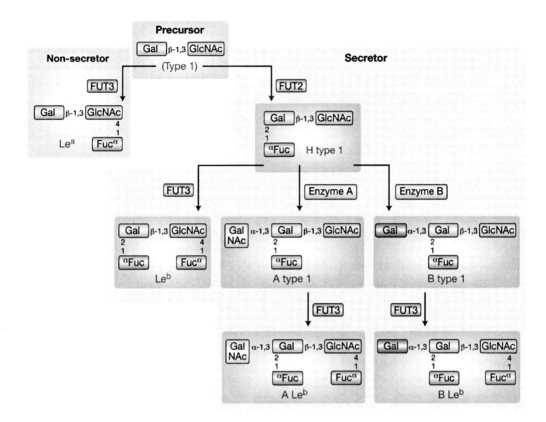

Figure 6.1 The biosynthesis pathways of the human ABH and Lewis histo-blood group antigens (HBGAs) based on the type 1 precursor. The synthesis of HBGA starts with a disaccharide precursor by the sequential addition of a monosaccharide. Each step of the synthesis is catalysed by a glycosyltransferase with specific substrates and linkages. FUT3, an α-1,3 or α-1,4 fucosyltransferase, is responsible for addition of a fucose residue to the disaccharide precursor (Galβ1–3GlcNAcβ) at an α-1,4 linkage, resulting in a trisaccharide called Lewis a (Le^a). FUT2, another fucosyltransferase, adds a fucose residue to the disaccharide precursor at an α–1,2 linkage, producing a trisaccharide designated H type 1. The FUT2 phenotypes are called secretor or secretor positive, whereas the FUT2-inactivating phenotypes are non-secretor or secretor negative. Further addition of another fucose residue to the H type 1 antigens by the FUT3 enzyme results in a tetrasaccharide antigen Lewis b (Le^b). Enzymes A and B are glycosyltransferases that add an N-acetylgalactosamine or a galactose at α-1,3 linkage of the H type 1 to produce the tetrasaccharide A type 1 and B type 1, respectively. These two antigens may be further developed into the pentasaccharide A and B antigens (ALe^b and BLe^b), respectively, by the addition of a fucose catalysed by the α–1,3/4 fucosyltransferase (FUT3). The entire pathways can occur on a type 2 precursor. The corresponding products Le^a and Le^b in type 1 are Le^x and Le^y for type 2, respectively. Abbreviations: Enzyme A, N-acetyl-galactosamine-transferase; enzyme B, galactosyltransferase; Fuc, L-fucose; FUT2, α-1,2-fucosyltransferase; FUT3, α-1, 3/4 fucosyltransferase; Gal, d-galactose; GalNAc, N-acetyl-galactosamine; GlcNAc, N-acetylglucosamine. Adapted with permission from reference (Tan and Jiang, 2007).

FUT3 enzyme produces a tetrasaccharide antigen named Lewis b (Le^b) and Lewis y (Le^y), respectively. On the other hand, both A and B enzymes are glycosyltransferases that catalyse the addition of an N-acetylgalactosamine or a galactose at α–1,3 linkage of the H type 1/2 trisaccharides to produce the tetrasaccharide A type 1/2 and B type 1/2, respectively. These two antigens may further develop into the pentasaccharide A and B antigens (ALe$^{b/y}$ and BLe$^{b/y}$), respectively, which are catalysed by the α-1,3/4-fucosyltransferase (Fig. 6.1).

The gene families encoding the various glycosyltransferases responsible for production of the ABO, secretor and Lewis blood types contain silent alleles, leading to null phenotypes of the loci.

For example, the FUT2-inactive mutations lead to the absence of A, B and H antigens. Individuals with FUT2 phenotypes are called secretors or secretor positive, while those with FUT2-inactive phenotypes are called non-secretors or secretor negative. Inactive mutations of the ABO family are common and show marked variation among different populations and races (Le Pendu, 2004). Even when functional genes encoding glycosyltransferases are expressed, the enzymes themselves can exhibit variable activities, and the relative amounts of displayed HBGA may vary among individuals within the same blood type. Another factor that affects the display of HBGA phenotypes are chemical modifications of the saccharides via linkages to functional groups such as acetyl or sialyl molecules. The highly polymorphic and complex HBGA phenotypes of the human population present a major challenge in studies of their role in calicivirus host interactions.

The diverse interaction between norovirus and HBGAs

Recognition of HBGAs by caliciviruses was first reported by Le Pendu and co-workers who discovered that RHDV interacts with the H type 2 antigen (Ruvoen-Clouet et al., 2000). Similar studies were then performed on human noroviruses, which demonstrated diverse and strain–specific interactions between noroviruses and HBGAs (Donaldson et al., 2008; Harrington et al., 2002; Huang et al., 2003, 2005; Hutson et al., 2003; Lindesmith et al., 2008; Shirato et al., 2008; Tan et al., 2004b; Thorven et al., 2005). The binding specificity and affinity of human noroviruses to HBGAs were initially analysed by in vitro assays that used recombinant VLPs as a surrogate for native virions and saliva samples as a source of native HBGA receptors. Although easily obtained, saliva samples often contain a mixture of HBGAs or unknown carbohydrates that vary among individuals, and a large panel of saliva samples representing variable blood types is needed in the binding assays. An alternative to the use of saliva has been the employment of commercially available synthetic oligosaccharides. A synthetic oligosaccharide has the defined structure of a single HBGA, which allows precise binding specificities of a norovirus to be determined, but the sensitivity of the assay can vary

among laboratories (see below). Haemagglutination (H) and haemagglutination inhibition (HI) assays have been developed also to examine the blood group specificity of noroviruses (Hutson et al., 2003), but they require a source of fresh blood cells. In summary, several methods have been developed to analyse calicivirus/HBGA interactions in vitro and a combination of different methods for cross-reference may help the confirmation of binding properties of individual strains.

Extensive studies to evaluate the diverse interactions between HBGAs and human noroviruses have been performed by numerous laboratories (Cannon et al., 2009; Donaldson et al., 2008; Harrington et al., 2002; Huang et al., 2003, 2005; Hutson et al., 2003; Lindesmith et al., 2008; Shirato et al., 2008; Tan et al., 2004b; Thorven et al., 2005). Noroviruses have been divided into five major genogroups (GI through GV), and most human pathogens belong to GI and GII. The GI and GII noroviruses have been further subdivided into several genetic clusters, or genotypes (Zheng et al., 2006). Tables 6.1 and 6.2 summarize available data on the binding specificities of VLPs representing GI and GII, respectively. Most VLPs exhibit a specific binding profile when screened against a panel of HBGAs, and this profile varies among strains. A few strains bind to HBGA oligosaccharides but not to saliva and a few strains bind to neither oligosaccharides nor saliva. Norovirus strains with a close genetic relatedness in capsid sequence generally share similar HBGA binding patterns, but dramatic differences in HBGA binding specificities or affinities have been observed among strains within the same genotype. Furthermore, same or similar HBGA binding patterns have been detected in strains that belong to different genotypes and genogroups. Most GII genotype 4 (GII.4) strains recognize the H-related antigens, including H, A, B, Lewis b and Lewis y, but some GII.4 strains do not bind or bind poorly to these antigens. This type of strain variation suggests a typical carbohydrate–protein interaction observed for many lectins, in which subtle changes of the viral capsid by random mutations could significantly alter the HBGA binding specificity or affinity (Tan et al., 2008c, 2009b) (see below).

Earlier studies sorted noroviruses into two major HBGA binding groups according to their

Table 6.1 Interactions of some genogroup I noroviral VLPs with synthetic HBGAs and/or saliva with defined ABO, secretor and Lewis blood types[1]

Strains	Genotype	Year	Binding to synthetic oligosaccharides									Binding to saliva[2]			
			H1	H2	H3	A	B	Lea	Leb	Lex	Ley	O	A	B	N
Norwalk	GI-1	1968	+	–	+	+	–	–	+	–	+	+	+	–	–
Aichi124	GI-1	1989	+	+	+	+	–	–	+	/	/	+	+	–	/
West Chester	GI-1	2001	–	/	+	–	–	–	–	–	–	/	/	/	/
Funabashi258	GI-2	1996	+	–	+	+	–	+	–	/	/	+	+	–	/
SoV	GI-2	1999	–	/	+	–	–	+	–	–	–	/	/	/	/
DSV32	GI-3	1999	–	/	–	+	–	+	–	–	–	–	–	–	–
DSV3	GI-3	1999	–	–	–	–	–	–	–	–	–	/	/	/	/
Kashiwa645	GI-3	1999	–	+	–	+	–	+	–	/	/	+	+	–	/
VA115	GI-3	1997	–	–	–	–	–	–	–	–	–	–	–	–	–
Chiba407	GI-4	1987	–	–	–	+	–	+	+	/	/	+	+	–	/
Chiba	GI-4	2000	–	/	–	–	–	+	–	–	–	/	/	/	/
Boxer	GI-8	2002	–	–	–	–	–	–	+	–	+	+	+	+	+
WUG1	GI-8	2000	–	–	–	+	+	+	+	/	/	+	+	+	/

1 The original data were published in Donaldson *et al.* (2008), Huang *et al.* (2003, 2005), Lindesmith *et al.* (2008) and Shirato *et al.* (2008). '+' indicates an interaction, '–', no interaction and '/', unknown.

2 O, A, and B represent saliva donors with O, A, and B blood types of secretors, while N indicates a donor of non-secretor.

3 Binding of these VLPs to HBGAs were performed by different laboratories independently.

specific interactions with the ABH and Lewis antigens, in which strains binding to the A and/or B antigens were often not reactive with the Lewis antigens and vice versa. These observations led to the proposal of a model in which noroviruses could be grouped based on their recognition of individual side chains on the HBGAs (Huang *et al.*, 2005; Tan and Jiang, 2005a, 2007). However, as discussed above, exceptions to this model were found, for example, four norovirus GI strains were found to bind both A/B and Lewis antigens, while a different GII.4 strain bound both A and Lewis x antigens (Shirato *et al.*, 2008) (Tables 6.1 and 6.2). In addition, a small number of the A/B binders in GII.3 and GII.4 were found to bind sialyl Lewis x (Rydell *et al.*, 2009).

The reported variation among norovirus-HBGA binding patterns may be caused, in part, by differences in the efficiency of the assays used in individual laboratories. The carbohydrate binding assays can be affected by the concentration of the reagents used (VLPs, oligosaccharides, saliva, and detector antibodies), incubation temperatures and length of time, and buffer conditions (pH, ion compositions and concentrations). Examples of such variation include the HBGA binding specificities of the Hawaii virus (HV, GII.1), Desert shield virus (DSV, GI.3) and BUDS virus (GII.2) (Tables 6.1 and 6.2), that were determined in different laboratories (Donaldson *et al.*, 2008; Huang *et al.*, 2003, 2005; Shirato *et al.*, 2008). Standardization of reagents and assay protocols including verification of the sequences of the cloned capsid genes used to generate the VLPs in different laboratories may improve the consistency of HBGA binding data reported in the literature (see below).

The norovirus strains that do not bind to any tested HBGAs (Tables 6.1 and 6.2) may recognize other carbohydrates as receptors (see

Table 6.2 Interactions of some genogroup II noroviral VLPs with synthetic HBGAs and/or saliva with defined ABO, secretor and Lewis blood types[1]

Strains	Genotype	Year	Binding to synthetic oligosaccharides									Binding to saliva[2]			
			H1	H2	H3	A	B	Lea	Leb	Lex	Ley	O	A	B	N
HV[3]	GII-1	1971	–	–	–	+	+	–	+	–	–	–	–	–	––
HV[3]	GII-1	1971	–	/	–	+	–	–	–	–	–	/	/	/	/
HV[3]	GII-1	1971	–	–	–	–	–	–	–	/	/	–	–	–	/
Noda485	GII-1	2000	–	–	–	–	–	–	–	/	/	–	–	–	/
Weisbaden	GII-1	2001	–	/	–	–	–	–	–	–	–	/	/	/	/
SMV[3]	GII-2	1976	–	–	+	–	+	–	–	/	/	–	–	+	–
SMV[3]	GII-2	1976	–	/	+	–	–	–	–	–	–	/	/	/	/
BUDS[3]	GII-2	2002	–	/	–	–	–	–	–	–	–	/	/	/	/
BUDS[3]	GII-2	2002	–	–	–	+	+	–	–	–	–	–	+	–	–
Ina	GII-2	2002	–	/	–	–	–	–	–	–	–	/	/	/	/
MxV	GII-3	1998	–	–	–	+	+	–	+	–	–	+	+	+	–
Toronto virus	GII-3	1999	–	/	+	+	–	–	–	–	–	/	/	/	/
Kashiwa336	GII-3	2000	–	–	+	–	–	–	–	/	/	–	+	+	/
Mutsudo18	GII-3	2000	–	–	+	–	–	–	–	/	/	–	+	+	/
PiV	GII-3	2003	–	–	–	+	+	–	+	–	–	–	+	+	–
GII-4 1987	GII-4	1987	–	/	+	–	–	–	–	–	+	+	+	+	–
GII-4 1997	GII-4	1997	–	/	+	+	+	–	–	–	+	+	+	+	–
Narita104	GII-4	1997	+	+	+	+	+	–	+	/	/	+	+	+	–
VA387	GII-4	1998	+	–	+	+	+	–	+	–	+	+	+	+	–
M7	GII-4	1999	–	/	–	–	–	–	–	–	–	/	/	/	/
GII-4 2002a	GII-4	2002	–	/	–	+	–	+	–	+	–	+	+	+	+
GII-4 2002	GII-4	2004	–	/	+	–	–	–	–	–	+	+	+	+	–
GII-4 2004	GII-4	2004	–	–	–	–	–	–	–	–	–	–	–	–	–
GII-4 2005	GII-4	2005	–	–	–	–	–	–	–	–	–	–	–	–	–
GII-4 2006	GII-4	2006	–	/	+	+	+	–	–	–	–	/	/	/	/
MOH	GII-5	1999	–	–	–	+	+	–	–	–	–	–	+	+	–
Ichikawa754	GII-5	1998	–	–	–	+	+	–	–	/	/	–	+	+	/
Ueno7k	GII-6	1994	–	+	+	–	+	+	+	/	/	+	+	+	/
Snbu445	GII-6	2000	–	–	+	–	–	–	–	/	/	+	+	+	/
Osaka 10–25	GII-7	1999	–	–	+	–	+	+	+	/	/	+	+	+	/
VA207	GII-9	1997	–	–	–	–	–	–	+	+	+	+	+	+	+
Chitta/Aichi 76	GII-12	1996	–	–	–	–	–	–	–	/	/	–	–	+	/
OIF	GII-13	2003	–	–	–	–	–	+	–	–	–	+	–	–	+
Kashiwa47	GII-14	1997	–	–	–	–	–	–	–	–	–	–	–	–	/

next section). Alternatively, these strains may be able to attach to the host cell *in vivo* with a low affinity, while the *in vitro* assay was not sensitive enough to show such an interaction. In summary, the available data provide evidence of diverse interactions between human noroviruses and HBGAs. Furthermore, the associations between norovirus infection and HBGA phenotypes are probably strain-specific rather genotype- or genogroup-specific because large variation within genogroups and genotypes exists. As pointed above, this is probably due to the nature of the carbohydrate–protein interaction between HBGAs and noroviruses. Nevertheless, understanding the mechanism of this binding diversity should facilitate future studies of the classification, evolution and epidemiology of noroviruses. Importantly, in an outbreak investigation that evaluates the association of HBGA phenotype with susceptibility to norovirus illness, care should be taken to confirm that only one norovirus strain is present in the clinical specimen.

Recognition of non-HBGA carbohydrates by human noroviruses

Human norovirus binding to HBGA carbohydrates is well established, but studies have explored also the recognition of other cell surface carbohydrates. Several human norovirus strains were reported to bind heparan sulphate (Tamura *et al.*, 2004). Heparan sulphate is a linear polysaccharide that is found on the surface of most cells as a component of certain proteoglycans. The specificity of norovirus binding to heparan sulphate was confirmed by blocking the binding with sulphated glycosaminoglycan and suramin, a highly sulphated derivative of urea. It was also supported by the demonstration of a marked reduction in binding to cells that had been treated with enzymes that specifically digest heparan sulphate. The enzymatic treatment resulted in 90% reduction of binding to undifferentiated Caco-2 cells but only 50% reduction in binding to differentiated Caco-2 cells, suggesting that additional molecules on the differentiated Caco-2 cells might be involved in norovirus binding. Human noroviruses show variable reactivities in their binding to heparan sulphate, similar to the variation observed with HBGAs. It was noted that VLPs derived from norovirus GII strains bound to cell surface heparan sulphate much stronger than those derived from GI strains (Tamura *et al.*, 2004). However, the biological significance of such binding in virus replication and clinical illness remains to be defined.

Recognition of carbohydrates by other caliciviruses

In addition to human noroviruses, several caliciviruses have been found also to recognize carbohydrates. Rabbit haemorrhagic disease virus in the genus *Lagovirus* was the first calicivirus studied that recognized the H type 2 antigen (Ruvoen-Clouet *et al.*, 2000). Evidence for this recognition specificity included (1) demonstration of virus binding to synthetic H type 2 blood group oligosaccharides; (2) inhibition of agglutination of human erythrocytes by saliva from secretor individuals but not from non-secretors; and 3) inhibition of attachment to the H type 2 antigens on rabbit epithelial cells by the H type 2-specific lectin (UEA-I) and H type 2 trisaccharide.

Tulane virus is a new calicivirus isolated from the stool of a rhesus monkey, and it is cultivable *in vitro* in a number of monkey cell lines (Farkas *et al.*, 2008a). Although phylogenic analyses indicate that Tulane virus is most closely related with members of the genus *Norovirus*, it is likely to represent a new genus in the calicivirus family (Farkas *et al.*, 2008b). Following purification of the virus propagated in cell culture, it was shown that the purified virions recognized type

---

Notes to Table 6.2

1. The original data were published in Donaldson *et al.* (2008), Huang *et al.* (2003, 2005), Lindesmith *et al.* (2008) and Shirato *et al.* (2008). '+' indicates an interaction, '–', no interaction and '/', unknown.

2. O, A, and B represent saliva donors with O, A, and B blood types of secretors, while N indicates a donor of non-secretor.

3. Binding of these VLPs to HBGAs were performed by different laboratories independently.

B oligosaccharides in carbohydrate binding assays as well as those in human saliva. The type B antigen was detected in the permissive monkey cell lines, which will allow further analysis of the role of this antigen as a functional receptor (Farkas *et al.*, 2008a).

Feline calicivirus of the genus *Vesivirus* and murine norovirus of the genus *Norovirus* recognize 2,6 linked sialyl acids on the host cell surface (Stuart and Brown, 2007; Taube *et al.*, 2009). Removal of the sialic acid with neuraminidase led to a marked reduction in viral binding and infectivity, suggesting a role for sialic acid in the initial steps of virus infection. In studies of murine norovirus, the sialic acid moiety was apparently linked to GD1, a ganglioside that characteristically bears a terminal sialic acid (Taube *et al.*, 2009).

The recognition of carbohydrates is apparently a shared feature among the caliciviruses, although additional studies of the clinical relevance and species specificity are needed. Because feline calicivirus, murine norovirus and Tulane virus are capable of replication in cell culture, they may prove useful as surrogates for the human caliciviruses. In particular, Tulane virus is the first cultivable enteric calicivirus identified that recognizes HBGAs, making it an attractive model for the non-cultivable human noroviruses.

Association of HBGA-binding patterns with clinical infection of human noroviruses and RHDV

The linkage between the HBGA-binding patterns and clinical infection has been successfully established for the prototype norovirus, Norwalk virus (GI.1), and a few other human noroviruses (Hutson *et al.*, 2002, 2005; Lindesmith *et al.*, 2003; Tan *et al.*, 2008b; Thorven *et al.*, 2005). The Norwalk virus VLPs bind strongly to saliva from individuals with blood types O and A, but show weak or no binding to type B saliva (Table 6.1) (Huang *et al.*, 2003, 2005). Norwalk VLPs are completely non-reactive with saliva from non-secretor individuals. Accordingly, volunteer studies showed that the infection rate with Norwalk virus was high in types O and A individuals, low in type B individuals, and negative in non-secretors (Hutson *et al.*, 2002, 2005; Lindesmith *et al.*, 2003). Similarly, some non-secretors were found

to be resistant to GII.4 strains in outbreak investigations: analyses of these GII.4 viruses found that they did not bind to the saliva of non-secretors (Hutson *et al.*, 2002, 2005; Lindesmith *et al.*, 2003; Tan *et al.*, 2008b; Thorven *et al.*, 2005).

Inconsistencies have been reported in the literatures concerning the association of HBGA phenotype with susceptibility to human norovirus illness. For example, saliva binding assays showed that the Snow Mountain virus (SMV, GII.2) bound only to type B saliva, but an association could not be made between a particular blood type, including type B and secretor status, and illness in volunteers following oral challenge with SMV (Lindesmith *et al.*, 2005). In addition, certain outbreak investigations have found no association between blood type and clinical illness (Rockx *et al.*, 2005). Because some noroviruses recognize non-HBGA carbohydrates such as heparan sulphate (Tamura *et al.*, 2004), additional studies are needed to determine if these outbreak strains have different carbohydrate binding specificities.

Certain RHDV strains are highly virulent and can achieve mortality rates up to 90% in a rabbit population (Guillon *et al.*, 2009; Meyers *et al.*, 1991; Ruvoen-Clouet *et al.*, 2000). A recent surveillance of wild rabbits (*Oryctolagus cuniculus*) in areas that had experienced an RHDV epidemic showed that the frequency of rabbits with a diminished HBGA H type 2 expression increased markedly compared to that in an RHDV-free area (Guillon *et al.*, 2009). A genetic analysis of the alleles in rabbits responsible for the expression of the H type 2 antigens found that one variant allele, *Sec1*, was elevated in the survivors and this allele was associated with decreased H type 2 antigen expression on cells. It was proposed that RHDV infection resulted in the selection of rabbits with this allele, providing the first evidence of natural selection for a particular HBGA binding pattern in a calicivirus host.

A functional receptor for FCV entry

The junctional adhesion molecule 1 (JAM-1) was the first membrane protein identified to be a receptor for FCV (Makino *et al.*, 2006). JAM-1 is a member of the immunoglobulin (Ig) superfamily expressed by various cells, such as endothelial and epithelial cells, and is specifically localized at apical tight junctions. The evidence for feline JAM-1 (fJAM-1) as a functional receptor of FCV included (1) demonstration of direct binding of FCV to fJAM-1; 2) the transfection of non-permissive cells with a fJAM-1 gene rendered the cells permissive to FCV, and (3) anti-fJAM-1 antibodies reduced replication of FCV in permissive cells (Makino et al., 2006). It has been reported also that FCV recognizes 2,6-linked sialic acid (Stuart and Brown, 2007), which is abundant in the respiratory tract and the primary site of FCV infection.

The JAM-1 molecule and other members of the Ig superfamily have been identified also as a cellular receptor for reoviruses (Barton et al., 2001), rhinoviruses (Bella et al., 1998, 1999; Bella and Rossmann, 2000), coxsackieviruses, polioviruses and adenoviruses (Bergelson et al., 1997a,b). However, in a recent study of MNV-1 using permissive murine macrophage and dendritic cells, evidence was not found for the utilization of JAM-1 as a receptor (Wobus, 2007). In another study, Takeda and co-workers (Tamura et al., 2000) reported that a 105-kDa membrane protein present on a number of mammalian cells interacted with norovirus VLPs. Attempts to identify this membrane protein were unsuccessful.

The observation that both sialic acid and JAM-1 are essential for FCV infection (Makino et al., 2006; Ossiboff and Parker, 2007; Stuart and Brown, 2007) raises interesting questions regarding the roles of the two molecules in viral attachment and entry. While the answers to these questions are not known at this time, the scenario of JAM-1 and sialic acid elucidated in reoviruses (Barton et al., 2001) may provide an interesting lead for FCV. Reoviruses display a serotype-dependent tropism and all strains infect cells expressing JAM-1. A model proposed by Barton and co-workers (Barton et al., 2001) suggests that tissue-specific sialic acids are the determinants of tropism involving the initial attachment of the viruses to cell surface, while JAM-1 mediates internalization of the viruses into the host cells. Similar functional relationships between sialic acid and fJAM-1 might occur in FCV. Furthermore, this relationship might also extend to the human noroviruses, in which the cell surface carbohydrates, including HBGAs and heparan sulphate, are the sites of initial attachment, while an additional molecule, most likely a membrane protein, serves as a receptor or co-receptor for subsequent penetration/entry of the viruses into host cells. Research in exploring this direction is warranted.

Structural basis of receptor–calicivirus interaction

Efforts were made to elucidate the receptor-binding interface of caliciviruses soon after the discovery of binding to HBGAs. An early study attempted co-crystallization of Norwalk virus VLPs with an oligosaccharide, resulting in a low resolution image of the complex (Hutson et al., 2004b). Efforts were then shifted to the protruding (P) domain of the capsid after the demonstration that HBGA-binding sites were located in this region of the capsid protein (Tan et al., 2004a; Tan and Jiang, 2005b). Expression of the P domain alone in E. coli yields P protein dimers that can self-assemble into 'P particles'. The ability to rapidly synthesize P dimers and P particles (and mutagenized forms) in high yield has facilitated co-crystallization studies of the P domain in complex with HBGAs. The crystal structures of the HBGA binding interfaces of two human noroviruses, the prototype Norwalk virus (GI.1) and VA387 (GII.4), have now been resolved (Bu et al., 2008; Cao et al., 2007; Choi et al., 2008). The structure of the P particles of VA387 has been determined also by cryo-EM (Tan et al., 2008a). In the meantime, the 3D structure of the fJAM-1–FCV complex has been reconstructed also by cryo-EM and the interaction interface on both fJAM-1 and FCV was analysed (Bhella et al., 2008). These advances have provided valuable insight into the mechanisms of calicivirus interactions with their receptors.

Interaction between HBGA and Norwalk virus

The Norwalk virus (GI.1) recognizes the human A and H HBGAs. The atomic structures of the Norwalk virus P dimers in complex with the A and H HBGAs have been solved by two laboratories (Bu et al., 2008; Choi et al., 2008). The HBGA binding interface is located at the P2 subdomain within a single P monomer, corresponding to the outermost surface of the capsid. Despite differences in their carbohydrate sequence and linkage, both A and H antigens bind to the same site on the capsid protein and project outward from the capsid surface (Fig. 6.2, left). The A trisaccharide binds to the P domain through interactions with six amino acids (Ser380, Pro378, Ser377, Asp327, His329, and W375). Only the two terminal saccharides (α-Fuc and α-GalNAc) of the A trisaccharide contribute to the interaction, in which four hydrogen bonds and a cation–pi interaction form between five residues (Asp327, His329, and W375, Ser377, Pro378) of the P domain and the terminal α-GalNAc, and one hydrogen bond links between the Ser380 and the α-Fucose (Bu *et al.*, 2008; Choi *et al.*, 2008) (Fig. 6.2E).

Similarly, the two terminal sugars (β-Gal and α-Fuc) of the H pentasaccharide bind to the same site of the Norwalk virus P domain, in which seven hydrogen bonds and a cation–pi interaction forms between the β-Gal-α-Fuc disaccharide and seven residues (Pro378, Ser377, Asp327, His329, Asp344, Gln342, and Trp375) of the P domain (Choi *et al.*, 2008). Further mutagenesis study has confirmed the HBGA binding interface with additional three amino acids (Ser338, Ala430, and Tyr431) near the interface that contributes also to the binding specificity of Norwalk virus (Tan *et al.*, 2009a).

The confirmed structures of the HBGA binding interfaces support the model of norovirus/HBGA interaction based on earlier studies using *in vitro* binding and blocking assays (Huang *et al.*, 2003; Huang *et al.*, 2005; Tan and Jiang, 2005a, 2007). According to the model, the binding specificity of norovirus was determined by one or two sugar binding sites within each HBGA binding interface. Each of the sites interacts with one terminal sugar of the HBGAs. The terminal A and H epitopes (α-GalNAc and α-Fuc) are responsible for binding of the A trisaccharide to Norwalk virus, while the β-Gal and H epitope (α-Fuc) are involved in binding to the H pentasaccharide. The interaction of the same interface with different carbohydrates in different binding modes provides a compelling explanation at the molecular level for the diversity and flexibility of the noroviruses in their interactions with human hosts.

The HBGA-binding interface of human GII.4 norovirus strain VA387

The HBGA binding interface was resolved for VA387, representing the major norovirus genotype associated with human disease, GII.4. This strain recognizes the A, B and H antigens and thus, has a broader HBGA recognition spectrum than the Norwalk virus (binds to the A and H but not B antigens). Crystal structures of the P dimers of VA387 in complex with the A and B trisaccharides showed the location of the HBGA binding interface on the P2 subdomain of the capsid (Cao *et al.*, 2007), similar to that of the Norwalk virus. However, the detailed receptor–binding interface of VA387 was distinct from that of the Norwalk virus, including different positions, amino acid compositions and binding modes to HBGA carbohydrates. The HBGA binding site of VA387 was positioned directly at the interface between the two P monomers and thus it was composed of residues from both P monomers. This feature indicates that the dimerization of the P domain is required for binding of VA387 to HBGAs.

Like the Norwalk virus, the two terminal sugars of the A and B HBGAs play major roles in VA387 binding (Cao *et al.*, 2007), although the third sugar (β-Gal) may also have weak interaction with the virus. The weak binding of β-Gal with the P domain should be interpreted with caution in this experimental system because its linkage *in vivo* to the backbone of the HGBA may affect its exposure. There are 12 hydrogen bonds between the trisaccharide and nine amino acids (Thr344, Arg345, Ala346, Lys348, Asp374, Cys440, Ser441, Gly442 and Asp391) of the P domain (Cao *et al.*, 2007) (Fig. 6.2, right) and none of these residues are shared with Norwalk virus. Of note, three residues (Cys440, S441 and G442) from the P1 subdomain of another monomer contribute to the binding interface of VA387, while residues from the P2 subdomain of only one monomer are involved in binding of the Norwalk virus. A mutagenesis study has confirmed the binding interface and identified additional amino acid residues near the interface involved in binding of VA387 (Tan *et al.*, 2008c).

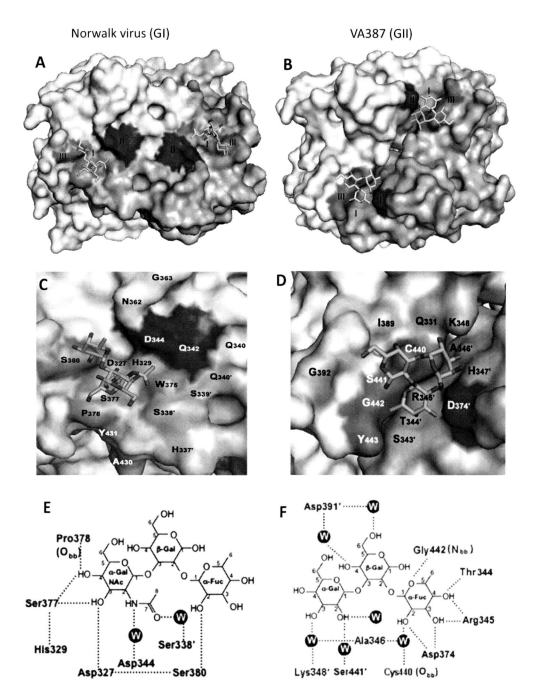

Figure 6.2 The crystal structures of the HBGA-binding interfaces of Norwalk virus (GI.1) and VA387 (GII.4). The surface models of the P dimers (top views) with indications of the HBGA-binding interfaces are shown in (A) and (B) with one monomer shown in darker grey. Enlargements of the HBGA-binding interfaces are shown in (C) and (D). The corresponding individual amino acids are labelled, with the prime symbol indicating a residue of another protomer. The three major components of the binding interfaces are labelled as site I, II, and III, respectively, while the oligosaccharides binding to the interface are also shown in stick mode 1. (E) and (F) are schematic diagrams of hydrogen bonding network (dash lines) between the amino acids of the P dimers of Norwalk virus (E), or VA387 (F) and the A- or B-type trisaccharides. The water-bridged hydrogen bonds are indicated by W. Adapted with permission from reference (Tan *et al.*, 2009b). The original data were published in Bu *et al.* (2008), Cao *et al.* (2007) and Choi *et al.* (2008).

Knowledge of the three-dimensional structure of the binding interface between certain HBGA carbohydrates and representative noroviruses has been critical in understanding these interactions at the molecular level. The ability to abolish binding to all relevant HBGAs with a single amino acid substitution in the capsid proteins studied indicates that the mapped binding sites are indeed the primary sites of interaction for Norwalk virus and VA387. However, it should be noted that these results were based on the binding of free, synthetic oligosaccharides to the crystallized P dimers. As such, this may not fully reflect the complexity of *in vivo* conditions where multiple species of HBGAs may be present, and many of the HBGAs are covalently linked to a protein or lipid backbone.

Human HBGA as a factor in norovirus evolution

The HBGA-binding interfaces of both Norwalk virus and VA387 can be divided into three major regions, representing the bottom (site I) and the walls (sites II and III) of the interface or pocket (Tan *et al.*, 2009b) (Fig. 6.2). Each of these sites is composed of one or a cluster of several scattered but sterically related amino acids (Figs 6.2 and 6.3), including Asp327 and His329 (site I), Gln342 and Asp344 (site II), and Trp375, Ser377, and Ser380 (site III) for Norwalk virus and Ser343 to His347 (site I), Asp374 (site II), and Ser441 and Gly442 (site III) for VA387. In addition to Norwalk virus and VA387, the involvement of the three sites in HBGA interactions has been confirmed by mutagenesis studies of other strains in both genogroups, including Boxer (GI.8), MOH (GII.5) and VA207 (GII.9) (Tan *et al.*, 2009b). Moreover, the specific interactions of these residues with HBGAs has been demonstrated also in the recently resolved crystal structures of the capsid P domain of VA207 (a Lewis binding strain) in complex with the Lewis HBGAs (Lewis y and sialyl Lewis x) (Chen, Tan, Jiang and Li, unpublished data).

Although the carbohydrate binding sites are not conserved between the two major genogroups of human noroviruses, they are highly conserved among strains within each of the two genogroups (Cao *et al.*, 2007; Choi *et al.*, 2008; Tan *et al.*, 2009b) (Figs 6.2 and 6.3). Sites I and III are more conserved than site II for the GI strains, while all three sites are highly conserved among GII strains with one exception. The three binding sites of GII.13 strains, including site III that is within the P1 subdomain of the capsid, show sequence variation (Tan *et al.*, 2009a) (Fig. 6.3B). The overall sequence identities of the P2 subdomains are 31–56% for strains within each of the two genogroups. These data indicate that HBGA recognition is an important selection factor in norovirus evolution and support the previous division of the two genetic lineages into two distinct genogroups of noroviruses.

These distinct lineages may have evolved from a common ancestor that did not have the ability to recognize HBGAs. Functional convergence of strains with the same HBGA targets subsequently resulted in acquisition of analogous HBGA binding interfaces in the two genogroups that share an overall structural similarity, despite their distinct locations and amino acid compositions. On the other hand, divergent evolution may have contributed to the observed overall differences between and within distinct lineages. Thus, both divergent and convergent evolution, as well as the polymorphic human HBGAs, probably contribute to the diversity of noroviruses.

Another factor believed to be involved in norovirus evolution is human herd immunity that may drive the epidemics of GII.4 strains (Lindesmith *et al.*, 2008), similar to the influenza viruses. Evidence supporting this hypothesis came from the finding of significant changes in the primary sequences of the P domain of emerging GII.4 strains compared to pre-existing circulating GII.4 strains. Changes in HBGA binding patterns and antigenicity among GII.4 variants have been reported (Lindesmith *et al.*, 2008). However, the observation that noroviruses usually induce a short-term immunity argues against this hypothesis. An alternative explanation on the predominance of GII.4 may be the 'spectrum hypothesis', in which the majority of GII.4 strains recognize the H-related antigens that represent 80% of the general population. Genotypes other than GII.4 may have a narrower spectrum of HBGA recognition, which is supported by the HBGA selection hypothesis (Tan *et al.*, 2009b). It has been proposed by Chan *et al.* (2006) that there may be a potential replicative advantage of

A

Components:	Site I	Site II	Site III
(Mutation study Norwalk virus)	D_{327} H_{329} * *	Q_{342} D_{344} * *	W_{375} S_{377} S_{380} * * *
GI1 Norwalk:	325GCDWHIN	340QTQYDVD	373LSWISPPSH
GI1 KY89:	325GCDWHIN	340QTQYDVD	373LSWVSPPSH
GI2 SOV:	328KCDWHMR	348MRSVSVQ	384IEWISQPST
GI3 HLL:	330ECDWHME	349IHQINVK	385LGWVSPVSD
GI3 DSV:	330DCDWHMS	349EYQILIK	384LSWISPVSD
GI3 VA115:	330ECDWHME	349IKQINVK	385LGWVSPASD
GI4 Chiba:	328SCDWHIE	348IVTNSVK	383IQWTSPPSD
GI5 Musgrove:	328TSDWHIE	348ILLRDIQ	383IQWTSQPSN
GI6 Wiscon:	329GCDWHVN	346SQSVTFA	382LGWISAPSD
GI7 Winchester:	330ACDWHVF	345EGSHVCT	383LAWVS-PST
GI8 Boxer:	330NCDLHMT	346STGDPSG	390LTWVSNRTG
(Mutation study Boxer)	H_{334} *	G_{348} D_{349} P_{350} ***	W_{392} *
GII4 VA387:	327VGKIQGM	346AHKATVS	385PVGVIQDGN

B

Components:	Site I	Site II	Site III
(Mutation study VA387)	343S.........H347 ****	D_{374} *	440C......Y443 ****
GII4 VA387:	338TREDGSTRAHKA	370DTNNDFQ	440GCSGYPN
GII1 Hawaii:	337RNPNNTCRAHDG	370WEESDLD	440LKGGTSD
GII2 SMV:	339RDK..ANRGHDA	378WQTDDLK	440LKGGYGN
GII3 Mexico:	351RNPDSTTRAHEA	382TESDDLD	447PSSGGRS
GII4 Lordsdale:	337TRADGSTRAHKA	370DTNNDFE	440GCSGYPN
GII5 MOH:	337RNR..ANRAHDA	372WNTNDVE	438LKGGFGN
GII6 Florida269:	340RDV..ATRAHEA	388D-SDDFN	450SAGGYGS
GII7 Leeds:	340RNK..ATRAQEV	375E-SQDFE	439PSSGGHE
GII8 Amsterdam:	340RSSDNATRAHEA	372P-STDFS	437GAGGFTD
GII9 VA207:	339RGPGDATRAHEA	371TSSNDFE	437GASGHTN
GII10 Erfurt:	347EGDLPANRAHEA	381WETQDVS	435SYSGALT
GII11 SW918:	350RNTDGQTRAHEA	381VESTDFH	446PSSGGVV
GII12 Wortley:	339RDHDNACRAHDA	371WEEDDVH	435LKGGVAD
GII13 Fayettevil:	339DNVNVST.GEAK	374SITEHVH	443GLQGQDA
GII14 M7:	338RDN..ATRAHDA	370SSSDDFD	436SAGGHTD
GII15 J23:	352GAGQNSNRAHFA	384FDTTDFQ	452FKGGYGE
GII16 Tiffin:	341TGTNPANRAHDA	373WDTEDLL	440LKGGHGD
GII17 CS-E1:	343GSNPNTTRAHEA	374STSTDFQ	440CAGGVSD
(Mutation study VA207)	R_{346}	D_{374}	G_{440}
(Mutation study MOH)	R_{347}	D_{376}	G_{441}
GI1 Norwalk:	331NMTQFGHSSQTQY	364IGSGNYV	427GPGAYNL
GI8 Boxer:	337FVKINPTELSTGD	375NNELDQF	441TVSNPKV
(Mutation study Boxer)	T_{347}	377E........F381	N_{444}

Figure 6.3 Sequence alignments of the HBGA-binding interfaces of various GI and GII noroviruses. Sequences of the three major components (site I, II, and III) of the HBGA-binding interfaces of 11 genogroup I (GI) (A) and 18 genogroup II (GII) (B) noroviruses, representing each of the 8 GI and 17 GII genetic types, respectively, are aligned based on the two known binding interfaces of Norwalk virus (GI) and VA387 (GII). Star symbols indicate the residues that have been experimentally shown to be required for binding to HBGAs. The three strains that have no detectable binding to examined HBGAs are underlined. The accession numbers of the sequence are: M87661 (Norwalk virus), L23828 (KY 89), L07418 (SOV), AF414403 (HLL), U04469 (DSV), AY038598 (VA115), AB042808 (Chiba), AJ277614 (Musgrove), AY502008 (Wiscon), AJ277609 (Winchester), AF538679 (Boxer), AY038600 (VA387), U07611 (Hawaii), AY134748 (SMV), U22498 (Mexico), X86557 (Lordsdale), AF397156 (MOH), AF414407 (Florida269), AJ277608 (Leeds), AF195848 (Amsterdam), AAK84676 (VA207), AF427118 (Erfurt), AB074893 (SW918), AJ277618 (Wortley), AY113106 (Fayettevil), AY130761 (M7), AY130762 (J23), AY502010 (Tiffin), AY502009 (CS-E1). Adapted with permission from (Tan et al., 2009b).

the GII noroviruses (with the majority of GII.4 noroviruses) over the GI noroviruses. Since the increase in GII.4 activity was detected and monitored only after 2000, continued surveillance is necessary in order to understand the underlying mechanisms, and establish the roles of herd immunity and HBGA recognition in norovirus evolution.

The findings of the conservation of the HBGA-binding interfaces within the two major genogroups of human noroviruses are significant in facilitating the rational design and development of therapeutics against noroviruses (Tan *et al.*, 2009a). For example, a single compound that inhibits the function of the conserved HBGA-binding interface may be capable of blocking infection of all strains with the shared receptor–binding interface and binding modes with human HBGAs. Thus, only a few compounds might be sufficient to block all noroviruses in the two genogroups studied here, each group sharing a similar binding interface that could be blocked by a common inhibitor.

Structural basis of FCV and fJAM-1 interactions

The discovery of fJAM-1 as the receptor for FCV (Bhella *et al.*, 2008; Makino *et al.*, 2006; Ossiboff and Parker, 2007) showed for the first time that a cellular membrane protein could facilitate entry of a calicivirus. This Ig-like protein is composed of an N-terminal signal peptide, the membrane-distal D1 and the membrane-proximal D2 Ig-like domains, a C-terminal transmembrane domain, and a short cytoplasmic tail (Kostrewa *et al.*, 2001; Prota *et al.*, 2003). The D1 domain was involved in the FCV–receptor interaction (Bhella *et al.*, 2008; Ossiboff and Parker, 2007). Structure-guided mutagenesis further suggested that three amino acids (Asp42, Lys43, and Ser97) within the D1 domain participate in the binding to FCV (Ossiboff and Parker, 2007). The 3D structures of FCV alone and fJAM-1-FCV in complex have been reconstructed by cryo-EM to 18 Å resolution (Bhella *et al.*, 2008) (fig. 6.4). Homology modelling and docking using the crystal structures of San Miguel sea lion virus (SMSV) (*Vesivirus* genus) (Chen *et al.*, 2006) capsid and the human JAM-1 (Prota *et al.*, 2003) showed that the D1 domain of fJAM-1 probably binds to the outer surface of the P2 subdomain of FCV capsid and the contact interface includes a hypervariable region of the P2 subdomain (Bhella *et al.*, 2008) (fig. 6.4). The significance of amino acid residues in the interaction suggested by mutagenesis (Asp42 and Lys43; (Ossiboff and Parker, 2007)) and sequence alignment (Ser91; (Makino *et al.*, 2006)) have been supported by their locations at the binding interface of the fJAM-1–FCV complex. Conformational changes in the viral capsid induced by fJAM-1 were observed, which might be an important step in the receptor-mediated uncoating process of FCV to release the genome into the host cells (Bhella *et al.*, 2008) (fig. 6.4).

Implications of the current understanding of calicivirus–host interaction

Nearly all caliciviruses examined thus far recognize carbohydrate molecules on the cell surface as potential ligands for virus attachment. Three of the four established genera (with the exception of *Sapovirus*, which has not been extensively studied) include virus strains that have been linked to specific interactions with variable carbohydrates. The HBGAs are the major players in these interactions, and a striking species-specificity in terms of types of the carbohydrates recognized has been noted. Several studies have shown that the recognition of these carbohydrate molecules is necessary for viral replication, and the observed variations might reflect differences in the determinants of the species- and/or tissue-tropisms of the viruses. In addition, as found in FCV, protein receptors may be required for the penetration or entry of the viruses into host cells, although direct evidence for such receptors or co-receptors remains lacking for other caliciviruses.

The interactions between carbohydrates and noroviruses follow the key–to-lock interactions that require a proper conformation of the carbohydrate to fit into the binding interface of the capsid (Bu *et al.*, 2008; Cao *et al.*, 2007; Choi *et al.*, 2008). The intermolecular bonding network between side chains of the carbohydrates and individual amino acids in the capsid interface facilitate the interaction. It has been noted that a given strain with a fixed binding interface can accommodate different carbohydrates as long as they meet the requirements of proper conformation to form the necessary bonding network. A typical example was the Norwalk virus, in which the binding pocket fit well to both β-Gal-α-Fuc and α-Gal-acetamido (*N*-acetylgalactosamine) (Bu *et al.*, 2008; Choi *et al.*, 2008). A further observation was that strains with distinct genetic identities share common HBGA targets. In other words, strains with different primary amino acid sequences in the capsid protein share similar conformations in their HBGA binding interfaces that recognize common HBGA targets (Tan *et al.*, 2009b). These findings provide insight into the mechanisms and dynamic changes of HBGA–norovirus interactions. This may serve as a model for other caliciviruses, such as Tulane virus and RHDV that bind also to HBGAs (Farkas *et al.*, 2008a; Ruvoen-Clouet *et al.*, 2000), and FCV and MNV that share the common target of sialic acid (Stuart and Brown, 2007; Taube *et al.*, 2009). A greater understanding of the molecular basis of carbohydrate–calicivirus interactions may facilitate the development of strategies for control and prevention of calicivirus-associated diseases by blocking or interrupting such interactions.

The interaction between noroviruses and HBGAs was highly strain-specific, probably reflecting typical protein–carbohydrate interactions that involve multivalent interactions. Norovirus strains with close genetic relatedness in the P domain of the capsid, in general, share similar HBGA binding patterns. For example, the majority of the GII.4 viruses share very conserved primary sequences in and around the HBGA binding interfaces (Tan *et al.*, 2008c, 2009b) and many of them recognize the common H-related antigens (Huang *et al.*, 2003; Huang *et al.*, 2005; Lindesmith *et al.*, 2008; Shirato *et al.*, 2008); and our unpublished data), representing ~80% of the blood types present in the general population. As discussed earlier, this may be a major factor in the predominance of GII.4 strains. However, non-HBGA-binding strains have been found within the GII.4 group (Lindesmith *et al.*, 2008). Since a number of GII.4 strains isolated from 2006 to 2008 bind to ABH antigens (Donaldson *et al.*, 2008; our unpublished data), it would be important to examine whether the non-HBGA binding GII.4 strains represent the minor natural variants. Future studies of HBGA binding with broader surveillance would help to answer this question.

Awareness of the strong strain-specificity of noroviruses in the recognition of HBGAs may help to clarify inconsistencies in the data generated from outbreak investigations that analyse an association between susceptibility to illness and HBGA phenotype. Some studies have reported an association (or lack of association) of HBGA types with infection of certain norovirus genogroups or genotypes based on surveillance of multiple outbreaks. Because the HBGA binding specificity is most likely virus strain-specific rather than genogroup- and genotype-specific (Tan and Jiang, 2008), knowledge of the HBGA binding patterns of the outbreak strains under investigation is critical in the analysis.

The finding of human HBGAs as an important factor in norovirus evolution may be significant for future disease control and prevention. If the noroviruses follow the same mechanism of evolution like the influenza viruses that are selected by the herd immunity, the same strategy of vaccine based on the antigenic types of future circulating strains may be necessary. However, the noticed short life of norovirus immunity apparently does not support this hypothesis and therefore, the influenza vaccine strategy may not be necessary or effective. On the other hand, the observation of major circulating GII.4 viruses binding to H-related antigens suggests that an alternative vaccine strategy to induce immune responses against the common epitopes of GII.4 strains may be an effective approach. Thus, further understanding the mechanism and dynamic changes of HBGA binding phenotypes of GII.4 variants in association with the host immune response are of most significance; a continuous accumulation of data in this field is highly demanded.

While HBGAs have been found to be an important factor in the tropism of noroviruses, there are undoubtedly additional factors involved in determining norovirus host range. The HBGAs are shared by many species but there is no direct evidence of cross-species transmission of noroviruses. In addition, human noroviruses bind to HBGAs on several human cell lines but none of these cell lines support norovirus replication. Noroviruses might utilize both a carbohydrate binding ligand and a fJAM-1-like receptor or co-receptor similar to the FCV model, although Wobus *et al.* did not find support for the utilization

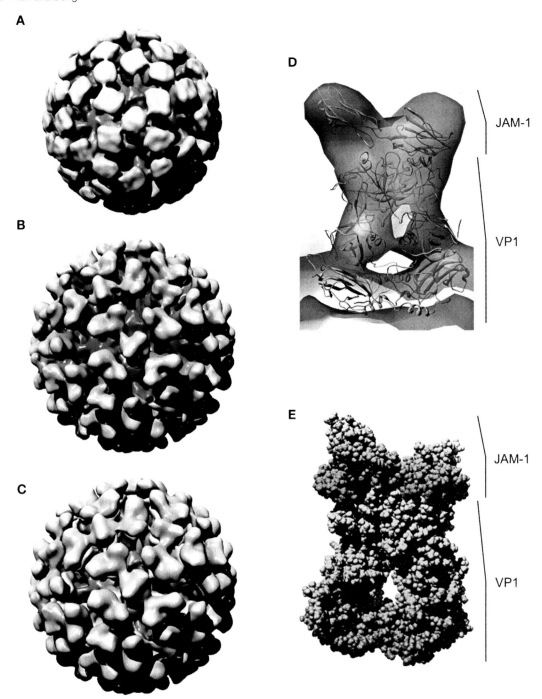

Figure 6.4 Structures of FCV virions and their interactions with fJAM-1. (A and B) Images of 3D reconstructions of unlabeled (A) and fJAM-1-labelled (B) FCV virions at 16 and 18-Å resolution, respectively, viewed along the icosahedral twofold symmetry axes. (C) A difference map was calculated by subtracting the unlabeled FCV reconstruction from the fJAM-1-labelled structure; concurrent rendering of the difference map and unlabeled reconstruction highlights additional density attached to the P2 domain of the labelled reconstruction. Difference density is also visible in the P1 domain, indicating changes in capsid conformation induced by fJAM-1 binding. (D) Docked homology models for FCV VP1 (ribbon) and fJAM-1 (ribbon) in the 18-Å resolution reconstruction for FCV labelled with fJAM-1 (transparent grey surface). (E) Atomic-resolution representation of the modelled structure for a FCV A-B dimer of VP1 (spheres) labelled with soluble fJAM-1 (spheres). Adapted with permission from (Bhella *et al.*, 2008).

of the JAM-1 molecule by MNV (Wobus, 2007). However, it is not unusual for viruses within the same family to utilize different cellular receptors. Continuing efforts to explore and characterize such receptors or co-receptors for the caliciviruses would be fundamentally important not only in elucidation of the mechanisms of pathogenesis, but also in the establishment of permissive cell culture systems and animal models for the study of fastidious viruses in the family.

Future research directions

Our understanding of calicivirus and host interactions has advanced significantly in the past decade, but many important questions remain. Caliciviruses are known for infecting a wide range of hosts with varying tissue tropisms, causing a broad spectrum of disease. Carbohydrates are likely to play an important role in determining host range and tissue specificity. Histo-blood group antigens and sialic acid serve as target molecules on the cell surface for attachment by human norovirus, MNV, FCV, RHDV and Tulane virus. It will be important to determine if additional carbohydrates are recognized by caliciviruses, and whether there are common recognition patterns among groups of viruses. The finding that FCV utilizes an additional membrane receptor potentially for penetration or entry indicates that an effort to explore such a protein receptor/co-receptor for other caliciviruses is necessary. The increasing number of epidemics of the GII.4 strains reported in many countries in recent years indicates an urgent need for research to understand the dynamic changes of the newly emerging GII.4 variants as it relates to host factor and evolution. Collaborative efforts to monitor such changes by laboratories in different geographical locations and countries would help to address this issue.

Conclusion

The discovery of a strong strain specificity among noroviruses in recognizing HBGAs and elucidation of the molecular mechanisms behind these variations have on one hand significantly advanced our understanding of the norovirus–host interaction and on the other hand presented new challenges in study of the role of HBGA in the epidemiology, immunology and classification of noroviruses. The continued accumulation of genetic and phenotypic data on such strain-specific variation obtained through collaborative research efforts with standardized protocols will be an important step towards this goal. In addition, determination of the structures of the receptor binding interfaces of additional strains representing different genogroups and genera of caliciviruses by crystallography is necessary. Finally, the readily available atomic structures of the HBGA binding interfaces of noroviruses representing the major receptor binding patterns should be explored for development of strategies against norovirus disease, such as through the computer-aided drug design (CADD) approach.

References

Barton, E.S., Forrest, J.C., Connolly, J.L., Chappell, J.D., Liu, Y., Schnell, F.J., Nusrat, A., Parkos, C.A., and Dermody, T.S. (2001). Junction adhesion molecule is a receptor for reovirus. Cell *104*, 441–451.

Bella, J., Kolatkar, P.R., Marlor, C.W., Greve, J.M., and Rossmann, M.G. (1998). The structure of the two amino-terminal domains of human ICAM-1 suggests how it functions as a rhinovirus receptor and as an LFA-1 integrin ligand. Proc. Natl. Acad. Sci. U.S.A. *95*, 4140–4145.

Bella, J., Kolatkar, P.R., Marlor, C.W., Greve, J.M., and Rossmann, M.G. (1999). The structure of the two amino-terminal domains of human intercellular adhesion molecule-1 suggests how it functions as a rhinovirus receptor. Virus Res. *62*, 107–117.

Bella, J., and Rossmann, M.G. (2000). ICAM-1 receptors and cold viruses. Pharm. Acta Helv. *74*, 291–297.

Bergelson, J.M., Cunningham, J.A., Droguett, G., Kurt-Jones, E.A., Krithivas, A., Hong, J.S., Horwitz, M.S., Crowell, R.L., and Finberg, R.W. (1997a). Isolation of a common receptor for Coxsackie B viruses and adenoviruses 2 and 5. Science *275*, 1320–1323.

Bergelson, J.M., Modlin, J.F., Wieland-Alter, W., Cunningham, J.A., Crowell, R.L., and Finberg, R.W. (1997b). Clinical coxsackievirus B isolates differ from laboratory strains in their interaction with two cell surface receptors. J. Infect. Dis. *175*, 697–700.

Bhella, D., Gatherer, D., Chaudhry, Y., Pink, R., and Goodfellow, I.G. (2008). Structural insights into calicivirus attachment and uncoating. J. Virol. *82*, 8051–8058.

Bu, W., Mamedova, A., Tan, M., Xia, M., Jiang, X., and Hegde, R.S. (2008). Structural basis for the receptor binding specificity of Norwalk virus. J. Virol. *82*, 5340–5347.

Cannon, J.L., Lindesmith, L.C., Donaldson, E.F., Saxe, L., Baric, R.S., and Vinje, J. (2009). Herd immunity to GII.4 noroviruses is supported by outbreak patient sera. J. Virol. *83*, 5363–5374.

Cao, S., Lou, Z., Tan, M., Chen, Y., Liu, Y., Zhang, Z., Zhang, X.C., Jiang, X., Li, X., and Rao, Z. (2007).

Structural basis for the recognition of blood group trisaccharides by norovirus. J. Virol. *81*, 5949–5957.

Chan, M.C., Sung, J.J., Lam, R.K., Chan, P.K., Lee, N.L., Lai, R.W., and Leung, W.K. (2006). Fecal viral load and norovirus-associated gastroenteritis. Emerg. Infect. Dis. *12*, 1278–1280.

Chen, R., Neill, J.D., Estes, M.K., and Prasad, B.V. (2006). X-ray structure of a native calicivirus: structural insights into antigenic diversity and host specificity. Proc. Natl. Acad. Sci. U.S.A. *103*, 8048–8053.

Choi, J.M., Hutson, A.M., Estes, M.K., and Prasad, B.V. (2008). Atomic resolution structural characterization of recognition of histo-blood group antigens by Norwalk virus. Proc. Natl. Acad. Sci. U.S.A. *105*, 9175–9180.

Donaldson, E.F., Lindesmith, L.C., Lobue, A.D., and Baric, R.S. (2008). Norovirus pathogenesis: mechanisms of persistence and immune evasion in human populations. Immunol. Rev. *225*, 190–211.

Estes, M.K., Prasad, B.V., and Atmar, R.L. (2006). Noroviruses everywhere: has something changed? Curr. Opin. Infect. Dis. *19*, 467–474.

Farkas, T., Sestak, K., and Jiang, X. (2008a). Tulane virus specifically binds to type B histo-blood group antigen. Scientific Program and Abstracts of 27th Annual Meeting for American Society for Virology, 271.

Farkas, T., Sestak, K., Wei, C., and Jiang, X. (2008b). Characterization of a rhesus monkey calicivirus representing a new genus of Caliciviridae. J. Virol. *82*, 5408–5416.

Guillon, P., Ruvoen-Clouet, N., Le Moullac-Vaidye, B., Marchandeau, S., and Le Pendu, J. (2009). Association between expression of the H histo-blood group antigen, alpha*1*, 2fucosyltransferases polymorphism of wild rabbits, and sensitivity to rabbit hemorrhagic disease virus. Glycobiology *19*, 21–28.

Harrington, P.R., Lindesmith, L., Yount, B., Moe, C.L., and Baric, R.S. (2002). Binding of Norwalk virus-like particles to ABH histo-blood group antigens is blocked by antisera from infected human volunteers or experimentally vaccinated mice. J. Virol. *76*, 12335–12343.

Huang, P., Farkas, T., Marionneau, S., Zhong, W., Ruvoen-Clouet, N., Morrow, A.L., Altaye, M., Pickering, L.K., Newburg, D.S., LePendu, J., et al. (2003). Noroviruses bind to human ABO, Lewis, and secretor histo-blood group antigens: identification of 4 distinct strain-specific patterns. J. Infect. Dis. *188*, 19–31.

Huang, P., Farkas, T., Zhong, W., Tan, M., Thornton, S., Morrow, A.L., and Jiang, X. (2005). Norovirus and histo-blood group antigens: demonstration of a wide spectrum of strain specificities and classification of two major binding groups among multiple binding patterns. J. Virol. *79*, 6714–6722.

Hutson, A.M., Airaud, F., LePendu, J., Estes, M.K., and Atmar, R.L. (2005). Norwalk virus infection associates with secretor status genotyped from sera. J. Med. Virol. *77*, 116–120.

Hutson, A.M., Atmar, R.L., and Estes, M.K. (2004a). Norovirus disease: changing epidemiology and host susceptibility factors. Trends Microbiol. *12*, 279–287.

Hutson, A.M., Atmar, R.L., Graham, D.Y., and Estes, M.K. (2002). Norwalk virus infection and disease is associated with ABO histo-blood group type. J. Infect. Dis. *185*, 1335–1337.

Hutson, A.M., Atmar, R.L., Marcus, D.M., and Estes, M.K. (2003). Norwalk virus-like particle hemagglutination by binding to h histo-blood group antigens. J. Virol. *77*, 405–415.

Hutson, A.M., Charkravarty, S., Atmar, R.L., Prasad, B.V., and Estes, M. (2004b). Loss of carbohydrate binding with point mutations of Norwalk virus virus-like particles. Second International Calicivirus Conference, Dijon, France, 6–10 November 2004.

Jiang, X., Huang, P., Zhong, W., Tan, M., Farkas, T., Morrow, A.L., Newburg, D.S., Ruiz-Palacios, G.M., and Pickering, L.K. (2004). Human milk contains elements that block binding of noroviruses to human histo-blood group antigens in saliva. J. Infect. Dis. *190*, 1850–1859.

Kostrewa, D., Brockhaus, M., D'Arcy, A., Dale, G.E., Nelboeck, P., Schmid, G., Mueller, F., Bazzoni, G., Dejana, E., Bartfai, T., et al. (2001). X-ray structure of junctional adhesion molecule: structural basis for homophilic adhesion via a novel dimerization motif. EMBO J. *20*, 4391–4398.

Le Pendu, J. (2004). Histo-blood group antigen and human milk oligosaccharides: genetic polymorphism and risk of infectious diseases. Adv. Exp. Med. Biol. *554*, 135–143.

Lindesmith, L., Moe, C., Lependu, J., Frelinger, J.A., Treanor, J., and Baric, R.S. (2005). Cellular and humoral immunity following Snow Mountain virus challenge. J. Virol. *79*, 2900–2909.

Lindesmith, L., Moe, C., Marionneau, S., Ruvoen, N., Jiang, X., Lindblad, L., Stewart, P., LePendu, J., and Baric, R. (2003). Human susceptibility and resistance to Norwalk virus infection. Nat. Med. *9*, 548–553.

Lindesmith, L.C., Donaldson, E.F., Lobue, A.D., Cannon, J.L., Zheng, D.P., Vinje, J., and Baric, R.S. (2008). Mechanisms of GII.4 norovirus persistence in human populations. PLoS Med. *5*, e31.

Makino, A., Shimojima, M., Miyazawa, T., Kato, K., Tohya, Y., and Akashi, H. (2006). Junctional adhesion molecule 1 is a functional receptor for feline calicivirus. J. Virol. *80*, 4482–4490.

Meyers, G., Wirblich, C., and Thiel, H.J. (1991). Rabbit hemorrhagic disease virus – molecular cloning and nucleotide sequencing of a calicivirus genome. Virology *184*, 664–676.

Morrow, A.L., Ruiz-Palacios, G.M., Altaye, M., Jiang, X., Guerrero, M.L., Meinzen-Derr, J.K., Farkas, T., Chaturvedi, P., Pickering, L.K., and Newburg, D.S. (2004a). Human milk oligosaccharide blood group epitopes and innate immune protection against campylobacter and calicivirus diarrhea in breastfed infants. Adv. Exp. Med. Biol. *554*, 443–446.

Morrow, A.L., Ruiz-Palacios, G.M., Altaye, M., Jiang, X., Guerrero, M.L., Meinzen-Derr, J.K., Farkas, T., Chaturvedi, P., Pickering, L.K., and Newburg, D.S. (2004b). Human milk oligosaccharides are associated with protection against diarrhea in breast-fed infants. J. Pediatr. *145*, 297–303.

Morrow, A.L., Ruiz-Palacios, G.M., Jiang, X., and Newburg, D.S. (2005). Human-milk glycans that inhibit pathogen binding protect breast-feeding infants against infectious diarrhea. J. Nutr. *135*, 1304–1307.

Newburg, D.S. (1999). Human milk glycoconjugates that inhibit pathogens. Curr. Med. Chem. *6*, 117–127.

Newburg, D.S. (2005). Innate immunity and human milk. J. Nutr. *135*, 1308–1312.

Newburg, D.S., Peterson, J.A., Ruiz-Palacios, G.M., Matson, D.O., Morrow, A.L., Shults, J., Guerrero, M.L., Chaturvedi, P., Newburg, S.O., Scallan, C.D., et al. (1998). Role of human-milk lactadherin in protection against symptomatic rotavirus infection. Lancet *351*, 1160–1164.

Newburg, D.S., and Walker, W.A. (2007). Protection of the neonate by the innate immune system of developing gut and of human milk. Pediatr. Res. *61*, 2–8.

Ossiboff, R.J., and Parker, J.S. (2007). Identification of regions and residues in feline junctional adhesion molecule required for feline calicivirus binding and infection. J. Virol. *81*, 13608–13621.

Prota, A.E., Campbell, J.A., Schelling, P., Forrest, J.C., Watson, M.J., Peters, T.R., Aurrand-Lions, M., Imhof, B.A., Dermody, T.S., and Stehle, T. (2003). Crystal structure of human junctional adhesion molecule 1: implications for reovirus binding. Proc. Natl. Acad. Sci. U.S.A. *100*, 5366–5371.

Rademacher, C., Krishna, N.R., Palcic, M., Parra, F., and Peters, T. (2008). NMR experiments reveal the molecular basis of receptor recognition by a calicivirus. J. Am. Chem. Soc. *130*, 3669–3675.

Ravn, V., and Dabelsteen, E. (2000). Tissue distribution of histo-blood group antigens. Apmis *108*, 1–28.

Rockx, B.H., Vennema, H., Hoebe, C.J., Duizer, E., and Koopmans, M.P. (2005). Association of histo-blood group antigens and susceptibility to norovirus infections. J. Infect. Dis. *191*, 749–754.

Ruvoen-Clouet, N., Ganiere, J.P., Andre-Fontaine, G., Blanchard, D., and Le Pendu, J. (2000). Binding of rabbit hemorrhagic disease virus to antigens of the ABH histo-blood group family. J. Virol. *74*, 11950–11954.

Rydell, G.E., Nilsson, J., Rodriguez-Diaz, J., Ruvoen-Clouet, N., Svensson, L., Le Pendu, J., and Larson, G. (2009). Human noroviruses recognize sialyl Lewis x neoglycoprotein. Glycobiology *19*, 309–320.

Saravanan, C., Cao, Z., Kumar, J., Qiu, J., Plaut, A.G., Newburg, D.S., and Panjwani, N. (2008). Milk components inhibit Acanthamoeba-induced cytopathic effect. Invest. Ophthalmol. Vis. Sci. *49*, 1010–1015.

Shirato, H., Ogawa, S., Ito, H., Sato, T., Kameyama, A., Narimatsu, H., Xiaofan, Z., Miyamura, T., Wakita, T., Ishii, K., et al. (2008). Noroviruses distinguish between type 1 and type 2 histo-blood group antigens for binding. J. Virol. *82*, 10756–10767.

Stuart, A.D., and Brown, T.D. (2007). Alpha2, 6-linked sialic acid acts as a receptor for Feline calicivirus. J. Gen. Virol. *88*, 177–186.

Tamura, M., Natori, K., Kobayashi, M., Miyamura, T., and Takeda, N. (2000). Interaction of recombinant norwalk virus particles with the 105-kilodalton cellular binding protein, a candidate receptor molecule for virus attachment. J. Virol. *74*, 11589–11597.

Tamura, M., Natori, K., Kobayashi, M., Miyamura, T., and Takeda, N. (2004). Genogroup II noroviruses efficiently bind to heparan sulfate proteoglycan associated with the cellular membrane. J. Virol. *78*, 3817–3826.

Tan, M., Fang, P., Chachiyo, T., Xia, M., Huang, P., Fang, Z., Jiang, W., and Jiang, X. (2008a). Noroviral P particle: structure, function and applications in virus–host interaction. Virology *382*, 115–123.

Tan, M., Hegde, R.S., and Jiang, X. (2004a). The P domain of norovirus capsid protein forms dimer and binds to histo-blood group antigen receptors. J. Virol. *78*, 6233–6242.

Tan, M., Huang, P., Meller, J., Zhong, W., Farkas, T., and Jiang, X. (2003). Mutations within the P2 domain of norovirus capsid affect binding to human histo-blood group antigens: evidence for a binding pocket. J. Virol. *77*, 12562–12571.

Tan, M., and Jiang, X. (2005a). Norovirus and its histo-blood group antigen receptors: an answer to a historical puzzle. Trends Microbiol. *13*, 285–293.

Tan, M., and Jiang, X. (2005b). The p domain of norovirus capsid protein forms a subviral particle that binds to histo-blood group antigen receptors. J. Virol. *79*, 14017–14030.

Tan, M., and Jiang, X. (2007). Norovirus–host interaction: implications for disease control and prevention. Expert Rev. Mol. Med. *9*, 1–22.

Tan, M., and Jiang, X. (2008). Association of histo-blood group antigens with susceptibility to norovirus infection may be strain-specific rather than genogroup dependent. J. Infect. Dis. *198*, 940–941.

Tan, M., Jin, M., Xie, H., Duan, Z., Jiang, X., and Fang, Z. (2008b). Outbreak studies of a GII-3 and a GII-4 norovirus revealed an association between HBGA phenotypes and viral infection. J. Med. Virol. *80*, 1296–1301.

Tan, M., Xia, M., Cao, S., Huang, P., Farkas, T., Meller, J., Hegde, R.S., Li, X., Rao, Z., and Jiang, X. (2008c). Elucidation of strain–specific interaction of a GII-4 norovirus with HBGA receptors by site-directed mutagenesis study. Virology *379*, 324–334.

Tan, M., Xia, M., Chen, Y., Bu, W., Hegde, R.S., Meller, J., Li, X., and Jiang, X. (2009a). Conservation of carbohydrate binding interfaces: evidence of human HBGA selection in norovirus evolution. PLoS ONE *4*, e5058.

Tan, M., Zhong, W., Song, D., Thornton, S., and Jiang, X. (2004b). E. coli-expressed recombinant norovirus capsid proteins maintain authentic antigenicity and receptor binding capability. J. Med. Virol. *74*, 641–649.

Taube, S., Perry, J.W., Yetming, K., Patel, S.P., Auble, H., Shu, L., Nawar, H.F., Lee, C.H., Connell, T.D., Shayman, J.A., et al. (2009). Ganglioside-linked terminal sialic acid moieties on murine macrophages function as attachment receptors for Murine Noroviruses (MNV). J. Virol. *83*, 4092–4101.

Thorven, M., Grahn, A., Hedlund, K.O., Johansson, H., Wahlfrid, C., Larson, G., and Svensson, L. (2005). A homozygous nonsense mutation (428G – >A) in the human secretor (FUT2) gene provides resistance to symptomatic norovirus (GGII) infections. J. Virol. *79*, 15351–15355.

Wobus, C.E. (2007). Mechanisms of Murine Norovirus 1 (MNV-1) Entry. Program and Abstracts of Third International Calicivirus Conference, S7–2.

Zheng, D.P., Ando, T., Fankhauser, R.L., Beard, R.S., Glass, R.I., and Monroe, S.S. (2006). Norovirus classification and proposed strain nomenclature. Virology 346, 312–323.

Calicivirus Reverse Genetics and Replicon Systems

7

Kyeong-Ok Chang and Yunjeong Kim

Abstract

Reverse genetics and replicon systems have become important tools in the elucidation of the calicivirus replication and pathogenicity. Reverse genetics systems are available for feline calicivirus, porcine enteric calicivirus, murine norovirus, rabbit haemorrhagic disease virus and a rhesus monkey calicivirus. For uncultivable caliciviruses, such as human norovirus, cell-based replicon systems have been established. Norovirus replicon systems are used to screen potential antivirals and therapeutic options against norovirus infection. Replicon systems with reporter genes such as those encoding green fluorescent protein or luciferase allow quantitative analysis of cellular and viral factors that promote virus replication. Further studies with reverse genetics and replicon systems could yield important information for cell culture adaptation of human noroviruses which is crucial for development of efficient vaccines and antivirals.

Introduction

A reverse genetics system is an important tool for the study of viral replication and pathogenicity because deliberate genetic changes can be introduced into a viral genome. In most cases, the establishment of a reverse genetics system is dependent on the availability of a cell culture system for the target virus. This observation is true for caliciviruses: reverse genetics systems for cell culture-adapted feline calicivirus (FCV), porcine enteric calicivirus (PEC), murine norovirus (MNV), rabbit haemorrhagic disease virus

(RHDV), and rhesus monkey calicivirus (Tulane virus) are available (Table 7.1), while those for uncultivable caliciviruses such as human noroviruses are not. Cell-based RNA replicon systems are an important tool also for both cultivable and non-cultivable viruses because authentic viral RNA replication can be studied in the absence of infectious virus. Furthermore, the introduction of reporter genes such as those encoding green fluorescent protein (GFP) or luciferase into the replicon can allow a quantitative approach to the mapping of essential replication elements and the screening of antiviral compounds. For uncultivable human noroviruses, a replicon system with a transient or a stable replication scheme including replicon-bearing cells has been established (Table 7.1) and has served as a useful tool for virus research. The purpose of this chapter is to review reverse genetics and replicon systems that have been developed for members of the family *Caliciviridae*, with an emphasis on the development of a replicon system for Norwalk virus, the prototype norovirus strain.

Calicivirus reverse genetics systems

Caliciviruses possess a positive-sense single-stranded RNA genome that is covalently linked to a small viral genome-linked protein (VPg) at the 5' end and polyadenylated at the 3' end. As with other positive strand RNA viruses, the genomic RNA of the virion is infectious when introduced into permissive cells (Black *et al.*, 1978; Chang *et al.*, 2002; Oglesby *et al.*, 1971). A

Table 7.1 Reverse genetics or replicon systems for various caliciviruses

Reference	Genus	Virus strain	Cell lines	RNA transcripts/ DNA based (promoter)	Recovery of virus
a	*Vesivirus*	FCV Urbana	CRFK/BHK21	RNA/DNA (T7)	Yes
b	*Vesivirus*	FCV 2024	CRFK/BHK21	RNA/DNA (T7)	Yes
c	*Sapovirus*	Cowden PEC	LLC-PK cells	RNA	Yes
d	*Norovirus*	MNV-1	Various cells	DNA (Pol II)	Yes
e	*Norovirus*	MNV-1	RAW264.7	DNA (T7)	Yes
f	*Lagovirus*	JX/CHA/97 RHDV	Liver (in *vivo*)/RK13	RNA/DNA (CMV)	Yes
g	Unclassified	Tulane virus	LLC-MK2	RNA	Yes
h	*Norovirus*	Norwalk virus	HEK 293T	DNA (T7)	No
i	*Norovirus*	U201	HEK 293T	DNA (T7)	No
j	*Norovirus*	Norwalk virus	BHK21/Huh-7	RNA	No
k	*Norovirus*	Norwalk virus	BHK21/Vero	DNA (T7)	No

a   Sosnovtsev and Green (1995), Sosnovtsev *et al.* (1997).

b   MVA-T7 was used for DNA based transfection (Thumfart and Meyers, 2002).

c   Bile acids or intestinal contents (IC) were required for virus replication (Chang *et al.*, 2005).

d   DNA was introduced by baculovirus and DNA Pol II promoter with a hepatitis delta virus ribozyme at 3′ end; various cells including HepG2, BHK-21, COS-7, or HEK293T were used (Ward *et al.*, 2007).

e   Fowlpox-T7 was used for DNA-based transfection, a hepatitis delta virus ribozyme was engineered at 3′ end (Chaudhry *et al.*, 2007).

f   Liu *et al.* (2006, 2008a,b).

g   Wei *et al.* (2008).

h   Replicon system with MVA-T7, a hepatitis delta virus ribozyme was engineered at 3′ end (Asanaka *et al.*, 2005).

i   Replicon system with MVA-T7, a hepatitis delta virus ribozyme was engineered at 3′ end (Katayama *et al.*, 2006).

j   Replicon-bearing cells (Chang *et al.*, 2006).

k   Replicon system with MVA-T7 (Chang *et al.*, 2008).

number of approaches has been used to develop reverse genetics systems for the caliciviruses. These approaches all involve the transcription of full-length RNA molecules from a cDNA clone corresponding to the complete viral genome. Variations among the systems relate to the choice of promoter used to drive RNA transcription from the full-length cDNA clone, and whether the RNA is synthesized by transcription outside the cell, or within the cell (Table 7.1).

FCV reverse genetics system (genus *Vesivirus*)

FCV causes a respiratory illness in cats. The symptoms of natural or experimental infection of FCV in cats are fever and lingual or oral ulceration conjunctivitis as well as upper respiratory tract signs such as sneezing, and rhinitis (Pesavento *et al.*, 2008). However, in many cases, FCV infection causes asymptomatic or mild illness, and can lead to persistent infection.

FCV grows efficiently in a feline kidney cell line [Crandell–Rees feline kidney (CRFK) cells]. The first calicivirus reverse genetics system was established by Sosnovtsev and Green in 1995 for the Urbana strain of FCV (Sosnovtsev and Green, 1995). They demonstrated that a full-length cDNA clone of the FCV genome constructed immediately downstream of the T7 promoter produced progeny viruses after being introduced into cells as RNA transcripts or as plasmid DNA in the presence of modified vaccinia virus expressing T7 polymerase (Sosnovtsev and Green, 1995; Sosnovtsev et al., 1997). The FCV VPg, involved in the initiation of translation of viral protein and essential for the infectivity of native viral RNA (Daughenbaugh et al., 2003; Schaffer et al., 1980), was not required for infectivity of the transcripts. Instead, the addition of a cap structure analogue [m7G(5')ppp(5')G] during in vitro transcription of the synthetic RNA was sufficient for successful virus recovery (Sosnovtsev and Green, 1995). Progeny viruses could be recovered from the transfection in permissive CRFK cells as well as non-permissive cell lines such as BHK-21 cells (Sosnovtsev and Green, 1995; Sosnovtsev et al., 1997). Using a FCV infectious clone, Neill et al. demonstrated the generation of chimeric viruses containing capsid protein domain exchanges from antigenically distinct FCV strains (Neill et al., 2000). Sosnovtsev et al. identified essential cleavage sites of the FCV open reading frame one (ORF1) non-structural polyprotein and showed that alteration of these sites prevented virus recovery (Sosnovtsev et al., 2006; Sosnovtsev et al., 2002). The same group also demonstrated that FCV virion protein 2 (VP2) was essential for productive replication and the synthesis and maturation of infectious virions. They also showed that the ORF3 sequence overlaps a cis-acting RNA signal at the genomic 3' end (Sosnovtsev et al., 2005). Infectious virus particles were rescued from a full-length FCV cDNA clone encoding a non-functional VP2 when VP2 was provided in trans from a eukaryotic expression plasmid (Sosnovtsev et al., 2005). Thumfart and Meyers demonstrated a similar reverse genetics system using a vaccine FCV strain (Thumfart and Meyers, 2002), however in their system, they inserted a green fluorescent protein (GFP) gene in the structural protein region. They showed that the genomic RNA was still able to replicate autonomously and that the GFP was packaged into virus particles when the intact structural protein was provided in trans. Without a cell culture system for human caliciviruses, FCV and its reverse genetics system have provided an excellent system for the study of the basic replication strategy of caliciviruses in cell culture.

PEC reverse genetics system (genus *Sapovirus*)

The Cowden PEC strain is a sapovirus that can replicate in LLC-PK cells, a continuous cell line of porcine kidney origin (Flynn and Saif, 1988; Guo et al., 1999; Parwani et al., 1991). This strain provides an excellent model system because it grows in cell culture, causes gastroenteric pathogenicity in experimentally inoculated animals and viruses are shed 1–7 post-inoculation day (PID) (Flynn et al., 1988; Guo et al., 2001). Using fluorescein-conjugated specific antibody and immunofluorescence, viral antigens were mainly observed in the villous epithelial cells in the small intestine, primarily the duodenum and jejunum. Histopathological findings showed villus atrophy in the duodenum and/or jejunum, i.e. short villi and long crypts were observed in the duodenum and jejunum between 3 and 7 PID. However, lesions were not observed in the histological sections of the ileum, colon, liver, lungs or kidney. Scanning electron microscopy (EM) confirmed the histopathological findings showing villous atrophy with shortening, blunting, and fusion, or absence of villi in the duodenum and jejunum.

The transfection of viral RNA (tissue culture-adapted Cowden PEC) into LLC-PK cells and incubation in the presence of intestinal contents (IC) or bile acids resulted in the recovery of progeny viruses (Chang et al., 2002), suggesting the possibility of generating infectious clones of Cowden PEC with IC or bile acids. A full-length cDNA copy of the tissue culture-adapted Cowden PEC genome was cloned into a plasmid vector directly downstream from the T7 RNA Pol promoter (Chang et al., 2005). The transfection of the capped RNA transcripts derived from this clone into LLC-PK cells yielded infectious progeny viruses only in the presence of IC or bile acids

(Chang *et al.*, 2005). This was consistent with the observation that virus replication requires IC or a bile acid-mediated signalling pathway (Chang *et al.*, 2005). The recovered PEC was administered to gnotobiotic pigs by the oral route to examine the pathogenicity. It induced a limited amount of virus shedding after 7 PID in pigs, and mild (6 PID) or no diarrhoea was observed in those pigs (Chang *et al.*, 2005). These results suggested that the recovered virus retained its ability to infect pigs and it showed an attenuated phenotype because the genome was cloned from the tissue culture-adapted Cowden PEC (Chang *et al.*, 2005; Guo *et al.*, 2001).

Reverse genetics system of murine norovirus (MNV) in the genus *Norovirus*

The first MNV (MNV-1) was originally found in severely immunocompromised mice lacking the recombination-activating gene 2 (RAG2) and signal transducer and activator of transcription 1 (STAT-1) (RAG2/STAT1−/− mice) with a high mortality (Karst *et al.*, 2003). This strain could also infect wild-type mice with per oral (p.o.) or intranasal (i.n.) inoculation. It has been established that MNVs are present widely in laboratory mouse colonies without apparent clinical symptoms (Hsu *et al.*, 2006; Wobus *et al.*, 2006). MNV-1 has a tissue tropism of macrophage-like cells *in vivo* and *in vitro* (Wobus *et al.*, 2004b). The replication of MNV-1 in cell culture was inhibited by interferon (IFN)-αβ receptor and STAT1, reflecting the importance of cellular innate immunity in the replication of MNVs *in vivo* and *in vitro* (Karst *et al.*, 2003; Wobus *et al.*, 2004a).

Successful isolation of MNV in a continuous cell line (murine macrophage-like RAW264.7) led to the generation of reverse genetics systems (Chaudhry *et al.*, 2007; Karst *et al.*, 2003; Ward *et al.*, 2007; Wobus *et al.*, 2004a). An interesting difference compared with the FCV and PEC systems was that transfection of the capped, synthetic RNA transcripts derived from the full length clones of MNV into various cell lines, including RAW264.7, did not result in the recovery of infectious progeny (Chaudhry *et al.*, 2007; Ward *et al.*, 2007). Ward *et al.* (Ward *et al.*, 2007) reported recovery of infectious MNV after baculovirus delivery of viral cDNA to human

hepatoma cells under the control of an inducible DNA Pol II promoter. Once recovered in the hepatoma cells, viruses could be amplified in permissive RAW264.7 cells (Ward *et al.*, 2007). The system was further simplified by the transfection of a Pol II promoter-driven infectious MNV cDNA clone into HEK293T cells and recovery of infectious virus (Ward *et al.*, 2007). In this study, the authors demonstrated that ablation of the protease-Pol (ProPol) cleavage site abolished the recovery of MNV (Ward *et al.*, 2007). Chaudhry *et al.* (Chaudhry *et al.*, 2007) developed a reverse-genetics system for MNV using a recombinant fowlpox virus (FWPV) expressing T7 RNA Pol. Interestingly, it was not successful when MVA-T7 was used as the source of T7 Pol in cells, and the authors concluded that MVA-T7 infection may exert a negative effect on MNV replication (Chaudhry *et al.*, 2007). In contrast, FWPV infection had no apparent deleterious effect and allowed the recovery of infectious MNV from cells previously transfected with MNV full-length cDNA constructs (Chaudhry *et al.*, 2007). Using the reverse genetics system, Bailey *et al.* (Bailey *et al.*, 2008) identified a virus-encoded molecular determinant of norovirus virulence in the major capsid protein (VP1). Following passage of MNV in RAW264.7 cells, two amino acid changes had been identified between the virulent (wild type) and cell-culture adapted viruses, with one of these located at residue 296 (Lys to Glu) in the VP1 capsid protein (Wobus *et al.*, 2004a). The introduction of the wild-type lysine at position 296 into the capsid protein (VP1) encoded in the infectious MNV clone was sufficient to restore virulence *in vivo* (Bailey *et al.*, 2008).

Reverse genetics of rabbit haemorrhagic disease virus (RHDV) in the genus *Lagovirus*

Rabbit haemorrhagic disease virus is a lagovirus, and primarily infects the liver causing liver necrosis, haemorrhages, and high mortality in rabbits. The first report of RHD outbreak was from China in 1985, and since then it has been reported worldwide include the US. In earlier studies, most studies of RHDV infection were done in the host animal due to lack of a stable cell culture system. Liu *et al.* (Liu *et al.*, 2006b) first reported a reverse genetics system using an animal model. They generated the full-length infectious clone of strain JX/CHA/97 of RHDV, and demonstrated recovery of infectious RHDV after the RNA transcripts from the full-length cDNA clones were mixed with Lipofectin and injected directly into the liver of rabbits (Liu *et al.*, 2006b). The virus recovery was also demonstrated in RK13 cells (rabbit kidney cells) transfected with the RNA transcripts, and the rescued viruses were adapted in the cells (Liu *et al.*, 2006a; Liu *et al.*, 2008b). The same group also reported a DNA-based reverse genetics system for RHDV under the control of the eukaryotic human cytomegalovirus promoter in RK13 cells (Liu *et al.*, 2008a). The authors demonstrated that the transfection with a cDNA clone of RHDV lacking VP2 in RK13 cells also resulted in the generation of infectious virions, suggesting VP2 may not be essential for RHDV infectivity (Liu *et al.*, 2008a).

Reverse genetics of Tulane virus, a newly discovered and unclassified calicivirus

Tulane virus, isolated from stool samples of rhesus macaques (*Macaca mulatta*), is a calicivirus with a significant genetic diversity, thus may represent a new genus (Farkas *et al.*, 2008). Although pathogenicity is not known, this virus can be cultured in LLC-MK2 cells (a monkey kidney cell line) (Farkas *et al.*, 2008). Wei *et al.* (2008) recently generated a reverse genetics system for Tulane virus, showing recovery of infectious virus after transfection of RNA transcripts from a full-length clone. The authors also demonstrated essential roles of the 5' end of the genome and both ORF2 and -3 in the recovery of virus using this system.

Calicivirus replicon systems

Although reverse genetic systems for cultivable viruses such as FCV, MNV, PEC, RHDV and Tulane virus are available and provide important tools for studying viruses, those for human enteric caliciviruses (both norovirus and sapovirus) have not been demonstrated. However, a replicon system for human noroviruses with either transient or stable expression of viral RNA has been reported and provided an important tool for the study of virus replication.

Transient calicivirus replication systems

It was demonstrated that transfection of a full-length cDNA clone of the Norwalk virus genome (under the control of T7 promoter) into MVA-T7 infected cells allowed the expression of viral proteins and subsequent Norwalk virus RNA replication (Asanaka *et al.*, 2005). The Norwalk virus subgenomic RNA (sgRNA) was transcribed from genomic RNA by the use of Norwalk virus non-structural proteins expressed from genomic RNA, and was subsequently translated into Norwalk virus capsid protein (VP1) (Asanaka *et al.*, 2005). Furthermore, viral genomic RNA was packaged into virus particles generated in mammalian cells, and the caesium chloride density gradient profile of purified virus particles containing Norwalk virus RNA was similar to that of Norwalk virus purified from stools (Asanaka *et al.*, 2005). Katayama *et al.* (2006) also demonstrated a similar replication system with U201 norovirus strain (a human norovirus genogroup II strain). The authors also used MVA-T7 in HEK 293T cells and demonstrated the expression of proteins from sgRNA and incorporation of viral RNA in particles (Katayama *et al.*, 2006). However, both systems required a helper virus to initiate the transcription of virus RNA, which may affect the replication of norovirus. Interestingly, Guix *et al.* (2007) demonstrated that Norwalk virus RNA isolated from stool samples of volunteers was infectious when transfected into various cells, leading to viral replication evidenced by the expression of viral antigens, RNA replication and release of viral particles into the medium. As shown with other caliciviruses, pre-treatment of RNA with proteinase K completely abolished RNA infectivity, suggesting a key role of the RNA–VPg complex in Norwalk virus replication (Guix *et al.*, 2007). The demonstration of infectious viral RNA in various cells suggests that at least limited replication of Norwalk virus occurs in cells, but the passage of the recovered virus was not demonstrated (Guix *et al.*, 2007).

The replicon systems for the FCV genome with GFP were developed also (Chang *et al.*, 2008; Thumfart and Meyers, 2002). Thumfart and Meyers (2002) demonstrated that an FCV replicon with the insertion of GFP in the ORF2 of the infectious cDNA clone was infectious and it could be packaged into virus particles when the structural proteins were provided *in trans*. The virus particle containing the replicon could infect cells producing GFP (Thumfart and Meyers, 2002). A similar FCV replicon with GFP (pQ-GFP) was constructed and the expression of GFP was evident after the transfection of pQ-GFP in MVA-T7 infected cells (various cell lines) (Chang *et al.*, 2008). Interestingly, co-transfection of pQ-GFP and pCI-LC, which provides leader of the capsid (LC) *in trans*, significantly increased the number of cells expressing GFP, suggesting that LC may promote the replication of FCV (Chang *et al.*, 2008).

Cell-based replicon system of Norwalk virus

Stable replicon-bearing cell lines for the human noroviruses

A cell-based replicon system was generated to establish a practical genetic system for the study of Norwalk virus replication (Chang and George, 2007; Chang *et al.*, 2006, 2008) using a cloned cDNA consensus sequence of the RNA genome of Norwalk virus (plasmid NV101)

(Fernandez-Vega *et al.*, 2004) engineered to encode the neomycin resistance gene as a selective marker within ORF2 (Fig. 7.1), designated as pNV-Neo. The antibiotic resistant gene replaced the most part of VP1 (in ORF2), therefore disrupting expression of an intact VP1. However, pNV-Neo contains the predicted subgenomic promoter, an intact ORF3, and the genomic 3' end (Fig. 7.1B). The RNA transcripts derived from pNV-Neo were transfected into various cells including BHK-21 cells and Huh-7 cells. Interestingly, viable cell colonies were selected only in BHK-21 (G3 cells) in the presence of G418 (0.5 mg/ml) (Chang *et al.*, 2006). The replicon-bearing cells based on Huh-7 cells (HG23 cells) were generated by the transfection of RNA extracted from G3 cells, because transfection of RNA transcripts did not yield any colony in the presence of the G418. The replicon RNA may be linked with VPg, thus improved the efficiency of virus replication in the cells (Chang *et al.*, 2006). The replicon RNA, full length genomic RNA and sgRNA species, both sense and anti-sense, in G3 and HG23 cells was confirmed by Northern blot analysis (Chang *et al.*, 2006). The antisera against Norwalk virus ProPol or helicase (NTPase) were used for detecting Norwalk virus proteins in

A Norwalk virus

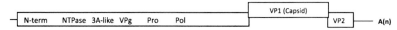

B pNV101, pNV-Neo, pNV-GFP, pNV-RL

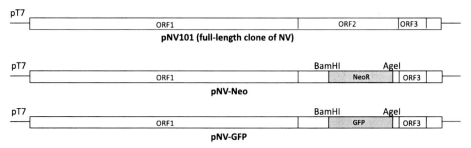

Figure 7.1 Genomic organization of human norovirus and recombinant plasmids based on the full-length genome of Norwalk virus. (A). Schematic diagram of the genome organization of Norwalk virus. (B) Recombinant plasmids of pNV101 (plasmid containing full-length genome of Norwalk virus under T7 promoter): pNV-Neo (with neomycin resistant gene: NeoR) and pNV-GFP. pNV-Neo and pNV-GFP were generated using unique sites of *Bam*HI and *Age*I in ORF2 of the Norwalk virus genome, and NeoR or GFP gene was cloned in place of VP1.

G3 and HG23 cells using immunofluorescence assay (IFA) and Western blot analysis. Interestingly, both ProPol and NTPase were detected in discrete areas near and around the nucleus in the cells, suggesting they are the part of the replicase complexes in cells. Mature Pol (57 kDa) and Pro (19 kDa) or NTPase (40 kDa) were detected in cells by Western blot analysis suggesting authentic proteolytic processing of the non-structural polyprotein encoded in ORF1 in replicon-bearing cells (Chang *et al.*, 2006). Interestingly, during the experiment for the selection of colonies containing Norwalk virus replicon in BHK21 cells, only a limited number of colonies were formed (Chang *et al.*, 2006). These results suggested that few cells could apparently support virus replication, and it was possible that severe growth restrictions were present in the cells. This observation may be related to the inability to grow human noroviruses in all cell lines examined thus far (Duizer *et al.*, 2004).

The role of adaptive mutations in the Norwalk virus replicon

In hepatitis C virus (HCV) replicon-bearing cells, it has been observed that within selected cells, HCV RNAs had acquired adaptive mutations that increased the efficiency of colony formation (Blight *et al.*, 2000; Blight *et al.*, 2003; Krieger *et al.*, 2001). The adaptive mutations that increased the replication of HCV were located in NS3 and NS5A, and these changes were proposed to affect the possible interactions between cellular and viral proteins important for RNA replication (Krieger *et al.*, 2001). It is possible that such adaptive mutations occurred in the Norwalk virus genome in replicon-bearing cells, but sequence analysis of the replicon RNA purified from G3 or HG 23 cells in different passage numbers failed to identify them. In G3 cells, there were two nucleotide changes that resulted in two amino acid changes: one in ORF1 (N-terminal protein) and one in ORF3 (VP2) (Chang *et al.*, 2006). Additional mutations were detected in the Norwalk virus genome in HG23 cells that included four HG23 cell-specific mutations in ORF1 (one amino acid each in the N-terminal protein, NTPase, Pro and Pol) and three additional mutations (two amino acids and one silent mutation) in the ORF3 region. However, there were no significant changes in the expression of Norwalk virus RNA or proteins in both G3 and HG23 cell lines with various passage numbers (up to 100 passages), suggesting little evidence for adaptive mutations that increase viral replication.

Infectivity of RNA isolated from Norwalk virus replicon-bearing cells

A colony-forming assay was developed to assess the infectivity of RNA transcripts derived from pNV-Neo and RNA purified from replicon-bearing cells (Chang *et al.*, 2006). Transfection of the capped RNA transcripts or RNA replicon extracted from HG23 cells into fresh BHK21 cells produced viable cell colonies in the presence of G418, suggesting those RNAs are infectious. An enzyme-based immunostaining technique that used Norwalk virus ProPol-specific antiserum confirmed the expression of Norwalk virus proteins in the cells of the colonies. The transfection of BHK21 cells with replicon RNA isolated from HG23 cells resulted in a greater number of selected colonies than the transfection of BHK21 cells with capped RNA transcripts (pNV-Neo) (Chang *et al.*, 2006). However, treatment of the replicon RNA with proteinase K prior to transfection abolished the selection of colonies in the presence of G418. This result suggested that the infectivity of replicon RNA purified from cells was enhanced by the presence of the VPg protein.

Interferon and the Norwalk virus replicon

Certain evasion mechanisms of intracellular host defence used by the virus during infection were identified using a similar cell-based replicon system for HCV (Blight *et al.*, 2000; Foy *et al.*, 2003; Gale and Foy, 2005). The NS3/4 protein of HCV was demonstrated to block interferon regulatory factor (IRF)–3 activation, thereby inhibiting the production of IFN-β (Foy *et al.*, 2003). In addition, the HCV NS5A protein, an inhibitor of double-stranded RNA-dependent protein kinase (PKR), was an important site where adaptive mutations occurred that allowed for higher levels of replication in HCV replicon-bearing cells (Blight *et al.*, 2000; Gale and Foy, 2005). In Norwalk virus replicon-bearing cells, several mutations in the genome during the passaging of cells were identified, but none of them was related to adaptive mutations which allowed higher levels of replication (Chang *et al.*, 2006). It was demonstrated that the presence of the HCV replicon in the cells inhibited cellular innate immune responses (Blight *et al.*, 2000; Foy *et al.*, 2003). However, unlike HCV, the Norwalk virus replicon in the cells failed to inhibit the ability of the cell to mount an innate immune response stimulated by Sendai virus infection as measured with the reporter plasmids under the control of the interferon-sensitive responsive element (pISRE-TA-luc) and the promoter of IFNβ (pIFNβ-TA-luc) (Chang *et al.*, 2006). In addition to the lack of adaptive mutations, this observation suggested that Norwalk virus may lack an active evasion strategy to counteract host cellular defences involving the IFN pathways. It has been reported that innate immunity plays an important role in the control of MNV infection (Karst *et al.*, 2003; Wobus *et al.*, 2004a). Mumphrey *et al.* (2007) showed that pre-treatment of IFN-α in RAW264.7 cells prior to infection with MNV resulted in the reduction of viral replication. The pretreatment of IFN-α in CRFK cells also reduced the replication of FCV transiently (unpublished observation from our laboratory). For Norwalk virus, IFNs efficiently reduced virus replication in replicon-bearing cells (G3 and HG23) (Chang and George, 2007; Chang *et al.*, 2006). Both IFN-α or γ inhibited Norwalk virus protein expression in a dose-dependent manner in HG23 cells (Chang and George, 2007; Chang *et al.*, 2006). The effective dose of IFN-α or -γ for reducing the replication of Norwalk virus by 50% (ED_{50}) compared to non-treated (mock) control cells at 72 h was calculated to be approximately 2 units/ml or 20 units/ml, respectively (Chang and George, 2007; Chang *et al.*, 2006). So far, there has been no report of a direct mechanism which may be used by caliciviruses to counteract the innate immune response in cells. It has been shown that viruses known to lack anti-innate immunity mechanisms (such as IFN-sensitive viruses) often do not grow well and show severe growth restriction in cells even in the presence of virus receptors (Durbin *et al.*, 1996; Garcia-Sastre *et al.*, 1998a,b). It is possible that the lack of such mechanisms by Norwalk virus might be related to its fastidious characteristics in cell culture.

Effect of ribavirin on the Norwalk virus replicon

Replicon-bearing cells can be used as a platform to screen potential antivirals. Ribavirin (1-β-D-ribofuranosyl-1,2,4-triazole-3-carboxamide), a synthetic guanosine analogue, is well known for antiviral actions against a range of DNA and RNA viruses, including HCV and respiratory syncytial virus (Graci and Cameron, 2006; Sidwell *et al.*, 1972). The effects of ribavirin in the replication of Norwalk virus was examined in HG23 cells. The RNA genome or proteins of Norwalk virus were monitored for virus replication. The expression of Norwalk virus RNA or protein was significantly reduced by the treatment of ribavirin at concentrations above 10 μM (Fig. 7.2) (Chang and George, 2007). The ED_{50} was calculated to be approximately 40 μM at 72 h of ribavirin treatment (Fig. 7.2), which was comparable to the ED_{50} of other viruses (Markland *et al.*, 2000). The replication of MNV in RAW264.7 cells was also significantly reduced by the treatment with ribavirin at a concentration above 50 μM (Chang and George, 2007). The antiviral effect of ribavirin was associated with the depletion of GTP in the cells, thereby reduced virus replication (Chang and George, 2007). The co-treatment with ribavirin and IFN-α in HG23 cells demonstrated that there was an additive effect by the co-treatment (Chang and George, 2007), suggesting that IFNs and ribavirin may represent possible therapeutic options for norovirus gastroenteritis.

Norwalk virus replicons expressing reporter genes

Norwalk virus replicon systems with GFP or renilla luciferase (RL) were generated. Plasmids similar to pNV-Neo based on plasmid NV101, pNV-GFP (Fig. 7.1B) or pNV-RL encoding GFP or RL, respectively, were generated. As it was shown with pNV-Neo, an initial transfection study with these plasmids demonstrated that only a few cells seemed to support virus replication (Table 7.2) (Chang *et al.*, 2008). Without adapted mutation as described above, it was hypothesized that there may be endogenous or exogenous factors (such as bile acids in Cowden PEC) which may enhance virus replication. It has been shown that vesiviruses encode a unique protein in the subgenomic region designated as LC. Because

A

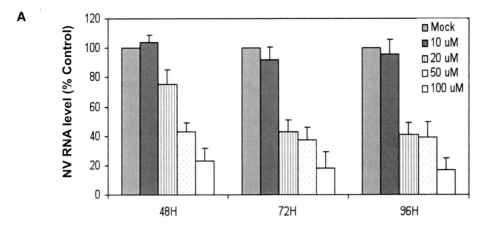

B

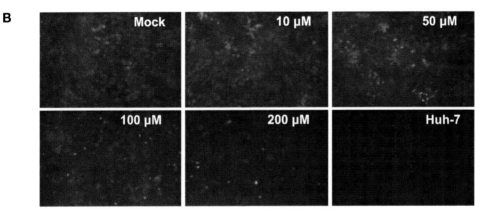

Figure 7.2 Use of Norwalk virus replicon-bearing cells to assay potential antivirals against noroviruses. In this figure, the effect of ribavirin was studied. For antiviral assay, Norwalk virus replicon-bearing cells (HG23) were treated with various concentrations of ribavirin for up to 96 h. (A). Virus replication was assessed by detecting Norwalk virus genome using real time qRT-PCR. The reduction of viral genome by ribavirin was calculated by the comparison to that with mock-treatment. (B). Virus replication was assessed by the expression of neomycin phosphotranferase II using immunofluorescent assay. Parental Huh-7 cells served as a negative control. (taken from Chang and George, 2007, with permission).

vesiviruses grow efficiently in cell culture, it was hypothesized that LC might be associated with the efficient replication of vesiviruses in cell culture and promote the replication of human norovirus in cells. Studies of Norwalk virus replication were performed while expressing a high level of LC *in trans* using a plasmid encoding LC under the control of the CMV promoter (pCI-LC) (Chang *et al.*, 2008). Virus replication assays were performed by transfecting pNV-GFP or pNV-LC alone or with pCI-LC in MVA-T7 infected cells, and then virus replication was assessed as the expression of GFP or LC.

The transfection of pNV-GFP or pNV-RL into Vero cells did not produce GFP-positive cells or RL expression above the level of the control MVA-T7 infected cells as measured by flow cytometry analysis or luciferase assay, respectively (Table 7.2). However, when pNV-GFP or pNV-RL was co-transfected with pCI-LC into MVA-T7 infected cells, the number of cells expressing GFP or RL was significantly increased (Table 7.2) (Chang *et al.*, 2008). The negative control using pNV-GFPΔGDD or pNV-RLΔGDD (mutant plasmids containing Norwalk virus genome with deleted Pol motif GDD) with or without pCI-LC confirmed that the expression of GFP or RL was mediated by viral Pol. Because it was well documented that the replication of noroviruses was significantly susceptible to type I IFN, the

Table 7.2 Expression of renilla luciferase after transfection of pNV-RL or pNV-RLΔGDD alone or with pCI-LC, pCI-V or pCI-LC+pCI-V into MVA-T7 infected Vero cells (Taken from Chang et al., 2008, with permission)

Plasmid	Fold induction in renilla luciferase (mean number ± standard deviation)[a]
pCI	1
pNV-GFP	0.95 ± 0.1
pNV-RL	1.5 ± 0.7
pNV-RLΔGDD	0.90 ± 0.3
pNV-RL/pCI-LC	7.1 ± 2.1**
pNV-RL/pCI-V	4.8 ± 2.4*
pNV-RL/pCI-LC+pCI-V	20.2 ± 8.7**
pNV-RLΔGDD/pCI-LC+pCI-V	1.1 ± 0.2

a   Cell lysates were prepared after 20 h following transfection for the renilla luciferase activity. Standard deviations were calculated with at least three independent measurements.

*$P < 0.05$; **$P < 0.01$ compared with pNV-RL group.

role of IFN in the replication of Norwalk virus was investigated in the replicon system. The plasmid expressing the V protein (pCI-V) of SV5 which down-regulates IFN system by degrading STAT1 (Young et al., 2003) was used for this study. Like LC, V protein also promoted the replication of Norwalk virus in this system but with much less efficiency than LC (Chang et al., 2008). This suggests that cells lacking innate immunity may better support the replication of Norwalk virus. Interestingly, there were synergistic effects with LC and V on promoting Norwalk virus replication: co-transfection of three plasmid, pNV-GFP (or pNV-RL), pCI-LC and pCI-V yielded significantly higher replication than co-transfection of pNV-GFP (or pNV-RL) and pCI-LC or pCI-V (Table 7.2) (Chang et al., 2008). However, it is not clear how LC promoted the replication of Norwalk virus in cells. It is possible that the LC may interfere with the innate immune system in cells as V protein does, or LC may interact with viral proteins and/or RNA to promote the replication of Norwalk virus. Nonetheless, the promotion of Norwalk virus replication in the presence of LC or/and V protein may provide a clue to the isolation of fastidious human noroviruses in cells.

Mutagenesis studies to identify cis-acting elements involved in Norwalk virus replication

Because the co-transfection with three plasmids, pNV-GFP, pCI-LC and pCI-V yielded significant numbers of GFP expressing cells, this method was used to identify a potential cis-acting element at the start of the subgenomic region and 3′ end of the Norwalk virus genome. There are four untranslated bases (GUAA) at the beginning of the sgRNA of Norwalk virus, and a series of substitution mutants were generated to examine each base in virus replication (Chang et al., 2008). The first base (G) was essential for virus replication because when the base was mutated to U, A, or C, few GFP-positive cells were seen (Chang et al., 2008). For the second base (U), the mutation yielded mixed results: while the mutation to A or C abolished the replication, the mutation to G resulted in comparable numbers of GFP-positive cells to that of the parental pNV-GFP. The third (A) or fourth base (A) did not appear to be essential for virus replication (Chang et al., 2008). A series of deletion mutations targeting the regions between the stop codon of GFP and the end of the genome was generated for mapping RNA elements at the 3′ end of the Norwalk virus genome essential for virus replication. Co-transfection study with each mutant plasmid demonstrated that the 3′ untranslated region (UTR) and the last part of the RNA sequences encoding ORF3 were essential, which contains *cis*-acting elements for virus replication, consistent with a previous study conducted with FCV (Sosnovtsev *et al.*, 2005). These results demonstrated that the Norwalk virus replicon system with reporter genes provided an important tool for the study of virus replication in the absence of cell culture system.

Conclusion

Reverse genetics and replicon systems of caliciviruses have provided an important tool for the study of virus replication. For uncultivable Norwalk virus, replicon-bearing cells and replicon with reporter genes were established and have provided important tools to the study of antivirals and virus replication in cells such as identification of cellular or viral factors which promote virus replication and *cis*-acting elements in viral genomes. Further studies with reverse genetics and replicon systems should yield important information for isolating human noroviruses, which is crucial to develop control measures such as vaccines and antivirals against norovirus infections.

Acknowledgements

This work was partly supported by NIH grants, P20 RR016443-07 and U01 AI081891-01. We thank David George for critical reading of this manuscript.

References

Asanaka, M., Atmar, R.L., Ruvolo, V., Crawford, S.E., Neill, F.H., and Estes, M.K. (2005). Replication and packaging of Norwalk virus RNA in cultured mammalian cells. Proc. Natl. Acad. Sci. U.S.A. *102*, 10327–10332.

Bailey, D., Thackray, L.B., and Goodfellow, I.G. (2008). A single amino acid substitution in the murine norovirus capsid protein is sufficient for attenuation in vivo. J. Virol. *82*, 7725–7728.

Black, D.N., Burroughs, J.N., Harris, T.J., and Brown, F. (1978). The structure and replication of calicivirus RNA. Nature *274*, 614–615.

Blight, K.J., Kolykhalov, A.A., and Rice, C.M. (2000). Efficient initiation of HCV RNA replication in cell culture. Science *290*, 1972–1974.

Blight, K.J., McKeating, J.A., Marcotrigiano, J., and Rice, C.M. (2003). Efficient replication of hepatitis C

virus genotype 1a RNAs in cell culture. J. Virol. *77*, 3181–3190.

Chang, K.O., and George, D.W. (2007). Interferons and ribavirin effectively inhibit Norwalk virus replication in replicon-bearing cells. J. Virol. *81*, 12111–12118.

Chang, K.O., George, D.W., Patton, J.B., Green, K.Y., and Sosnovtsev, S.V. (2008). Leader of the capsid protein in feline calicivirus promotes the replication of Norwalk virus in cell culture. J. Virol. *82*, 9306–9317.

Chang, K.O., Kim, Y., Green, K.Y., and Saif, L.J. (2002). Cell-culture propagation of porcine enteric calicivirus mediated by intestinal contents is dependent on the cyclic AMP signaling pathway. Virology *304*, 302Chang, K.O., Sosnovtsev, S.S., Belliot, G., Wang, Q., Saif, L.J., and Green, K.Y. (2005). Reverse genetics system for porcine enteric calicivirus, a prototype sapovirus in the Caliciviridae. J. Virol. *79*, 1409–1416.

Chang, K.O., Sosnovtsev, S.V., Belliot, G., King, A.D., and Green, K.Y. (2006). Stable expression of a Norwalk virus RNA replicon in a human hepatoma cell line. Virology *353*, 463–473.

Chaudhry, Y., Skinner, M.A., and Goodfellow, I.G. (2007). Recovery of genetically defined murine norovirus in tissue culture by using a fowlpox virus expressing T7 RNA polymerase. J. Gen. Virol. *88*, 2091–2100.

Daughenbaugh, K.F., Fraser, C.S., Hershey, J.W., and Hardy, M.E. (2003). The genome-linked protein VPg of the Norwalk virus binds eIF3, suggesting its role in translation initiation complex recruitment. EMBO J. *22*, 2852–2859.

Duizer, E., Schwab, K.J., Neill, F.H., Atmar, R.L., Koopmans, M.P., and Estes, M.K. (2004). Laboratory efforts to cultivate noroviruses. J. Gen. Virol. *85*, 79–87.

Durbin, J.E., Hackenmiller, R., Simon, M.C., and Levy, D.E. (1996). Targeted disruption of the mouse Stat1 gene results in compromised innate immunity to viral disease. Cell *84*, 443–450.

Farkas, T., Sestak, K., Wei, C., and Jiang, X. (2008). Characterization of a rhesus monkey calicivirus representing a new genus of Caliciviridae. J. Virol. *82*, 5408–5416.

Fernandez-Vega, V., Sosnovtsev, S.V., Belliot, G., King, A.D., Mitra, T., Gorbalenya, A., and Green, K.Y. (2004). Norwalk virus N-terminal nonstructural protein is associated with disassembly of the Golgi complex in transfected cells. J. Virol. *78*, 4827–4837.

Flynn, W.T., and Saif, L.J. (1988). Serial propagation of porcine enteric calicivirus-like virus in primary porcine kidney cell cultures. J. Clin. Microbiol. *26*, 206–212.

Flynn, W.T., Saif, L.J., and Moorhead, P.D. (1988). Pathogenesis of porcine enteric calicivirus-like virus in four-day-old gnotobiotic pigs. Am. J. Vet. Res. *49*, 819–825.

Foy, E., Li, K., Wang, C., Sumpter, R., Jr., Ikeda, M., Lemon, S.M., and Gale, M., Jr. (2003). Regulation of interferon regulatory factor-3 by the hepatitis C virus serine protease. Science *300*, 1145–1148.

Gale, M., Jr., and Foy, E.M. (2005). Evasion of intracellular host defence by hepatitis C virus. Nature *436*, 939–945.

Garcia-Sastre, A., Durbin, R.K., Zheng, H., Palese, P., Gertner, R., Levy, D.E., and Durbin, J.E. (1998a). The role of interferon in influenza virus tissue tropism. J. Virol. 72, 8550–8558.

Garcia-Sastre, A., Egorov, A., Matassov, D., Brandt, S., Levy, D.E., Durbin, J.E., Palese, P., and Muster, T. (1998b). Influenza A virus lacking the NS1 gene replicates in interferon-deficient systems. Virology 252, 324–330.

Graci, J.D., and Cameron, C.E. (2006). Mechanisms of action of ribavirin against distinct viruses. Rev. Med. Virol. 16, 37–48.

Guix, S., Asanaka, M., Katayama, K., Crawford, S.E., Neill, F.H., Atmar, R.L., and Estes, M.K. (2007). Norwalk virus RNA is infectious in mammalian cells. J. Virol. 81, 12238–12248.

Guo, M., Chang, K.O., Hardy, M.E., Zhang, Q., Parwani, A.V., and Saif, L.J. (1999). Molecular characterization of a porcine enteric calicivirus genetically related to Sapporo-like human caliciviruses. J. Virol. 73, 9625–9631.

Guo, M., Hayes, J., Cho, K.O., Parwani, A.V., Lucas, L.M., and Saif, L.J. (2001). Comparative pathogenesis of tissue culture-adapted and wild-type Cowden porcine enteric calicivirus (PEC) in gnotobiotic pigs and induction of diarrhea by intravenous inoculation of wild-type PEC. J. Virol. 75, 9239–9251.

Hsu, C.C., Riley, L.K., Wills, H.M., and Livingston, R.S. (2006). Persistent infection with and serologic cross-reactivity of three novel murine noroviruses. Comp. Med. 56, 247–251.

Karst, S.M., Wobus, C.E., Lay, M., Davidson, J., and Virgin, H.W.t. (2003). STAT1-dependent innate immunity to a Norwalk-like virus. Science 299, 1575–1578.

Katayama, K., Hansman, G.S., Oka, T., Ogawa, S., and Takeda, N. (2006). Investigation of norovirus replication in a human cell line. Arch. Virol. 151, 1291–1308.

Krieger, N., Lohmann, V., and Bartenschlager, R. (2001). Enhancement of hepatitis C virus RNA replication by cell culture-adaptive mutations. J. Virol. 75, 4614–4624.

Liu, G., Ni, Z., Yun, T., Yu, B., Chen, L., Zhao, W., Hua, J., and Chen, J. (2008a). A DNA-launched reverse genetics system for rabbit hemorrhagic disease virus reveals that the VP2 protein is not essential for virus infectivity. J. Gen. Virol. 89, 3080–3085.

Liu, G., Ni, Z., Yun, T., Zhang, Y., Du, Q., Sheng, Z., Liang, H., Hua, J., Li, S., and Chen, J. (2006a). Rescued virus from infectious cDNA clone of rabbit hemorrhagic disease virus is adapted to RK13 cells line Chinese Sci. Bull. 51, 1698–1702.

Liu, G., Zhang, Y., Ni, Z., Yun, T., Sheng, Z., Liang, H., Hua, J., Li, S., Du, Q., and Chen, J. (2006b). Recovery of infectious rabbit hemorrhagic disease virus from rabbits after direct inoculation with in vitro-transcribed RNA. J. Virol. 80, 6597–6602.

Liu, G.Q., Ni, Z., Yun, T., Yu, B., Zhu, J.M., Hua, J.G., and Chen, J.P. (2008b). Rabbit hemorrhagic disease virus poly(A) tail is not essential for the infectivity of the virus and can be restored in vivo. Arch. Virol. 153, 939–944.

Markland, W., McQuaid, T.J., Jain, J., and Kwong, A.D. (2000). Broad-spectrum antiviral activity of the IMP dehydrogenase inhibitor VX-497: a comparison with ribavirin and demonstration of antiviral additivity with alpha interferon. Antimicrob. Agents Chemother. 44, 859–866.

Mumphrey, S.M., Changotra, H., Moore, T.N., Heimann-Nichols, E.R., Wobus, C.E., Reilly, M.J., Moghadamfalahi, M., Shukla, D., and Karst, S.M. (2007). Murine norovirus 1 infection is associated with histopathological changes in immunocompetent hosts, but clinical disease is prevented by STAT1-dependent interferon responses. J. Virol. 81, 3251–3263.

Neill, J.D., Sosnovtsev, S.V., and Green, K.Y. (2000). Recovery and altered neutralization specificities of chimeric viruses containing capsid protein domain exchanges from antigenically distinct strains of feline calicivirus. J. Virol. 74, 1079–1084.

Oglesby, A.S., Schaffer, F.L., and Madin, S.H. (1971). Biochemical and biophysical properties of vesicular exanthema of swine virus. Virology 44, 329–341.

Parwani, A.V., Flynn, W.T., Gadfield, K.L., and Saif, L.J. (1991). Serial propagation of porcine enteric calicivirus in a continuous cell line. Effect of medium supplementation with intestinal contents or enzymes. Arch. Virol. 120, 115–122.

Pesavento, P.A., Chang, K.O., and Parker, J.S. (2008). Molecular Virology of feline calicivirus. Vet Clin North Am. Small Anim. Pract. 38, 775–786.

Schaffer, F.L., Ehresmann, D.W., Fretz, M.K., and Soergel, M.I. (1980). A protein, VPg, covalently linked to 36S calicivirus RNA. J. Gen. Virol. 47, 215–220.

Sidwell, R.W., Huffman, J.H., Khare, G.P., Allen, L.B., Witkowski, J.T., and Robins, R.K. (1972). Broad-spectrum antiviral activity of Virazole: 1-beta-D-ribofuranosyl-1, 2, 4-triazole-3-carboxamide. Science 177, 705–706.

Sosnovtsev, S., and Green, K.Y. (1995). RNA transcripts derived from a cloned full-length copy of the feline calicivirus genome do not require VpG for infectivity. Virology 210, 383–390.

Sosnovtsev, S., Sosnovtseva, S., and Green, K.Y., eds. (1997). Recovery of feline calicivirus from plasmid DNA containing a full-length copy of the genome. (Weybridge: European Society for Veterinary Virology and the Central Veterinary Laboratory).

Sosnovtsev, S.V., Belliot, G., Chang, K.O., Onwudiwe, O., and Green, K.Y. (2005). Feline calicivirus VP2 is essential for the production of infectious virions. J. Virol. 79, 4012–4024.

Sosnovtsev, S.V., Belliot, G., Chang, K.O., Prikhodko, V.G., Thackray, L.B., Wobus, C.E., Karst, S.M., Virgin, H.W., and Green, K.Y. (2006). Cleavage map and proteolytic processing of the murine norovirus nonstructural polyprotein in infected cells. J. Virol. 80, 7816–7831.

Sosnovtsev, S.V., Garfield, M., and Green, K.Y. (2002). Processing map and essential cleavage sites of the nonstructural polyprotein encoded by ORF1 of the feline calicivirus genome. J. Virol. 76, 7060–7072.

Thumfart, J.O., and Meyers, G. (2002). Feline calicivirus: recovery of wild-type and recombinant viruses after transfection of cRNA or cDNA constructs. J. Virol. *76*, 6398–6407.

Ward, V.K., McCormick, C.J., Clarke, I.N., Salim, O., Wobus, C.E., Thackray, L.B., Virgin, H.W.t., and Lambden, P.R. (2007). Recovery of infectious murine norovirus using pol II-driven expression of full-length cDNA. Proc. Natl. Acad. Sci. U.S.A. *104*, 11050–11055.

Wei, C., Farkas, T., Sestak, K., and Jiang, X. (2008). Recovery of infectious virus by transfection of in vitro-generated RNA from tulane calicivirus cDNA. J. Virol. *82*, 11429–11436.

Wobus, C.E., Karst, S.M., Thackray, L.B., Chang, K.O., Sosnovtsev, S.V., Belliot, G., Krug, A., Mackenzie, J.M., Green, K.Y., and Virgin, H.W. (2004a). Replication of norovirus in cell culture reveals a tropism for dendritic cells and macrophages. PLoS Biol. *2*, e432.

Wobus, C.E., Karst, S.M., Thackray, L.B., Chang, K.O., Sosnovtsev, S.V., Belliot, G., Krug, A., Mackenzie, J.M., Green, K.Y., and Virgin, H.W.t. (2004b). Replication of Norovirus in Cell Culture Reveals a Tropism for Dendritic Cells and Macrophages. PLoS Biol. *2*, e432.

Wobus, C.E., Thackray, L.B., and Virgin, H.W.t. (2006). Murine norovirus: a model system to study norovirus biology and pathogenesis. J. Virol. *80*, 5104–5112.

Young, D.F., Andrejeva, L., Livingstone, A., Goodbourn, S., Lamb, R.A., Collins, P.L., Elliott, R.M., and Randall, R.E. (2003). Virus replication in engineered human cells that do not respond to interferons. J. Virol. *77*, 2174–2181.

Feline Calicivirus

Christine Luttermann and Gregor Meyers

Abstract

Feline calicivirus (FCV) represents an important pathogen of cats that has been studied extensively on the molecular level. FCV was the first calicivirus for which milestones like a reverse genetics system or the identification of a verified virus receptor were reached. Recently, great efforts were made to investigate unusual mechanisms of translation initiation driven by the RNA bound protein VPg or an RNA structure named TURBS.

Introduction

Feline calicivirus (FCV), along with San Miguel sea lion virus (SMSV) and vesicular exanthema of swine virus (VESV), was among the first members of the family *Caliciviridae* characterized by physicochemical and molecular techniques (Black *et al.*, 1978; Burroughs and Brown, 1978; Burroughs *et al.*, 1978a; Burroughs *et al.*, 1978b; Ehresmann and Schaffer, 1977; Ehresmann and Schaffer, 1979; Schaffer *et al.*, 1980a,b; Studdert, 1978). The reason for this choice in the early phase of calicivirus research was the fact that these three viruses could be easily propagated in tissue culture cells, in contrast to many other family members, especially the human viruses. The development of modern molecular tools led to the intensification of research, especially in the study of FCV. In many areas, investigations of FCV together with rabbit haemorrhagic disease virus (RHDV) paved the way for the analysis of other caliciviruses. Even though the human caliciviruses have become the major target of research in the past few years and molecular analysis of murine norovirus (MNV), a cell culture compatible

surrogate for human noroviruses, is booming, FCV still represents the best-characterized member of the *Caliciviridae* thus far. The advantages of FCV as a surrogate system for other caliciviruses are that it is easily propagated in tissue culture, it produces high yields of virus particles and a reverse genetics system is available (Sosnovtsev and Green, 1995; Thumfart and Meyers, 2002a). The following chapter summarizes published molecular data on FCV that, despite the many differences among individual family members, provides a framework for understanding the biology of other caliciviruses.

Classification

Originally, caliciviruses were regarded as members of the *Picornaviridae* but when the results of molecular analyses provided an increasing bulk of evidence that there were major differences between both virus groups, a reclassification resulted in formation of a new virus family, *Caliciviridae* (Studdert, 1978; Schaffer *et al.*, 1980a). Currently, the family *Caliciviridae* consists of four accepted genera and FCV represents a member of the genus *Vesivirus* (Green *et al.*, 2000). Examples of other members in this genus are VESV SMSV. A typical characteristic of vesiviruses and providing the basis for their classification is their genome organization that is unique among caliciviruses (see below). Even though the genus name is derived from VESV (in recognition of its historical place in the family), FCV represents by far the best-studied member of the genus and is often regarded as the preferred prototype of the vesiviruses.

Disease and control

The following paragraph intends to provide a brief overview of FCV induced disease and FCV control. More information on these aspects can be found in a recent review by Radford *et al.* and the references therein (Radford *et al.*, 2007).

FCV infects cats via the nasal, oral or conjunctival route and replicates predominantly in oral and respiratory tissues (Radford *et al.*, 2007). The infection is associated with a wide range of clinical symptoms ranging from unapparent infections to relatively mild oral and upper respiratory tract symptoms. The most characteristic lesion is oral ulceration. Additionally, frequent signs of FCV-induced disease include conjunctivitis and respiratory disorders accompanied by ocular and nasal discharge. Less often though more severe respiratory infections are observed that can be lethal in young kittens. Other signs of disease induced by FCV are acute arthritis, sometimes accompanied by lameness and icterus. Upon infection of pregnant animals the virus can be transmitted to the foetus and lead to death of the embryo. Even though there is evidence that the severity of the symptoms induced by FCV infection is dependent on the individual virus strain, an allocation of different FCV strains to biotypes has not been done so far.

Recently, highly virulent strains of FCV were identified in the US and UK (Radford *et al.*, 2007; Ossiboff *et al.*, 2007; Foley *et al.*, 2006; Pesavento *et al.*, 2004; Hurley *et al.*, 2004; Coyne *et al.*, 2006b; Hurley *et al.*, 2004; Pedersen *et al.*, 2000). These viruses induce systemic disease with haemorrhagic fever, resulting in high rates of mortality (~50%). The symptoms observed in adult cats are more severe than those seen in young animals (Hurley and Sykes, 2003). This syndrome is designated 'virulent systemic disease' (VSD) and is reminiscent of rabbit haemorrhagic disease (RHD) that is induced by the calicivirus RHDV (Hurley and Sykes, 2003; Xu and Chen, 1989). It is noteworthy that RHD is not seen in young rabbits and that a harmless ancestor of RHDV circulated for a long time in the rabbit population before suddenly a highly pathogenic variant arose in the late 1980s (Capucci *et al.*, 1996). The molecular mechanisms responsible for the observed change in virulence are not known for either RHDV or FCV.

Most cats shed FCV in oropharyngal secretions for approximately 30 days post infection. After this period, the virus is usually eliminated from the host organism (Wardley and Povey, 1977). Rather rarely, a so-called carrier state is observed, in which virus shedding continues beyond the above-mentioned 30-day period (Wardley, 1976). In some cases, cats were reported to shed virus for life. Such animals can be regarded as a constant source of virus and are probably responsible for keeping the virus circulating in a rather small population.

Generally, FCV is widespread in the cat population (Coyne *et al.*, 2006a; Wardley *et al.*, 1974; Mochizuki *et al.*, 2000). As expected, the chance for infection of an animal in a given population correlates with the frequency of contacts with different other individuals, so that the actual frequency of antibody positive animals increases with group size and usually ranges between 10% and 40%. However, up to 90% prevalence has been reported in some cases (Coyne *et al.*, 2006a; Radford *et al.*, 2003). It is generally accepted that there is no reservoir for FCV in alternative hosts. Thus, the virus is kept within the population via spread from an acutely infected cat or a long-term carrier to other animals. Since, however, the virus is resistant to environmental influences, shed virus can persist in the environment for several days or even weeks, dependent on the temperature and other conditions (Clay *et al.*, 2006; Doultree *et al.*, 1999; Duizer *et al.*, 2004). Thus, indirect transmission of the virus can occur by contamination of the environment with secretions from infected animals.

The only reasonable means for prevention and control of FCV infections is vaccination. Several types of FCV vaccines prepared from FCV infected cell culture material are commercially available that are considered safe and effective with regard to preventing or at least reducing FCV-induced disease. These vaccines comprise either live attenuated FCV or inactivated viral antigen. However, it has to be stressed that these vaccines do not prevent productive infection and establishment of a carrier state is also possible upon infection of vaccinated animals (Pedersen and Hawkins, 1995). Moreover, vaccination does not prevent FCV-associated VSD (Coyne *et al.*, 2006b; Hurley *et al.*, 2004; Pedersen *et al.*,

2000). A further flaw of the available vaccines is that cross-reactivity to different FCV strains is only partially seen (Pedersen and Hawkins, 1995; Poulet *et al.*, 2005; Povey and Ingersoll, 1975). Taken together, there is an obvious need for improved FCV vaccines inducing sterile immunity against a broad range of virus isolates.

The virion

Virions of many members of the family *Caliciviridae* including FCV show a characteristic morphology with cup-shaped depressions from which the family got its name that is derived from the Latin word 'calix' for cup or chalice (Fig. 8.1) (Green *et al.*, 2000). The viral particle is non-enveloped and has a size of 30–40 nm in diameter, dependent on the preparation method. Cryo-electron microscopy (cryo-EM) analysis of FCV and comparison of the reconstructed images with the crystal structure of SMSV particles (Chen *et al.*, 2006) provided a rather well defined picture of the three dimensional structure of the virions (Bhella *et al.*, 2008). The capsid is composed of 90 dimers of the major capsid protein (VP1) assembled in a T=3 icosahedral shell. VP1 consists of two different domains termed shell (S) and protruding (P). The S domain makes up the inner shell of the capsid whereas the P domain forms arch-like structures protruding from the capsid surface. The architecture resulting from

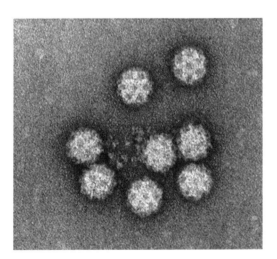

Figure 8.1 Electron microscopy picture of purified FCV particles (kindly provided by Frank Weiland, FLI Tübingen).

these protrusions is responsible for the typical calicivirus morphology.

Expression of VP1 in the absence of other viral proteins results in self-assembly of virus like particles (VLPs) (Geissler *et al.*, 2002; Di Martino *et al.*, 2007). This is a common feature of all caliciviruses. Different systems can be used for the expression. Using EM, the morphology of VLPs is indistinguishable from virions prepared from the supernatant of infected cells. An obvious frequently observed difference, however, is the absence of closed particles in many VLP preparations indicating that the final steps of virion formation are dependent on the presence of viral RNA and probably other viral proteins.

Based on the degree of amino acid conservation between different calicivirus capsid proteins, the capsid was divided into five regions designated B, C, D, E and F (from amino terminus to carboxy terminus; region A is identical to the so-called leader peptide of the capsid protein precursor and is cleaved off; see below) (Neill, 1992). Region B represents residues 1 to 275 of VP1 (121 to 396 of the precursor). It is moderately conserved and corresponds roughly to the above mentioned S domain building the inner shell of the particle (Bhella *et al.*, 2008; Prasad *et al.*, 1994, 1999). Regions C to F are equivalent to the P domain, which can be further subdivided into two subdomains P1 and P2 with P2 representing a ~130 amino acid insert into the P1 sequence (residues ~280 to 410 in the Norwalk virus crystal structure). The P2 domain is almost equivalent to the E region and contains two hypervariable regions possibly involved in receptor binding (see below) (Radford *et al.*, 1999b). Regions C, D and F are distinguished by their different degrees of sequence conservation with highly conserved sequences in D and considerably variable sequences in C and F. The last three regions are within the P1 domain of the capsid, but have not been linked to specific structural functions.

FCV infected cats mount a neutralizing antibody response within approximately one week (Radford *et al.*, 2007). As expected, all serum neutralizing activity analysed so far could be contributed to antibodies directed against VP1. A significant number of publications report on the identification of epitopes in VP1 that are recognized by sera from cats, rabbits or mouse monoclonal antibodies (Geissler *et al.*, 2002; Guiver *et al.*, 1992; Milton *et al.*, 1992; Neill *et al.*, 2000; Tohya *et al.*, 1991, 1997; Carter *et al.*, 1989; Seal *et al.*, 1993). A recombinant polypeptide corresponding to amino acids 287 to 396 of VP1 (408 to 517 of the precursor) was able to elicit an FCV neutralizing antibody response in rabbits (Guiver *et al.*, 1992). Within the same region (residues 301 to 337) the epitopes for two neutralizing mouse monoclonal antibodies were detected (Milton *et al.*, 1992). Analyses of sequences from neutralization resistant FCV strains identified residues (residues 320 to 344) that were critical in the formation of four linear B-cell epitopes. Additional residues (residues 372/373) outside this region have been mapped also as forming important conformational epitopes (Tohya *et al.*, 1997). In a methodical approach, a random expression library of VP1 from the vaccine strain F9 was established in a lambda phage and served for detection of linear B-cell epitopes (Radford *et al.*, 1999b). The library was screened with sera from cats that had been infected with the F9 virus or other FCV strains or that had been vaccinated with F9-based vaccines. Four regions containing linear B-cell epitopes [antigenic sites (ags) 1 to 4] were identified. Two of these regions (ags 2 and 3) are located within a stretch of 13 amino acids (residues 324 to 336) located in the amino terminal hypervariable sequence within region E. These results align well with data resulting from the analysis of neutralization resistant viruses (Tohya *et al.*, 1997). Interestingly, ags1 (in region D) and ags4 (in the conserved part of region E) are rather conserved among different isolates and thus represent the first identified antigenic regions of FCV VP1 that show considerable sequence conservation.

The overall homology of FCV capsid protein sequences is approximately 70%, but variation is of course higher in the hypervariable regions (Thumfart and Meyers, 2002a). The so far defined immunoreactive region of VP1 consists of a stretch of roughly 150 amino acids spanning the variable parts of regions C and E. Because the neutralizing epitopes are clustered in the hypervariable regions, it is not surprising that feline sera against FCV often show only limited cross-reactivity so that vaccination with a single FCV strain may not afford protection against a broad panel

of field strains (Sato *et al.*, 2002; Horimoto *et al.*, 2001; Glenn *et al.*, 1999; Radford *et al.*, 1997; Geissler *et al.*, 1997; Seal, 1994). Further evidence for this hypothesis was gained by a reverse genetics approach in which viruses with chimeric variants of VP1 were generated and analysed with regard to their antibody reactivity (Neill *et al.*, 2000). Even though domain exchange was not able to confer complete recognition by the parental antiserum in most cases, these experiments confirmed once more the importance of region E and especially the hypervariable stretches within this region for virus neutralization. Nevertheless, the identification of conserved antigenic sites (Povey and Ingersoll, 1975; Tohya *et al.*, 1991; Povey, 1974) and especially the description of a mouse monoclonal antibody directed against a conformational epitope, which consistently neutralizes different FCV isolates (Tohya *et al.*, 1991), are promising for future developments towards improved FCV vaccines.

A second protein termed VP2 is present in the FCV virion in much lower amounts than VP1 (Luttermann and Meyers, 2007; Sosnovtsev and Green, 2000). The function of this minor capsid protein is unclear. The number of VP2 molecules present in FCV particles was determined to be one to two or about eight in the two different publications. It was shown for other caliciviruses that assembly of VLPs is possible in the absence of VP2 (Green *et al.*, 1997; Guo *et al.*, 2001; Jiang *et al.*, 1995, 1999; Laurent *et al.*, 1994; Leite *et al.*, 1996; Sibilia *et al.*, 1995; Williams *et al.*, 1997) and co-expression of VP1 and VP2 does not lead to formation of VLPs of altered morphology, so that VP2 has no obvious assembly inducing or structure determining function. The only indication for VP2 to exhibit a structural role was obtained for Norwalk virus, for which VP2 was shown to enhance the stability of VLPs (Bertolotti-Ciarlet *et al.*, 2003). It was therefore to be investigated whether VP2 is an essential viral protein at all. Reverse genetics experiments showed that VP2 is necessary for formation of infectious FCV particles. However, the protein does not need to be expressed from the viral RNA but can be provided *in trans* (Sosnovtsev *et al.*, 2005). Interestingly, complete deletion of the VP2 coding sequence from the viral genome was deleterious for RNA replication indicating the presence of a *cis*-acting RNA element in the open reading frame (ORF) 3 (see below).

A full-length genomic RNA and a 3′ co-terminal subgenomic RNA (sgRNA) have each been found in association with calicivirus virions (Meyers *et al.*, 1991; Neill, 2002). Thus, two species of viral particles that contain either genomic RNA or sgRNA are released from infected cells. Because of the different lengths of the two RNA species, the amount of encapsidated RNA is different and consequently the density of the two types of particles is different so that they can be separated by ultracentrifugation techniques. However, the centrifugation analyses gave no indication for the presence of particles that contain both the genomic RNA and sgRNA, so that it is likely that one virus particle encapsidates only one molecule of RNA. The presence of both genomic RNA and sgRNA in virus particles can only be explained when both RNAs carry the encapsidation signal. Thus, encapsidation of FCV RNA is either promoted by a *cis* acting RNA element located within the 3′ terminal third of the genome or by the VPg that is found at the 5′ end of both types of viral RNAs (see below). For FCV, the purification of a fraction of empty capsids was also reported (Zhou *et al.*, 1994).

Host cell receptor

Specific recognition of a susceptible host cell is a basic prerequisite for successful virus propagation. The first indications of such a specific interaction were published for FCV in 1994 (Kreutz *et al.*, 1994). The authors used virus binding and binding competition assays to demonstrate specific interaction of FCV with Crandell-Reese feline kidney (CRFK) cells that are highly susceptible to FCV infection. These studies showed that treatment of cells with proteases resulted in loss of binding, suggesting that the receptor was probably a protein.

The definition of a virus receptor as a single cell surface molecule promoting virus binding and penetration has been challenged in the past two decades for many viruses by experimental results showing that infection of a cell is actually a multistep process involving several receptor or co-receptor molecules. Also for FCV, two different types of host cellular surface proteins have now been described as important for infection. The

first one is a glycoprotein of the immunoglobulin-like superfamily and belongs to a class of cellular adhesion proteins that are quite frequently found to be involved in virus entry. For FCV, feline junctional adhesion molecule 1 (fJAM-1) is important for infection (Makino *et al.*, 2006). This molecule is composed of an ectodomain with two immunoglobulin-like domains (D1 and D2), a transmembrane region and a rather short cytoplasmic tail (Prota *et al.*, 2003; Mandell and Parkos, 2005; Ebnet *et al.*, 2004). The evidence for fJAM-1 receptor function was obtained in a set of experiments typically conducted for identification of virus receptors (Makino *et al.*, 2006). The authors established a retroviral expression library with sequences of CRFK cells. After transduction with the retroviruses, P3U1 cells expressing putative FCV receptor fragments were identified via their ability to adhere to plates coated with fixed FCV that was specifically bound via a monoclonal antibody. Adherent cells forming colonies were expanded and analysis of the sequence introduced via the retroviral vector led to discovery of fJAM-1. Specific interaction of FCV with fJAM-1 was shown by virus binding to hamster cells expressing the feline protein. Virus binding and infection of cells was prevented by fJAM-1 specific antibodies. To obtain final proof for the receptor function of fJAM-1, the protein was expressed in non-permissive cells. FCV was known before to be able to replicate in non-permissive cells upon transfection of viral RNA (Makino *et al.*, 2006). Expression of fJAM-1 in hamster lung or 293T cells allowed infection of the cells and the completion of a full replication cycle.

In a recent paper, structural aspects of the interaction of FCV particles with fJAM-1 were analysed by cryo-EM studies with FCV particles incubated with soluble receptor molecules (Ossiboff and Parker, 2007). The results demonstrated that the putative receptor interacted with the regions of VP1 containing epitopes critical for virus neutralization and thus supported the idea of a specific virus/receptor interaction of the two partners. Taken together, the published data provided good evidence that fJAM-1 represents a cellular receptor for FCV that is crucial for the infection process.

JAM-1 is expressed in a variety of cell types including leucocytes, endothelial and epithelial cells and is found to be specifically associated with apical tight junctions (Mandell and Parkos, 2005; Ebnet et al., 2004). This distribution does not fit with the tropism of the virus in the natural host, since FCV is known to cause local infections of the respiratory tract with the exception of strains causing VSD. Along those lines it also has to be mentioned that some FCV strains are able to infect simian (Vero) cells and there is good evidence that this is achieved via simian JAM-1 (sJAM-1) (Makino et al., 2006). Human JAM-1 contains some differences to the simian protein that are thought to be responsible for the absence of FCV infection of human cells. However, in vivo the host range of FCV is characteristically restricted to cats. These data indicate that, even though fJAM-1 is essential for infection and a critical determinant of FCV host specificity, it alone is not sufficient to explain host and cell tropism of the virus. Accordingly, further determinants have to exist and rather recently it was published that α2,6-linked sialic acid acts as a second receptor for FCV (Stuart and Brown, 2007). The first evidence for involvement of carbohydrate moieties in FCV binding and infection was obtained by the observation of dramatically reduced infection rates in consequence of destruction of such molecules via periodate treatment. The importance of sialic acid was then shown by neuraminidase treatment of cells. The use of an enzyme derived from Vibrio cholerae, that cleaves both α2,6-linked and α2,3-linked sialic acid reduced infection rates by about 77%. Since treatment with sialidase S from Streptococcus pneumoniae, which cleaves α2,3-linked sialic acid only, was not able to hinder FCV infection, α2,6-linked sialic acid was identified as a specific binding partner for FCV. Further experimental work using different lectins supported this conclusion and experiments with 1-phenyl-2-decanoylamino-3-morpholino-1-propanol (PDMP), a specific inhibitor of glycolipid biosynthesis, showed that glycolipids are not important for FCV infection, whereas tunicamycin, an inhibitor of N-glycosylation, was able to reduce virus binding and infection. In conclusion, the published data show that FCV uses glycoprotein bound α2,6-linked sialic acid as a second cellular receptor. It is not clear so far, to which degree this glycoprotein influences FCV tropism, whether the combination of this receptor and fJAM-1 alone determine FCV tropism or whether additional factors are important. As with other viruses, the prerequisites for FCV permissiveness might be complex and will have to be further investigated in future experiments.

Virus entry

After attachment of a virus to its cellular receptor(s) penetration of the host cellular membrane barrier and uncoating of the viral genome have to occur to allow initiation of viral replication. Virus entry into the host cell can either occur at the plasma membrane via a receptor mediated penetration mechanism or is initiated by receptor-mediated endocytosis followed by penetration of an intracellular (vesicle) membrane. For FCV, it has been reported that receptor binding induces endocytosis (Kreutz and Seal, 1995; Stuart and Brown, 2006). In a first report, it was shown that addition of chloroquine, a lysosomotropic agent that prevents acidification of intracellular vesicles, inhibited the production of infectious FCV progeny virus when added to the culture before or during the early phase of infection (Kreutz and Seal, 1995). The effects of chloroquine were reversible, so that elimination of the drug allowed virus infection and replication to proceed. Moreover, chloroquine does not principally interfere with later stages of the virus replication since its addition to the culture at 2 hours post infection had little effect. These results indicated that entry of FCV into CRFK cells requires a low pH-dependent step. A recent report confirmed and expanded these results (Stuart and Brown, 2006). The authors used a set of reporter systems for different steps and pathways of endocytosis and a number of specific inhibitors. Chlorpromazine, a drug which blocks endocytosis via clathrin-coated pits, inhibited FCV infection. The same effect was obtained by ammonium chloride, chloroquine and bafilomycin A1, all of which block the acidification of endosomes. All these drugs did not prevent virus binding to CRFK cells. Since also nystatin, an inhibitor of lipid raft/caveolae endocytosis, had no negative effect on FCV infection, the authors were able to conclude that FCV endocytosis is achieved via clathrin-coated pits and requires endosome acidification. This was also confirmed by the study of dominant negative mutants of rab5 and eps15, two essential components of the clathrin-dependent endocytosis pathway (Stuart and Brown, 2007). Expression of these mutant proteins in CRFK cells was able to reduce FCV infection by approximately 80–90%. These values are very similar to the effects seen with the inhibitors and provide final evidence for the importance of clathrin dependent endocytosis for FCV infection.

Also Na+/H+ ion channel inhibitors were able to block FCV infection whereas a Ca^2+ channel blocker had no effect (Stuart and Brown, 2007). In contrast to the inhibitors of vesicle acidification, the ion channel inhibitors and the microfilament disrupting agent cytochalasin D were able to block FCV propagation when added to the culture post infection. Since the latter drugs had to be present during the process of virus replication, it is likely that they do not interfere with virus entry, but affect a later step of the replication cycle like protein translation or RNA replication. This can also be concluded from the fact that these inhibitors were able to block virus replication even after transfection of viral RNA which stands in marked contrast to the results obtained with the inhibitors of virus entry that had no effect on virus propagation initiated by RNA transfection.

It was shown for some non-enveloped RNA viruses like poliovirus and rotavirus that the host cell membranes become permeabilized for certain drugs during virus infection. The same was observed for FCV when α-sarcin was added to the infection mixture (Stuart and Brown, 2006). The effect was multiplicity of infection dependent, but saturable. Similar to the observations made for virus infection, permeabilization was also prevented by inhibitors of endosome acidification, which shows a direct connection between the drug and virus entry. The mechanism underlying this finding is not clear and will probably only be understood when the details of FCV membrane penetration are clarified.

So far, all available data support the idea that FCV enters the host cell via receptor-mediated endocytosis and subsequent acid induced membrane penetration, but the details are not known. There are only hints from the cryo-EM analyses that binding of fJAM-1 to the FCV particle induces considerable conformational changes (Ossiboff and Parker, 2007). These receptor-mediated

changes could be regarded as a first step in the process of membrane penetration and virus uncoating. Receptor binding could prime the capsid protein for an acid induced gross conformational change finally leading to the formation of a pore in the endosomal membrane. However, such considerations represent speculation so far and the detailed protein structural analyses have yet to be conducted in order to understand the process of FCV infection.

The viral RNA

The FCV genome is a single-stranded RNA of positive polarity (Carter *et al.*, 1992a; Green *et al.*, 2000; Sosnovtsev and Green, 1995; Thumfart and Meyers, 2002a). It has a length of approximately 7.5 kb and is polyadenylated at the 3′ terminus. The 5′ end carries a covalently bound protein that, by analogy with picornaviruses, is named VPg for 'viral protein, genome linked', even though it obviously serves functions that differ from those of picornaviral VPgs (see below) (Herbert *et al.*, 1997).

Dependent on the genus they belong to, calicivirus genomes contain either two or three ORFs (Green *et al.*, 2000), though recently a fourth ORF (ORF4) was identified in the MNV genome (Thackesy *et al.*, 2007). FCV and other vesivirus genomes have three major ORFs (Fig. 8.2). ORF1 encompasses about two thirds of the genome and encodes for the non-structural proteins in the form of a polyprotein that is processed by a protease. The second ORF starts around genomic position 5300, only a few nucleotides downstream of the ORF1 stop codon. It has a length of approximately 2 kb and encodes for the major capsid protein precursor that is processed into the leader peptide (LC) and VP1 (Carter, 1990; Carter *et al.*, 1992b; Neill *et al.*, 1991; Sosnovtsev *et al.*, 1998). ORF2 overlaps the 3′ terminal ORF3 by four nucleotides. FCV ORF3 encompasses approximately 320 nucleotides and encodes for the minor capsid protein, VP2.

The coding regions of the FCV genome are flanked by short non-translated regions (NTR) that encompass only 19 nucleotides at the 5′ end whereas the 3′ NTR consists of about 50 residues [without poly(A)tail]. The overall homology of the genome sequences of different FCV isolates is about 80%. The region with the highest degree

Viral RNA

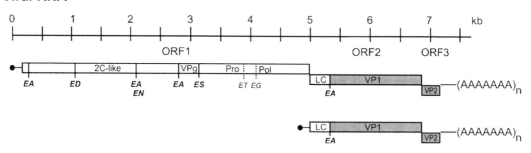

ORF1-encoded proteins

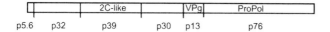

Figure 8.2 Genome organization of FCV. On top a scale in kilobases (kb) is shown to estimate the location of the RNA coding region. The presentation of the RNA includes the 5′ terminal VPg (filled black circle), the 5′ and 3′ non-translated regions (black lines), the poly(A) tail and the location of the 3 ORFs shown as boxes. In the ORFs the location of known sequence motifs coding for e.g. protease (Pro) or polymerase (Pol), leader of the capsid protein (LC), major capsid protein (VP1) or minor capsid protein (VP2) is shown together with the cleavage sites for the viral protease demonstrated by a vertical line and the two amino acids flanking the site (bold face). Two non-essential cleavage sites of so far unknown function located in the ProPol region are also shown. White bars indicate non-structural protein coding regions, grey bars the capsid protein coding regions. Below the genomic RNA, a schematic representation of the sgRNA is shown. At the bottom, the ORF1 encoded non-structural polyprotein is indicated with the sizes of the cleavage products.

of conservation is the 5′ NTR followed by the 3′ NTR and ORF3. The lowest homology is found in ORF2, which is mainly due to the hypervariable regions of the capsid protein (Horimoto et al., 2001; Neill et al., 2000; Radford et al., 1999a, 2003; Seal et al., 1993; Seal, 1994).

In infected cells and viral particles a sgRNA species of about 2.4 kb is found that is 3′ co-terminal with the genome, also carrying a poly(A) tail (Carter, 1990; Neill and Mengeling, 1988). VPg is linked to the 5′ end of this sgRNA that encompasses ORF2 and ORF3 of the viral genome and thus encodes for the capsid protein precursor and VP2. The 5′ terminal sequence of the sgRNA is very similar to the sequence found at the 5′ end of the genome so that this sequence motif can be regarded as important, but so far no function has been assigned to this element. It may function as a cis acting element in RNA replication. This would explain the presence of sgRNA minus strand in infected cells which indicates independent replication of the sgRNA (Carter, 1990). However, this function is most likely not connected to or at least not alone responsible for the initiation of

sgRNA transcription templated by the genomic RNA minus strand since in analogy to RHDV the sequence complementary to the promoter for sgRNA transcription should be located in the genome upstream of the position where the sgRNA starts (Morales et al., 2004). For FCV the sgRNA promoter has not yet been identified. For the RHDV genome, this cis acting sequence important for transcription of the sgRNA was mapped to the 50 nucleotides upstream of the VP1-coding region within the 3′ terminal region of the polymerase gene (Morales et al., 2004). Even though the genome organization of lagoviruses and vesiviruses differ considerably in this area due to the presence of a continuous ORF coding for non-structural proteins and VP1 in the former, it can be hypothesized that the sgRNA promoter is located in similar positions in both genera. However, the sequences of these corresponding regions of the genome show little similarity, so that the molecular basis for its promoter function is still obscure. A recent article predicts the presence of secondary structural elements in the region immediately upstream of the sgRNA start site (Simmonds *et al.*, 2008). This finding will need to be analysed further in regard to its functional importance.

A systematic search for *cis*-acting elements along the entire FCV genome has not been reported. Sequence analysis did not provide indications for such elements except for the above mentioned conserved 5′ terminal sequence motif present in genomic RNA as well as sgRNA. The only proven *cis*-acting sequences identified by loss of function in a mutated FCV RNA are both found in the 3′ terminal part of the viral RNA. One is located close to the 3′ end of ORF3 (Sosnovtsev *et al.*, 2005). It was identified during experiments aimed at elucidation of the importance of VP2 for virus viability. Deletion of the ORF3 sequence resulted in loss of virus viability even when VP2 was provided *in trans*. A variety of mutants with smaller deletions or blocked VP2 expression by introduction of translational stop codons were also not viable but could be complemented by full-length VP2. Because mutants defective *in trans*-complementation assays were not able to express VP1 at levels that could be detected in immunofluorescence assays, and sgRNA could not be detected in Northern blots, it is likely that ORF3 contains a *cis* acting element important for replication and/or transcription. Further analyses are necessary to determine the role of this proposed element.

The second *cis* acting sequence identified by experimental approaches is the so-called termination upstream ribosomal binding site (TURBS) region located in the 3′ terminal part of ORF2 (Luttermann and Meyers, 2007). The TURBS is important for VP2 expression and will be described in more detail below.

Further *cis*-acting RNA elements can be hypothesized in the 3′ NTR that should be involved in promoting initiation of negative strand RNA synthesis. For noroviruses, formation of a stem–loop structure of 47 nucleotides was predicted for the 3′ NTR. Proteins from HeLa cell extracts, such as La and PTB, were found to form stable complexes with this region (see below). Similar putative secondary structures were also found in other calicivirus RNAs including FCV (Seal *et al.*, 1994; Simmonds *et al.*, 2008). In addition, the specific binding of the poly(A) binding protein (PABP) was shown to occur, when the 3′ NTR

was elongated by a poly(A) tail of 24 nucleotides (Gutierrez-Escolano *et al.*, 2003). Since La, PTB, and PABP were found to represent important *trans*-acting factors required for translation and replication of other RNA viruses, the identified RNA–protein interactions may play a role in calicivirus replication and/or translation. For norovirus it was shown also that the 5′ end of the RNA interacts with cellular proteins La, PCBP-2, PTB and hnRNP L, but again a functional assay supporting the importance of this interaction for viral replication is not yet available (Gutierrez-Escolano *et al.*, 2000).

Genome organization and encoded proteins

As already mentioned above, the genomic RNA of FCV and other vesiviruses contain three ORFs. ORF1 codes for a polyprotein that gives rise to the non-structural proteins (Fig. 8.2) (Sosnovtsev *et al.*, 2002). Production of region-specific antisera by immunization of guinea pigs with bacterially expressed polyprotein fragments and amino terminal sequencing of selected mature viral proteins resulted in establishment of a map for ORF1. According to the published results, the ORF1-encoded polyprotein encompasses six mature viral proteins designated NS1 (p5.6), NS2 (p32), NS3 (p39), NS4 (p30), NS5 (p13 or VPg) and NS6–7 (p76 or ProPol).

The first protein from the amino terminus has a length of about 5.6 kDa and is named p5.6. This polypeptide was not detected by immunoprecipitation of proteins from infected cells with an antiserum directed against the respective part of the polyprotein, which only precipitated p32. Amino terminal sequencing of the latter protein showed, that it started with alanine 47 of the polyprotein, the P1′ residue of a predicted cleavage site for the viral protease. Accordingly, it was rather likely that p32 was generated by proteolytic cleavage and thus should be preceded by a small polypeptide of 46 amino acids obtained by translation initiation at the first AUG of ORF1. Mutagenesis of the predicted cleavage site in an infectious clone prevented recovery of viable FCV (Sosnovtsev *et al.*, 2002). To detect the amino terminal protein, a GST-tag was fused to the amino terminal end of the polyprotein. Using a GST-specific antiserum a band corresponding to the expected size for the fusion protein GST-FCVaa1–46 was identified after *in vitro* translation of RNA transcribed from the construct so that expression of the utmost part of ORF1 was formally demonstrated. At the moment, it is not clear whether the failure to detect p5.6 in FCV infected cells is due to the absence of p5.6 specific antibodies in the respective antiserum or to instability of the protein. In this respect it seems interesting that p16, the corresponding protein of RHDV, is also hardly detectable in extracts from infected rabbit hepatocytes (König *et al.*, 1998) or from BHK cells transiently expressing the RHDV ORF1 polyprotein (Meyers *et al.*, 2000). Since, however, a p16 band of reasonable intensity was detected after precipitation of *in vitro* translated protein (Wirblich *et al.*, 1996) it seems more likely that the protein is unstable within the cell than not properly recognized by the serum.

A function has not yet been assigned to p5.6. Since caliciviruses show rather close similarity to picornaviruses with regard to the nature and polyprotein arrangement of the non-structural proteins, some indication on functional aspects of calicivirus proteins can be obtained by comparison with picornaviruses. However, in the case of FCV p5.6 and RHDV p16 the picornavirus counterpart is 2A, a protein showing highly variable functions among the picornavirus group itself (Semler and Wimmer, 2002). Thus, the function of p5.6 is obscure at the moment and has to be determined in future analyses.

The second protein contained in the non-structural polyprotein is p32. This protein was readily detected in extracts of FCV infected cells upon immunoprecipitation and extends from amino acid 47 to 331 (Sosnovtsev *et al.*, 2002). Again, the function of this protein has not yet been determined. The p32 represents one of the most variable proteins among caliciviruses and there are no reports in the literature about identification of known protein motifs. In analogy to picornaviruses, it could correspond to 2B, a protein, that alone or as part of the precursor protein 2BC is involved in modification of intracellular membranes within the infected cells, including membrane rearrangement and vesicle formation, changes in membrane permeability and disassembly of the Golgi complex (Aldabe *et al.*, 1996; Aldabe and Carrasco, 1995; Cho *et al.*,

1994; Jecht *et al.*, 1998; Sandoval and Carrasco, 1997; Teterina *et al.*, 1997). As a matter of fact p32 represents an overall hydrophobic protein with several strongly hydrophobic clusters and is similar to the corresponding RHDV protein p23 (Meyers *et al.*, 2000; Wirblich *et al.*, 1996).

The 39-kDa protein is located immediately downstream of p32 in the polyprotein (Sosnovtsev *et al.*, 2002). The p39 protein contains a nucleotide binding motif and shares sequence similarity with RHDV p37 and the corresponding protein p41 of Southampton virus (Meyers *et al.*, 2000; Pfister and Wimmer, 2001; Wirblich *et al.*, 1996). For the last two viruses, nucleoside triphosphate binding and NTPase activity have been demonstrated experimentally (Pfister and Wimmer, 2001; Marin *et al.*, 2000). These findings are consistent with the observed homology with the 2C protein of the picornaviruses.

The protein following p39 in the polyprotein is p30 that corresponds to RHDV p29 and picornavirus 3A. It is a hydrophobic protein of unknown function. The sequence of p30 and its counterparts in other caliciviruses is quite divergent, representing the second most variable non-structural protein region of these viruses. A rather stable precursor of p30 and p13 (VPg) is found in infected cells (p43) (Sosnovtsev *et al.*, 2002) that is reminiscent of RHDV p41 (König *et al.*, 1998; Meyers *et al.*, 2000) and picornavirus 3AB (Semler and Wimmer, 2002). By analogy with the picornaviruses, p43 could function as a donor for VPg during RNA replication/mRNA transcription with p30 as a hydrophobic anchor tethering VPg to the membranous compartment to which the replication complex is bound.

As already indicated above, VPg with a molecular weight of about 13 kDa is located downstream of p30 in the polyprotein. This position fits very well with the genome organization known from picornaviruses (Semler and Wimmer, 2002). In contrast to the latter viruses, VPg is much larger in caliciviruses (Green *et al.*, 2000) and has been proposed to exert additional functions as described below. VPg is found in infected cells and virus particles covalently bound to the 5′ end of viral RNA (Carter *et al.*, 1992a; Herbert *et al.*, 1997; Sosnovtsev and Green, 2000). It is present at the ends of both the genomic RNA and the sgRNA, linked via a phosphodiester linkage between tyrosine 24 of VPg and the 5′ hydroxyl group of the ribose moiety of the terminal nucleotide (Mitra et al., 2004).

VPg is followed in the polyprotein by a large protein of 76 kDa that is not processed further. This polypeptide encompasses a protease and a polymerase domain and is therefore reminiscent of the 3CD ProPol precursor found in picornavirus infected cells. Similarly, a 72-kDa polypeptide with both enzymatic functions was described for RHDV. However, in contrast to picornavirus 3CD (Semler and Wimmer, 2002) and RHDV p72 (König et al., 1998; Meyers et al., 2000; Wirblich et al., 1996; Vazquez et al., 1998) the FCV p76 (ProPol) is not cleaved into protease and polymerase (Sosnovtseva et al., 1999). This absence of ProPol protein processing might be a characteristic of FCV and other vesiviruses, but also for sapoviruses (Robel et al., 2008). The absence of cleavage between protease and polymerase might point to functional differences since in picornaviruses 3CD has different activities and functions than 3C or 3D and the delayed processing of 3CD seems to be an important element for regulation of virus propagation with a special emphasis on the time point of the switch to RNA replication (Semler and Wimmer, 2002). It will be an interesting task for future work to determine the alternative way(s) of FCV to coordinate this important step of the virus life cycle (see also below).

The FCV ORF1 terminates with a translational stop codon at the end of the polymerase gene (position 5309–5311). Only a few nucleotides downstream thereof, ORF2 starts. In this region, the 'promoter' responsible for transcription of ORF2 should be located (see above).

As already mentioned, the sgRNA encompasses ORF2 and ORF3 coding for the viral capsid proteins VP1 and VP2, which have already been discussed in the context of the virus particle. A unique feature of FCV and the other members of the genus *Vesivirus* is the presence of an additional coding region preceding the VP1 gene. This part of ORF2 gives rise to the so-called leader peptide or leader of the capsid protein (LC) that represents the amino terminal part of the capsid protein precursor. Cleavage of this precursor by the viral protease separates LC from VP1. So

far, no function has been assigned to the leader peptide. It is obvious that this protein is not part of the viral capsid. Nevertheless, the expression of the protein and VP1 in equimolar amounts rather points at a function in virus structure formation than replication. However, it was reported recently that the expression of FCV LC enhances the replication of norovirus RNA (Chang et al., 2008). The authors observed a very low number of fluorescent cells upon transfection of cells with a norovirus replicon RNA in which GFP was inserted into the capsid region of the genome. However, when the FCV LC was provided in *trans* by co-transfection of a plasmid encoding FCV LC under control of a cytomegalovirus promoter, the number of cells expressing the norovirus replicon increased significantly. Further analyses are necessary to identify the mechanism by which LC promotes norovirus replication in this experimental system.

RNA replication

The basic functions of the ProPol protein can be easily summarized in a way that the protease domain is responsible for the processing of the viral proteins and the polymerase conducts the basic steps of genome replication and mRNA transcription. This view is of course oversimplified since it is known from other RNA viruses that polyprotein processing and RNA synthesis are concerted actions involving many viral and host cellular factors that act in a finely tuned interplay. This interplay of the different factors allows coordination of important steps in the viral life cycle like translation, RNA replication and particle assembly. The knowledge about this interplay is very poor for FCV and awaits further investigation. In a recent publication, profiles of protein/protein interactions were described for different FCV polypeptides (Kaiser et al., 2006). Based on yeast two-hybrid analyses and biochemical tests the authors were able to demonstrate such interactions for a variety of FCV proteins. Evidence was found for the existence of a complex containing p32, p39 (NTPase), p30 and the ProPol fusion protein. The experimental results suggested that p32 might be involved in the formation of a scaffold for this complex. The interaction of p32 with p30 would also lead to incorporation of VPg into the complex since p30 is part of the VPg precursor p43. The authors also found evidence for interaction between VPg and ProPol as well as homo-oligomerization of ProPol. By analogy to picornavirus RNA replication, one might expect that VPg, perhaps in the form of its precursor p43, would function as a primer for RNA replication. Noroviruses were the first caliciviruses for which VPg primed RNA synthesis was demonstrated (Rohayem *et al.*, 2006). The formation of at least dimers of the polymerase is an important point also in the models of picornavirus RNA replication (Semler and Wimmer, 2002), and, once again, a recent paper showed that the active form of the norovirus RdRp is a dimer (Hogbom *et al.*, 2009). Some of the identified interactions of FCV proteins are, however, unexpected. This is true for ProPol interacting with the ORF2 product (capsid protein precursor) and the minor capsid protein VP2 (Kaiser *et al.*, 2006). Even though the latter interaction seemed to be rather weak, further elucidation of the meaning of these interactions will be interesting for future work.

Important insights into the function of proteins can be obtained by the interpretation of the results of structural analyses. So far, no FCV non-structural protein has been analysed on the structural level. However, the 3D structures of some proteins from other caliciviruses have been determined and allow conclusions also with regard to their counterparts in FCV (Fullerton *et al.*, 2007; Zamyatkin *et al.*, 2008; Garriga *et al.*, 2007). This point is presented in another chapter of this book.

Strategy of gene expression

Caliciviruses belong to the Baltimore classification group IV that encompasses the positive-strand RNA viruses (Baltimore, 1971). As for other members of this group, the calicivirus genomic RNA is infectious, which means that the introduction of the RNA alone into susceptible cells is sufficient to start a complete cycle of virus replication ending with the release of infectious progeny virus. The basis for this feature is the fact that the viral genomic RNA functions as an mRNA that is translated within the cell to give rise to all replication-relevant viral proteins. These proteins are able to initiate all subsequent steps of genome replication and translation of large amounts of viral proteins and assembly and

release of virus particles. The Baltimore scheme of classification divides the positive-strand RNA viruses into groups A and B depending mainly on the absence or presence of sgRNAs within the infected cell, respectively. According to these stringent criteria, caliciviruses represent members of group B, even though lagoviruses, sapoviruses and according to a recent publication also noroviruses (McCormick *et al.*, 2008) are able to express their structural proteins not only from the subgenomic mRNA but also from the genome, a feature that is reminiscent of members of group A and not found in other group B viruses.

The principles of calicivirus gene expression are characterized by the features (i) translation of a polyprotein from the genomic RNA with subsequent proteolytic processing and (ii) transcription of a sgRNA that is translated to give rise to the structural proteins. These two strategies will be described in more detail below.

As already mentioned above, the FCV genome contains three ORFs. Only ORF1 is translated from the genomic RNA to give rise to a polyprotein of approximately 1760 aa that encompasses the non-structural virus proteins. A usual prerequisite for translation of RNAs in eukaryotic cells is the presence of a 5′ cap structure and a 3′ poly(A) tail. These two *cis*-acting elements are crucial for the assembly of the translation initiation complex, composed of the RNA, the ribosome, a whole set of translation initiation factors, the poly(A) binding protein PABP, initiator t-RNA Met-tRNA; and a variety of other molecules (see Kapp and Lorsch, 2004 for a recent review). As an initial step, the cap structure is recognized by the eIF4E component of initiation factor eIF4F that recruits other initiation factors and thereby mediates the contact between the RNA 5′ end and the small ribosomal subunit. One component of the complex is also PABP that bridges the initiation complex with the mRNA 3′ end, resulting in a quasi-circular structure of the RNA.

The FCV RNA contains a poly(A) tail but no 5′ cap structure so it is clear that translation initiation cannot follow the standard scheme. The problem of how to achieve (preferential) translation of their own RNAs is common to all viruses and has led to the evolution of a variety of special viral translation initiation mechanisms.

Many of these mechanisms have been discovered recently to be in principle also used by the host cells for special purposes (recently reviewed in Bushell and Sarnow, 2002; Jackson, 2000, 2005; Ryabova et al., 2006; Schneider and Mohr, 2003).

Picornaviruses and also several other positive-strand RNA viruses utilize an internal ribosomal entry site (IRES) for recruitment of ribosomes. This IRES represents a considerable part of the long 5′ terminal non-translated region of the virus genome and folds into a characteristic structure that is able to bind ribosomes independently of the RNAs 5′ end (reviewed in Hellen and Sarnow, 2001; Sarnow et al., 2005). In caliciviruses, the 5′ NTR is very short (only 19 nucleotides in FCV) so that the presence of a classical IRES can be excluded and another mechanism for translation initiation is needed. It was recognized quite early in calicivirus research that the genomic RNA is covalently linked to the already mentioned VPg (Burroughs and Brown, 1978; Herbert et al., 1997; Meyers et al., 1991; Schaffer et al., 1980b). Covalently linked polypeptides are also found in a variety of other viral RNAs but the protein linked to calicivirus RNAs is much larger, so that it was speculated to fulfil additional functions not found in picornavirus VPg. In fact, evidence already published more than ten years ago supported the idea that the calicivirus VPg is involved in translation initiation of the viral RNA since destruction of VPg by proteinase K treatment of viral RNA interfered with efficient translation of viral RNA (Herbert et al., 1997). Recently reported data show that VPg binds to canonical translation initiation factors (Chaudhry et al., 2006; Daughenbaugh et al., 2003; Goodfellow et al., 2005). The earliest publication on this issue reported that VPg of norovirus interacted with initiation factor eIF3, one of the most universally used initiation factors essential for almost all initiation processes analysed so far (Daughenbaugh et al., 2003). This was first found in yeast two-hybrid analyses that indicated binding of VPg to the eIF3d subunit of the initiation factor. This interaction was further proven by demonstration of binding of VPg to purified mammalian eIF3 in vitro and to eIF3 in cell lysates. In addition, these authors reported the detection of interaction of VPg with eIF4E and eIF2α in GST pull down assays. It was published soon afterwards that VPg from FCV and the human norovirus Lordsdale virus interact with eIF4E, the cap binding protein (Goodfellow et al., 2005). Similar results were also reported for MNV (Chaudhry et al., 2006). Although the structural basis for this interaction has not been elucidated so far, this finding fits very nicely with the idea of VPg working as a cap substitute, since the binding of eIF4E to VPg would in theory result in formation of a standard preinitiation complex at the 5′ end of the viral RNA. For FCV nearly the complete set of published data are in line with this model since depletion of eIF4E as well as addition of 4E-BP1, a protein that binds and temporarily inhibits eIF4E, were able to inhibit translation of FCV RNA. Moreover, the foot and mouth disease virus L protease-induced cleavage of eIF4G, an initiation factor that mediates the contact to the ribosome by simultaneous binding to the small ribosomal subunit and to eIF4E, abrogates translation of FCV RNA just as reported for capped RNA. A still open point concerns the binding site for VPg on eIF4E. Addition of an excess quantity of cap analogue does not inhibit FCV RNA translation which indicates that VPg and cap do not compete for the same binding site on eIF4E (Chaudhry et al., 2006). Whether separate binding sites are utilized for the VPg and cap or both bind to the same site without influencing each other has to be investigated in future experiments. There are additional questions concerning the initiation of translation in the noroviruses. The functional importance of the binding of the norovirus VPg to translation initiation factors has not been determined. For MNV and LDV, eIF4E binding was demonstrated but the translation of MNV RNA was not affected by depletion or inactivation of this initiation factor or the destruction of its partner eIF4G. Whether these results really reflect individual differences with regard to the basic mechanism of translation initiation or represent variations of a common principle needs to be investigated in future work. There is also a need for further analysis with regard to the process following establishment of the preinitiation complex. It is not clear whether the preinitiation complex reaches the translational start site in the FCV RNA via a very short scanning process or is assembled directly at the correct position.

The polyprotein is processed co- and post-translationally into the mature viral proteins. As far as can be concluded from the published data, there is only one protease involved in this process, namely the 3C-like cysteine protease that in case of FCV represents one domain of the p76 ProPol protein (Sosnovtsev et al., 1998, 1999, 2002). Features important for the catalytic activity of this protease are highly conserved among caliciviruses (Oka et al., 2007). The cleavage sites for the FCV protease in the polyprotein have been identified by N-terminal sequencing of cleavage products and/or site-directed mutagenesis at positions crucial for cleavage (Sosnovtsev et al., 2002). Fig. 8.2 shows a cleavage map that is able to allocate all polypeptides found within the infected cells to a defined part of the genome. All identified cleavage sites exhibit sequences that are in accordance with the so far known requirements of the 3C-like protease that usually cleaves after glutamic acid followed by alanine, glycine, serine, asparagine or aspartic acid. This finding differs from the situation in RHDV, where one cleavage site was found that does not fulfil the criteria for cleavage by the 3C-like viral protease and indeed seems to be processed by a cellular protease (Thumfart and Meyers, 2002b).

The description of the above given definition of a cleavage site based on the P1 and P1' residues makes obvious, that not all putative cleavage sites can be processed. Otherwise, many small cleavage products would be generated. So far, the requirements for a cleavage site beyond the nature of the flanking residues are not clear, but it is likely that structural features play a major role.

As in other positive strand RNA viruses, rather stable precursors composed of more than one mature viral protein [p60, p88, p120 (Sosnovtsev et al., 2002)] can be found in FCV-infected cells. It can be speculated that these processing intermediates have functions different from those of the final cleavage products but further investigation with regard to this point is necessary.

As noted above, members of the genus *Vesivirus* express a capsid protein precursor composed of VP1 and LC from the sgRNA (Green et al., 2000). This fusion protein is also processed by the viral 3C-like protease (Sosnovtsev et al., 1998). Cleavage occurs at a Glu/Ala site located at positions 124/125 of the ORF2 translation product. Cleavage of the major capsid protein precursor is essential for virus viability, since all mutations at the cleavage site that abrogated precursor processing were shown to be lethal whereas a variety of mutations preserving cleavage were tolerated. In this context it is interesting that the presence of the Glu residue at P1 is very important. Even conservative exchanges of Gln for Glu or Asp for Glu reduced cleavage efficiency considerably and prevented the recovery of infectious FCV (Sosnovtsev et al., 1998). The mechanistic basis for the latter result is somewhat obscure, since a reduced provision of mature VP1 could also be hypothesized to only reduce virus yield but not abrogate recovery of viable virus. It has been put forward by the authors that incorporation of unprocessed capsid precursor proteins might interfere with the assembly of the infectious virion. Further analyses are necessary to clarify the reason for the importance of (complete) processing of the capsid protein precursor for virus viability.

Expression of VP2

Even though the initiation of translation at the 5' ends of calicivirus RNA is not fully clarified from a mechanistic point of view, the basic principles seem to be rather clear with VPg serving as a cap substitute that directs translation initiation at the RNA 5' end in a near classical way. A more difficult question concerns the expression of the 3' terminal ORF present in all calicivirus genomic RNA and sgRNAs. Because of its position, this ORF (number 3 in FCV and other vesiviruses) could either be expressed from a second sgRNA or has to be translated via a special mechanism from the known viral RNAs. The first alternative was excluded, because of the absence of a second sgRNA from FCV infected cells. An early publication reported the identification of several nested sgRNAs in FCV infected cells (Carter, 1990), but this finding was questioned later by the same author (Carter, 1994). Our own analyses did not find evidence for the existence of more than one sgRNA. Similarly, elaborate studies with RNA from liver material of RHDV infected rabbits using, for example, Northern blot hybridization and size fractionation of RNA with subsequent *in vitro* translation and immunoprecipitation of viral proteins did not give any indication for a functional small mRNA coding for VP2 (Meyers, unpublished results). Thus, a special mechanism of translation initiation had to be responsible for VP2 expression.

In a standard scanning dependent process, the small ribosomal subunit starts scanning at the 5′ end and then migrates linearly in 3′ direction. Therefore, translation usually starts at the AUG codon closest to the 5′ end. In higher eukaryotes, the first AUG has to reside in a favourable sequence context; otherwise, at least some of the ribosomal subunits can initiate at an AUG further downstream, resulting in expression of different proteins from one mRNA (Kozak, 1991, 1999; Peri and Pandey, 2001; Curran and Kolakofsky, 1988; Ryabova *et al.*, 2006; Stacey *et al.*, 2000). As mentioned above, it is not clear whether scanning occurs at all during translation of calicivirus RNA, but even then expression of VP2 could hardly be due to leaky scanning since the FCV sgRNA contains more than 30 AUG codons upstream of the ORF3 start codon.

Several alternative mechanisms have been identified that lead to start of translation. Many of these were first detected in viruses since viruses have developed a variety of strategies to ensure translation of their RNAs in competition with cellular RNAs (reviewed in Jackson, 2000; Ryabova *et al.*, 2006; Schneider and Mohr, 2003; Bushell and Sarnow, 2002).

One possible alternative initiation mechanism is the transfer of an initiation complex assembled at the 5′ end to a downstream initiation site by a so-called shunt mechanism (Cuesta *et al.*, 2001; Ryabova *et al.*, 2002; Latorre *et al.*, 1998). Different types of this mechanism have been identified that rely on special sequences in the RNA and in some cases on the presence of specific proteins that facilitate shunting (Park *et al.*, 2001; Xi *et al.*, 2005).

As mentioned above, translation can be initiated without implication of the RNA 5′ end when an IRES is present, a principle that is found in a variety of cellular and especially viral RNAs (reviewed in Pestova *et al.*, 2001; Sarnow *et al.*, 2005; Hellen and Sarnow, 2001). Furthermore, ribosomes can be recruited by downstream translational start sites after having terminated translation of a preceding ORF (reviewed in Kozak, 1987, 2002; Ryabova *et al.*, 2006). The

biochemistry behind such a termination/reinitiation process, which occurs only rarely in eukaryotes, is not fully understood. A major concern in reinitiation is the difference between a translational initiation complex loaded with a bulk of initiation factors and a 'later' elongating ribosome that has lost the initiation factors and isn't able to start a new round of translation unless it is reloaded with the factors. It is therefore believed that reinitiation of translation is possible only when the post termination ribosome is tethered for a while to the mRNA to give time for initiation factor binding.

Analyses conducted in our lab showed that VP2 expression in FCV and RHDV is neither an IRES- nor a shunt-induced process but depends on a termination/reinitiation mechanism (Luttermann and Meyers, 2007, 2009; Meyers, 2003, 2007). Since eukaryotic translation is usually initiated close to the 5′ end and not designed for promoting a restart after termination, caliciviruses use a special mechanism to promote reinitiation. This mechanism depends on an upstream sequence termed the termination upstream binding site, TURBS (about 70 nucleotides), which is thought to bind the post-termination ribosome and therefore increase the chance of reinitiation (Meyers, 2003, 2007; Luttermann and Meyers, 2007, 2009).

The TURBS region contains three short sequence motifs (1, 2 and 2*) essential for VP2 translation as determined by deletion mapping (Luttermann and Meyers, 2007; Meyers, 2007) (Fig. 8.3). Motif 1 is found conserved among caliciviruses and is located at similar positions in the mRNAs of the different caliciviruses, upstream of the 3′ terminal ORFs. This motif is the core of a region, which is complementary to the loop region of helix 26 within 18S rRNA. Hybridization of motif 1 to 18S rRNA could tether the ribosome to the viral RNA. Published data indicate interaction of motif 1 with initiation factor eIF3 (Pöyry *et al.*, 2007), a process that could assist the hypothetical hybridization mediated contact or could represent an alternative mechanism for tethering of the post-termination ribosome. Analyses in a special yeast expression system provided final proof for the importance of motif 1 18S rRNA complementarity (Luttermann and Meyers, 2009). The *S. cerevisiae* strain NOY908 lacks all chromosomal

A

ORF2

```
          -70          -60         -50         -40         -30         -20         -10        -1
-ACTGCCUCCUACAUGGGAAUUCAAUUGGCAAAGAUUCGGCUUGCCUCAAACAUUAGGAGUUCAAUGACUAAAUUAUGA
         motif 2*  motif1                                        motif 2                AUGAAUUCA-
```

ORF3

B

```
18S rRNA:  3´ -cccaguacccuuauugc-  5´
   FCV   :  5´ -CCUACAUGGGAAUUCAA-  3´
```

C

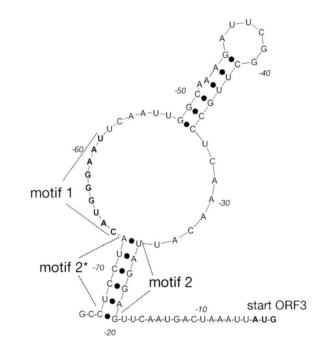

Figure 8.3 Organization and features of the TURBS region located at the 3′ end of ORF2. (A) Parts of the RNA sequences of ORF2 and ORF3 are shown with the essential motifs 1, 2 and 2* indicated (bold face, underlined). Above the RNA a number line is shown that gives the position of the nucleotides with respect to the end of ORF2. (B) The complementarity of motif 1 and the loop region of helix 26 of mammalian 18S rRNA is shown. Sequences given in bold face are able to hybridize. (C) Putative secondary structure of the TURBS region.

copies of the 35S rRNA gene and expresses the rRNA precursor from the episomally maintained plasmid pNOY373 (Wai et al., 2000). Transfection of such cells with a plasmid containing mutated rRNA sequences allows the expression of rRNA variants. Using this system, we were able to test the influence of (reciprocal) mutations in rRNA and TURBS motif 1 on expression of a 3′ terminal ORF coding for VP2/GFP, a truncated VP2 fused to green fluorescent protein (GFP). These analyses showed that changes in the yeast 18S rRNA sequence corresponding to the loop of helix 26 in mammals were able to significantly enhance translation of the 3′ ORF when the changes improved the complementarity of this sequence and motif 1. Moreover, we found that motif 1 mutations blocking or at least dramatically reducing translation of the 3′ terminal ORF could be compensated by reciprocal changes in the corresponding 18S rRNA sequence (Luttermann and Meyers, 2009). The highest level of VP2/GFP translation was achieved when the motif 1 sequence was adjusted to the yeast 18S rRNA sequence, most likely, because this manipulation raised the number of putative hybridization partners to a maximum, so that all translating ribosomes were able to interact with motif 1. Thus, the experiments showed that complementarity of motif 1 and 18S rRNA is crucial for VP2 translation and that the amount of ribosomes with complementary RNA sequences correlates with the expression efficiency, which very strongly supports the 18S rRNA hybridization theory for the motif 1 function.

In contrast to motif 1 the sequence of motif 2 is not conserved. This sequence was supposed to play a role in positioning of the ribosome relative to the start site of the 3′ terminal ORF (Luttermann and Meyers, 2007; Meyers, 2007). The effect of mutations in motif 2 could be divided into two classes, namely A to G or C to U changes that only reduced VP2 expression levels and other changes that severely impaired reinitiation. This finding is indicative for motif 2 working via hybridization to another RNA since G or U instead of A or C, respectively, can preserve this interaction via a non- Watson-Crick G-U pairing whereas the other changes lead to loss of pairing. Since, however, the sequence of motif 2 is not conserved among caliciviruses it is unlikely that it interacts with a presumably rather conserved host cellular RNA and thus should bind to another piece of viral RNA. Recently, we indeed were able to identify a sequence element in the FCV RNA that hybridizes to motif 2 (Luttermann and Meyers, 2009). This so-called motif 2* is located directly upstream of motif 1, so that motif 2*/2 interaction would lead to a secondary structure in which motif 1 is presented in a loop above of a 2*/2 stem (Fig. 8.3C). Mutational analyses again provided compelling evidence for the importance of the motif 2*/2 hybridization because lethal mutations in one partner could be compensated by reciprocal changes in the other sequence.

Taken together, the termination/reinitiation mechanism leading to translation of the FCV minor capsid protein VP2 is dependent on intra- and intermolecular base pairing. In our model (Fig. 8.4) motif 2 would interact with motif 2*

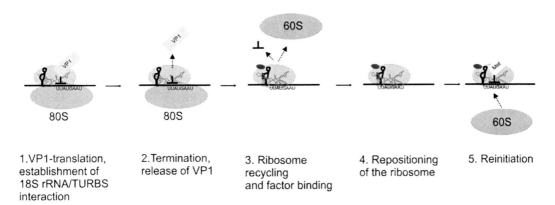

1. VP1-translation, establishment of 18S rRNA/TURBS interaction

2. Termination, release of VP1

3. Ribosome recycling and factor binding

4. Repositioning of the ribosome

5. Reinitiation

Figure 8.4 Hypothetical model of the reinitiation mechanism leading to expression of VP2. The different steps necessary for translation reinitiation are indicated. The viral RNA (black line) with the TURBS (black lined secondary structure) is indicated together with the ribosome (grey ellipses represent small and large subunit of the ribosome). The 18S rRNA contained in the small ribosomal subunit (grey line shown in hypothetical secondary structure) is indicated. The sequence at the start/stop site is given. The tRNA is indicated by an upside down T. The (initiation) factors except for eIF3 (dark grey ellipse) necessary to promote reinitiation are not known. It has to be stressed that the details of the process especially with regard to the timely coordination are mostly obscure so far.

and build a secondary structure as shown in Fig. 8.3C. This structure could help to present motif 1 in a way that allows binding and positioning of the ribosome for the reinitiation process. As part of this structure motif 1 is able to tether a certain proportion of the terminating ribosomes via hybridization with the helix 26 sequence of the 18S rRNA that is part of the 40S subunit of the ribosome. Thereby, time would be available to assemble an initiation complex containing the necessary factors. The tethering of this complex at the translation start site would bypass the requirement for a good initiation codon context that is observed in scanning dependent initiation, thereby allowing initiation on non-canonical start codons, as was shown to occur in FCV (Luttermann and Meyers, 2007). This model is in accordance with the recent finding that a residue of helix 26 (nt 1107 of mammalian 18S rRNA) interacts in initiation events with the mRNA. This interaction occurs at position −17 of the RNA as shown by cross-linking experiments (Pisarev *et al.*, 2008). This position corresponds to the 3′ end of motif 2 of the FCV TURBS that is located at position −16 relative to the ORF3 start codon (nt −20 with regard to the ORF2 stop codon). Thus, the position of the ribosome bound to the RNA for reinitiation seems to parallel the arrangement in standard initiation complexes.

Further aspects of the interaction of FCV with its host cell

The infection of a suitable cell with FCV leads to a fulminant cytopathic effect. This finding is not surprising for a virus that replicates quickly to very high titres. It can be hypothesized that the infected cell is soon exhausted of metabolites and energy and that this situation could lead to cell death. Moreover, the accumulation of a large number of virus particles in the cytoplasm, the host shutoff on the translational level (see below) and various stress signals can kill the infected cell. Such conditions are prone to induction of apoptosis and indeed signs of programmed cell death were identified by several groups upon FCV infection (Natoni *et al.*, 2006; Roberts *et al.*, 2003; Sosnovtsev *et al.*, 2003). It is not clear at the moment whether apoptosis leads to death of FCV infected cells or the cells die in consequence of,

for example, energy exhaustion before reaching the final stage of the suicide program.

The induction of apoptosis might also be related to another aspect of the interaction between FCV and the host cell. It has been reported that caspase-2 and to a lesser extent caspase-6 cleave the FCV major capsid protein VP1 in infected cells (Al-Molawi *et al.*, 2003). This process might result from or at least be enhanced by the observed FCV-induced activation of caspase-2, 3 and -7. The significance of this process, however, is not clear at the moment.

In addition to basic interactions like receptor binding, penetration, use of cellular resources for translation, genome replication and transport pathways, as well as induction of cell damage many viruses have evolved a second level of host cell interactions primarily designed for interference with host cellular defence mechanisms and reprogramming of metabolic pathways in order to improve the basis for virus replication. There are many open questions with regard to these aspects in FCV and caliciviruses in general. It is very likely that caliciviruses like many if not all other viruses have evolved mechanisms to interfere with the innate immune system of the host (cell), but, so far, no data have been published elucidating such mechanisms for family members. In contrast, at least preliminary results are available that indicate the existence of viral functions leading to a host cell shutoff. It has been reported that the overall translation of cellular mRNAs is significantly reduced in the course of an FCV infection. This finding can at least in part be explained by the observed cleavage of Poly(A) binding protein (PABP) executed by the 3C-like protease (Kuyumcu-Martinez *et al.*, 2004). The meaning of this process for FCV replication is not yet clear, because destruction of PABP should not only affect host cellular but also viral translation so that it probably does not result in a simple host shut-off. It might be that reduction of translation in consequence of PABP destruction is (also) important for induction of the switch from FCV gene expression to replication. In this context it is interesting that the effect on translation manifests between 4 and 5 hpi, a rather late time point in FCV propagation. Thus, it could well be that this reduction of translation activity coincides with increasing genome replication activity.

An interesting observation with regard to host cell shutoff is the cleavage of translation initiation factor eIF4G-I and -II in FCV infected cells (Willcocks *et al.*, 2004). It is not yet clear which protease mediates this cleavage. The cleavage occurs concomitantly with the reduction of host cellular translation, but at least most of the cleavage products retain the ability to bind eIF4E so that the connection between eIF4G processing and reduction of protein translation is not yet obvious. Moreover, it has to be analysed whether this process also affects FCV protein translation, to be able to evaluate whether it is more a strategy to push viral gene expression or a means to switch from the translation to the replication phase.

Viruses not only reprogramme the host cellular system to improve replication efficiency but also use cellular components, in particular cellular nucleic acid binding proteins, for regulation and/or support of virus replication. It has been published rather recently, that FCV RNA interacts specifically with polypyrimidine tract-binding protein (PTB) (Karakasiliotis *et al.*, 2006). Binding was found to occur to the 5′ terminal region of the viral genomic RNA and sgRNAs. Using RNA interference approaches to knock down PTB in infected cells the authors showed that PTB is required for efficient FCV replication. Interestingly, especially pronounced effects were only seen at temperatures below or above of 37°C leading to speculation about an RNA chaperon function of PTB that would be more important when the host activates its defence mechanisms. Elevation of temperature occurs in FCV-infected cats so that PTB might facilitate virus propagation in animals with high fever.

Conclusion

FCV is an important pathogen in cats and represents so far the best-studied calicivirus. Nevertheless, many fundamental aspects of its life cycle are still obscure and additional experimental and theoretical work will be necessary to obtain a detailed view of the molecular biology of this virus. FCV has many fascinating molecular features that we just have begun to elucidate. This work will continue despite the importance of human noroviruses and the increased research on these viruses in consequence of the identification of MNV, the first norovirus amenable to tissue culture propagation. The availability of infectious cDNA clones for two different FCV isolates (Sosnovtsev and Green, 1995; Thumfart and Meyers, 2002a) will greatly help in the ongoing and future work.

References

Al-Molawi, N., Beardmore, V.A., Carter, M.J., Kass, G.E., and Roberts, L.O. (2003). Caspase-mediated cleavage of the feline calicivirus capsid protein. J. Gen. Virol. *84*, 1237–1244.

Aldabe, R., Barco, A., and Carrasco, L. (1996). Membrane permeabilization by poliovirus proteins 2B and 2BC. J. Biol. Chem. *271*, 23134–23137.

Aldabe, R., and Carrasco, L. (1995). Induction of membrane proliferation by poliovirus proteins 2C and 2BC. Biochem. Biophys. Res. Commun. *206*, 64–76.

Baltimore, D. (1971). Expression of animal virus genomes. Bacteriol. Rev. *35*, 235–241.

Bertolotti-Ciarlet, A., Crawford, S.E., Hutson, A.M., and Estes, M.K. (2003). The 3′ end of Norwalk virus mRNA contains determinants that regulate the expression and stability of the viral capsid protein VP1: a novel function for the VP2 protein. J. Virol. *77*, 11603–11615.

Bhella, D., Gatherer, D., Chaudhry, Y., Pink, R., and Goodfellow, I.G. (2008). Structural insights into calicivirus attachment and uncoating. J. Virol. *82*, 8051–8058.

Black, D.N., Burroughs, J.N., Harris, T.J., and Brown, F. (1978). The structure and replication of calicivirus RNA. Nature *274*, 614–615.

Burroughs, J.N., and Brown, F. (1978). Presence of a covalently linked protein on calicivirus RNA. J. Gen. Virol. *41*, 443–446.

Burroughs, J.N., Doel, T.R., Smale, C.J., and Brown, F. (1978a). A model for vesicular exanthema virus, the prototype of the calicivirus group. J. Gen. Virol. *40*, 161–174.

Burroughs, N., Doel, T., and Brown, F. (1978b). Relationship of San Miguel sea lion virus to other members of the calicivirus group. InterVirology *10*, 51–59.

Bushell, M., and Sarnow, P. (2002). Hijacking the translation apparatus by RNA viruses. J. Cell Biol. *158*, 395–399.

Capucci, L., Fusi, P., Lavazza, A., Pacciarini, M.L., and Rossi, C. (1996). Detection and preliminary characterization of a new rabbit calicivirus related to rabbit hemorrhagic disease virus but nonpathogenic. J. Virol. *70*, 8614–8623.

Carter, M.J. (1994). Genomic organization and expresion of astroviruses and caliciviruses. Arch. Virol. [Suppl.] *9*, 429–439.

Carter, M.J. (1990). Transcription of feline calicivirus RNA. Arch. Virol. *114*, 143–152.

Carter, M.J., Milton, I.D., Meanger, J., Bennett, M., Gaskell, R.M., and Turner, P.C. (1992a). The complete nucleotide sequence of a feline calicivirus. Virology *190*, 443–448.

Carter, M.J., Milton, I.D., Turner, P.C., Meanger, J., Bennett, M., and Gaskell, R.M. (1992b). Identification and sequence determination of the capsid protein gene of feline calicivirus. Arch. Virol. *122*, 223–235.

Carter, M.J., Routledge, E.G., and Toms, G.L. (1989). Monoclonal antibodies to feline calicivirus. J. Gen. Virol. *70*, 2197–2200.

Chang, K.O., George, D.W., Patton, J.B., Green, K.Y., and Sosnovtsev, S.V. (2008). Leader of the capsid protein in feline calicivirus promotes replication of Norwalk virus in cell culture. J. Virol. *82*, 9306–9317.

Chaudhry, Y., Nayak, A., Bordeleau, M.E., Tanaka, J., Pelletier, J., Belsham, G.J., Roberts, L.O., and Goodfellow, I.G. (2006). Caliciviruses differ in their functional requirements for eIF4F components. J. Biol. Chem. *281*, 25315–25325.

Chen, R., Neill, J.D., Estes, M.K., and Prasad, B.V. (2006). X-ray structure of a native calicivirus: structural insights into antigenic diversity and host specificity. Proc. Natl. Acad. Sci. U.S.A. *103*, 8048–8053.

Cho, M.W., Teterina, N., Egger, D., Bienz, K., and Ehrenfeld, E. (1994). Membrane rearrangement and vesicle induction by recombinant poliovirus 2C and 2BC in human cells. Virology *202*, 129–145.

Clay, S., Maherchandani, S., Malik, Y.S., and Goyal, S.M. (2006). Survival on uncommon fomites of feline calicivirus, a surrogate of noroviruses. Am. J. Infect. Control *34*, 41–43.

Coyne, K.P., Dawson, S., Radford, A.D., Cripps, P.J., Porter, C.J., McCracken, C.M., and Gaskell, R.M. (2006a). Long-term analysis of feline calicivirus prevalence and viral shedding patterns in naturally infected colonies of domestic cats. Vet. Microbiol. *118*, 12–25.

Coyne, K.P., Jones, B.R., Kipar, A., Chantrey, J., Porter, C.J., Barber, P.J., Dawson, S., Gaskell, R.M., and Radford, A.D. (2006b). Lethal outbreak of disease associated with feline calicivirus infection in cats. Vet. Rec. *158*, 544–550.

Cuesta, R., Xi, Q., and Schneider, R.J. (2001). Preferential translation of adenovirus mRNAs in infected cells. Cold Spring Harb. Symp. Quant. Biol. *66*, 259–267.

Curran, J., and Kolakofsky, D. (1988). Ribosomal initiation from an ACG codon in the Sendai virus P/C mRNA. EMBO J. *7*, 245–251.

Daughenbaugh, K.F., Fraser, C.S., Hershey, J.W., and Hardy, M.E. (2003). The genome-linked protein VPg of the Norwalk virus binds eIF3, suggesting its role in translation initiation complex recruitment. EMBO J. *22*, 2852–2859.

Di Martino, B., Marsilio, F., and Roy, P. (2007). Assembly of feline calicivirus-like particle and its immunogenicity. Vet. Microbiol. *120*, 173–178.

Doultree, J.C., Druce, J.D., Birch, C.J., Bowden, D.S., and Marshall, J.A. (1999). Inactivation of feline calicivirus, a Norwalk virus surrogate. J. Hosp. Infect. *41*, 51–57.

Duizer, E., Bijkerk, P., Rockx, B., De, G.A., Twisk, F., and Koopmans, M. (2004). Inactivation of caliciviruses. Appl. Environ. Microbiol. *70*, 4538–4543.

Ebnet, K., Suzuki, A., Ohno, S., and Vestweber, D. (2004). Junctional adhesion molecules (JAMs): more molecules with dual functions? J. Cell Sci. *117*, 19–29.

Ehresmann, D.W., and Schaffer, F.L. (1977). RNA synthesized in calicivirus-infected cells is atypical of picornaviruses. J. Virol. *22*, 572–576.

Ehresmann, D.W., and Schaffer, F.L. (1979). Calicivirus intracellular RNA: fractionation of 18–22 s RNA and lack of typical 5'-methylated caps on 36 S and 22 S San Miguel sea lion virus RNAs. Virology *95*, 251–255.

Foley, J., Hurley, K., Pesavento, P.A., Poland, A., and Pedersen, N.C. (2006). Virulent systemic feline calicivirus infection: local cytokine modulation and contribution of viral mutants. J. Feline Med. Surg. *8*, 55–61.

Fullerton, S.W., Blaschke, M., Coutard, B., Gebhardt, J., Gorbalenya, A., Canard, B., Tucker, P.A., and Rohayem, J. (2007). Structural and functional characterization of sapovirus RNA-dependent RNA polymerase. J. Virol. *81*, 1858–1871.

Garriga, D., Navarro, A., Querol-Audi, J., Abaitua, F., Rodriguez, J.F., and Verdaguer, N. (2007). Activation mechanism of a noncanonical RNA-dependent RNA polymerase. Proc. Natl. Acad. Sci. U.S.A. *104*, 20540–20545.

Geissler, K., Schneider, K., Platzer, G., Truyen, B., Kaaden, O.R., and Truyen, U. (1997). Genetic and antigenic heterogeneity among feline calicivirus isolates from distinct disease manifestations. Virus Res. *48*, 193–206.

Geissler, K., Schneider, K., and Truyen, U. (2002). Mapping neutralizing and non-neutralizing epitopes on the capsid protein of feline calicivirus. J. Vet. Med. B Infect. Dis. Vet. Public Health *49*, 55–60.

Glenn, M., Radford, A.D., Turner, P.C., Carter, M., Lowery, D., DeSilver, D.A., Meanger, J., Baulch-Brown, C., Bennett, M., and Gaskell, R.M. (1999). Nucleotide sequence of UK and Australian isolates of feline calicivirus (FCV) and phylogenetic analysis of FCVs. Vet. Microbiol. *67*, 175–193.

Goodfellow, I., Chaudhry, Y., Gioldasi, I., Gerondopoulos, A., Natoni, A., Labrie, L., Laliberte, J.F., and Roberts, L. (2005). Calicivirus translation initiation requires an interaction between VPg and eIF 4 E. EMBO Rep. *6*, 968–972.

Green, K.Y., Ando, T., Balayan, M.S., Berke, T., Clarke, I.N., Estes, M.K., Matson, D.O., Nakata, S., Neill, J.D., Studdert, M.J., and Thiel, H.J. (2000). Family Caliciviridae. In Virus Taxonomy, M.H.V.Regenmortel, C.M.Fauquet, and D.H.L.Bishop, eds. (New York: Academic Press), pp. 725–734.

Green, K.Y., Kapikian, A.Z., Valdesuso, J., Sosnovtsev, S., Treanor, J.J., and Lew, J.F. (1997). Expression and self-assembly of recombinant capsid protein from the antigenically distinct Hawaii human calicivirus. J. Clin. Microbiol. *35*, 1909–1914.

Guiver, M., Littler, E., Caul, E.O., and Fox, A.J. (1992). The cloning, sequencing and expression of a major antigenic region from the feline calicivirus capsid protein. J. Gen. Virol. *73*, 2429–2433.

Guo, M., Qian, Y., Chang, K.O., and Saif, L.J. (2001). Expression and self-assembly in baculovirus of porcine enteric calicivirus capsids into virus-like particles and their use in an enzyme-linked immunosorbent assay for antibody detection in swine. J. Clin. Microbiol. 39, 1487–1493.

Gutierrez-Escolano, A.L., Brito, Z.U., del Angel, R.M., and Jiang, X. (2000). Interaction of cellular proteins with the 5' end of Norwalk virus genomic RNA. J. Virol. 74, 8558–8562.

Gutierrez-Escolano, A.L., Vazquez-Ochoa, M., Escobar-Herrera, J., and Hernandez-Acosta, J. (2003). La, PTB, and PAB proteins bind to the 3(') untranslated region of Norwalk virus genomic RNA. Biochem. Biophys. Res. Commun. 311, 759–766.

Hellen, C.U., and Sarnow, P. (2001). Internal ribosome entry sites in eukaryotic mRNA molecules. Genes Dev. 15, 1593–1612.

Herbert, T.P., Brierley, I., and Brown, T.D. (1997). Identification of a protein linked to the genomic and subgenomic mRNAs of feline calicivirus and its role in translation. J. Gen. Virol. 78, 1033–1040.

Hogbom, M., Jager, K., Robel, I., Unge, T., and Rohayem, J. (2009). The active form of the norovirus RNA-dependent RNA polymerase is a homodimer with cooperative activity. J. Gen. Virol. 90, 281–291.

Horimoto, T., Takeda, Y., Iwatsuki-Horimoto, K., Sugii, S., and Tajima, T. (2001). Capsid protein gene variation among feline calicivirus isolates. Virus Genes 23, 171–174.

Hurley, K.E., Pesavento, P.A., Pedersen, N.C., Poland, A.M., Wilson, E., and Foley, J.E. (2004). An outbreak of virulent systemic feline calicivirus disease. J. Am. Vet. Med. Assoc. 224, 241–249.

Hurley, K.F., and Sykes, J.E. (2003). Update on feline calicivirus: new trends. Vet. Clin. North Am. Small Anim Pract. 33, 759–772.

Jackson, R.J. (2000). A comparative view of initiation site selection mechanisms. In Translational Control of Gene Expression, N.Sonenberg, J.W.B.Hershey, and M.B.Mathews, eds. (Cold Spring Harbor, NY: Cold Spring Harbor Laboratory Press), pp. 127–183.

Jackson, R.J. (2005). Alternative mechanisms of initiating translation of mammalian mRNAs. Biochem. Soc. Trans. 33, 1231–1241.

Jecht, M., Probst, C., and Gauss-Muller, V. (1998). Membrane permeability induced by hepatitis A virus proteins 2B and 2BC and proteolytic processing of HAV 2BC. Virology 252, 218–227.

Jiang, X., Matson, D.O., Ruiz-Palacios, G.M., Hu, J., Treanor, J., and Pickering, L.K. (1995). Expression, self-assembly, and antigenicity of a snow mountain agent-like calicivirus capsid protein. J. Clin. Microbiol. 33, 1452–1455.

Jiang, X., Zhong, W., Kaplan, M., Pickering, L.K., and Matson, D.O. (1999). Expression and characterization of Sapporo-like human calicivirus capsid proteins in baculovirus. J. Virol. Methods 78, 81–91.

Kaiser, W.J., Chaudhry, Y., Sosnovtsev, S.V., and Goodfellow, I.G. (2006). Analysis of protein–protein interactions in the feline calicivirus replication complex. J. Gen. Virol. 87, 363–368.

Kapp, L.D., and Lorsch, J.R. (2004). The molecular mechanics of eukaryotic translation. Annu. Rev. Biochem. 73, 657–704.

Karakasiliotis, I., Chaudhry, Y., Roberts, L.O., and Goodfellow, I.G. (2006). Feline calicivirus replication: requirement for polypyrimidine tract-binding protein is temperature-dependent. J. Gen. Virol. 87, 3339–3347.

König, M., Thiel, H.J., and Meyers, G. (1998). Detection of viral proteins after infection of cultured hepatocytes with rabbit hemorrhagic disease virus. J. Virol. 72, 4492–4497.

Kozak, M. (1987). Effects of intercistronic length on the efficiency of reinitiation by eucaryotic ribosomes. Mol. Cell Biol. 7, 3438–3445.

Kozak, M. (1991). An analysis of vertebrate mRNA sequences: intimations of translational control. J. Cell Biol. 115, 887–903.

Kozak, M. (1999). Initiation of translation in prokaryotes and eukaryotes. Gene 234, 187–208.

Kozak, M. (2002). Pushing the limits of the scanning mechanism for initiation of translation. Gene 299, 1–34.

Kreutz, L.C., and Seal, B.S. (1995). The pathway of feline calicivirus entry. Virus Res. 35, 63–70.

Kreutz, L.C., Seal, B.S., and Mengeling, W.L. (1994). Early interaction of feline calicivirus with cells in culture. Arch. Virol. 136, 19–34.

Kuyumcu-Martinez, M., Belliot, G., Sosnovtsev, S.V., Chang, K.O., Green, K.Y., and Lloyd, R.E. (2004). Calicivirus 3C-like proteinase inhibits cellular translation by cleavage of poly(A)-binding protein. J. Virol. 78, 8172–8182.

Latorre, P., Kolakofsky, D., and Curran, J. (1998). Sendai virus Y proteins are initiated by a ribosomal shunt. Mol. Cell Biol. 18, 5021–5031.

Laurent, S., Vautherot, J.F., Madelaine, M.F., Le, G.G., and Rasschaert, D. (1994). Recombinant rabbit hemorrhagic disease virus capsid protein expressed in baculovirus self-assembles into viruslike particles and induces protection. J. Virol. 68, 6794–6798.

Leite, J.P., Ando, T., Noel, J.S., Jiang, B., Humphrey, C.D., Lew, J.F., Green, K.Y., Glass, R.I., and Monroe, S.S. (1996). Characterization of Toronto virus capsid protein expressed in baculovirus. Arch. Virol. 141, 865–875.

Luttermann, C., and Meyers, G. (2009). The importance of inter- and intromolecular base pairing for translation reinitiation on a eukaryotic bicistronic mRNA. Genes Dev. 23, 331–344.

Luttermann, C., and Meyers, G. (2007). A bipartite sequence motif induces translation reinitiation in feline calicivirus RNA. J. Biol. Chem. 282, 7056–7065.

Makino, A., Shimojima, M., Miyazawa, T., Kato, K., Tohya, Y., and Akashi, H. (2006). Junctional adhesion molecule 1 is a functional receptor for feline calicivirus. J. Virol. 80, 4482–4490.

Mandell, K.J., and Parkos, C.A. (2005). The JAM family of proteins. Adv. Drug Deliv. Rev. 57, 857–867.

Marin, M.S., Casais, R., Alonso, J.M., and Parra, F. (2000). ATP binding and ATPase activities associated with recombinant rabbit hemorrhagic disease virus 2C-like polypeptide. J. Virol. 74, 10846–10851.

McCormick, C.J., Salim, O., Lambden, P.R., and Clarke, I.N. (2008). Translation termination reinitiation between open reading frame 1 (ORF1) and ORF2 enables capsid expression in a bovine norovirus without the need for production of viral subgenomic RNA. J. Virol. 82, 8917–8921.

Meyers, G. (2007). Characterization of the sequence element directing translation reinitiation in RNA of the calicivirus rabbit hemorrhagic disease virus. J. Virol. 81, 9623–9632.

Meyers, G. (2003). Translation of the minor capsid protein of a calicivirus is initiated by a novel termination-dependent reinitiation mechanism. J. Biol. Chem. 278, 34051–34060.

Meyers, G., Wirblich, C., and Thiel, H.J. (1991). Genomic and subgenomic RNAs of rabbit hemorrhagic disease virus are both protein-linked and packaged into particles. Virology 184, 677–686.

Meyers, G., Wirblich, C., Thiel, H.J., and Thumfart, J.O. (2000). Rabbit hemorrhagic disease virus: genome organization and polyprotein processing of a calicivirus studied after transient expression of cDNA constructs. Virology 276, 349–363.

Milton, I.D., Turner, J., Teelan, A., Gaskell, R., Turner, P.C., and Carter, M.J. (1992). Location of monoclonal antibody binding sites in the capsid protein of feline calicivirus. J. Gen. Virol. 73, 2435–2439.

Mitra, T., Sosnovtsev, S.V., and Green, K.Y. (2004). Mutagenesis of tyrosine 24 in the VPg protein is lethal for feline calicivirus. J. Virol. 78, 4931–4935.

Mochizuki, M., Kawakami, K., Hashimoto, M., and Ishida, T. (2000). Recent epidemiological status of feline upper respiratory infections in Japan. J. Vet. Med. Sci. 62, 801–803.

Morales, M., Barcena, J., Ramirez, M.A., Boga, J.A., Parra, F., and Torres, J.M. (2004). Synthesis in vitro of rabbit hemorrhagic disease virus subgenomic RNA by internal initiation on (–)sense genomic RNA: mapping of a subgenomic promoter. J. Biol. Chem. 279, 17013–17018.

Natoni, A., Kass, G.E., Carter, M.J., and Roberts, L.O. (2006). The mitochondrial pathway of apoptosis is triggered during feline calicivirus infection. J. Gen. Virol. 87, 357–361.

Neill, J.D. (2002). The subgenomic RNA of feline calicivirus is packaged into viral particles during infection. Virus Res. 87, 89–93.

Neill, J.D. (1992). Nucleotide sequence of the capsid protein gene of two serotypes of San Miguel sea lion virus: identification of conserved and non-conserved amino acid sequences among calicivirus capsid proteins. Virus Res. 24, 211–222.

Neill, J.D., and Mengeling, W.L. (1988). Further characterization of the virus-specific RNAs in feline calicivirus infected cells. Virus Res. 11, 59–72.

Neill, J.D., Reardon, I.M., and Heinrikson, R.L. (1991). Nucleotide sequence and expression of the capsid protein gene of feline calicivirus. J. Virol. 65, 5440–5447.

Neill, J.D., Sosnovtsev, S.V., and Green, K.Y. (2000). Recovery and altered neutralization specificities of chimeric viruses containing capsid protein domain exchanges from antigenically distinct strains of feline calicivirus. J. Virol. 74, 1079–1084.

Oka, T., Yamamoto, M., Yokoyama, M., Ogawa, S., Hansman, G.S., Katayama, K., Miyashita, K., Takagi, H., Tohya, Y., Sato, H., and Takeda, N. (2007). Highly conserved configuration of catalytic amino acid residues among calicivirus-encoded proteases. J. Virol. 81, 6798–6806.

Ossiboff, R.J., and Parker, J.S. (2007). Identification of regions and residues in feline junctional adhesion molecule required for feline calicivirus binding and infection. J. Virol. 81, 13608–13621.

Ossiboff, R.J., Sheh, A., Shotton, J., Pesavento, P.A., and Parker, J.S. (2007). Feline caliciviruses (FCVs) isolated from cats with virulent systemic disease possess in vitro phenotypes distinct from those of other FCV isolates. J. Gen. Virol. 88, 506–517.

Park, H.S., Himmelbach, A., Browning, K.S., Hohn, T., and Ryabova, L.A. (2001). A plant viral 'reinitiation' factor interacts with the host translational machinery. Cell 106, 723–733.

Pedersen, N.C., Elliott, J.B., Glasgow, A., Poland, A., and Keel, K. (2000). An isolated epizootic of hemorrhagic-like fever in cats caused by a novel and highly virulent strain of feline calicivirus. Vet. Microbiol. 73, 281–300.

Pedersen, N.C., and Hawkins, K.F. (1995). Mechanisms for persistence of acute and chronic feline calicivirus infections in the face of vaccination. Vet. Microbiol. 47, 141–156.

Peri, S., and Pandey, A. (2001). A reassessment of the translation initiation codon in vertebrates. Trends Genet. 17, 685–687.

Pesavento, P.A., MacLachlan, N.J., llard-Telm, L., Grant, C.K., and Hurley, K.F. (2004). Pathologic, immuno-histochemical, and electron microscopic findings in naturally occurring virulent systemic feline calicivirus infection in cats. Vet. Pathol. 41, 257–263.

Pestova, T.V., Kolupaeva, V.G., Lomakin, I.B., Pilipenko, E.V., Shatsky, I.N., Agol, V.I., and Hellen, C.U. (2001). Molecular mechanisms of translation initiation in eukaryotes. Proc. Natl. Acad. Sci. U.S.A. 98, 7029–7036.

Pfister, T., and Wimmer, E. (2001). Polypeptide p41 of a Norwalk-like virus is a nucleic acid-independent nucleoside triphosphatase. J. Virol. 75, 1611–1619.

Pisarev, A.V., Kolupaeva, V.G., Yusupov, M.M., Hellen, C.U., and Pestova, T.V. (2008). Ribosomal position and contacts of mRNA in eukaryotic translation initiation complexes. EMBO J. 27, 1609–1621.

Poulet, H., Brunet, S., Leroy, V., and Chappuis, G. (2005). Immunisation with a combination of two complementary feline calicivirus strains induces a broad cross-protection against heterologous challenges. Vet. Microbiol. 106, 17–31.

Povey, C., and Ingersoll, J. (1975). Cross-protection among feline caliciviruses. Infect. Immun. 11, 877–885.

Povey, R.C. (1974). Serological relationships among feline caliciviruses. Infect. Immun. 10, 1307–1314.

Pöyry, T.A., Kaminski, A., Connell, E.J., Fraser, C.S., and Jackson, R.J. (2007). The mechanism of an exceptional case of reinitiation after translation of a long ORF reveals why such events do not generally occur in mammalian mRNA translation. Genes Dev. 21, 3149–3162.

Prasad, B.V., Hardy, M.E., Dokland, T., Bella, J., Rossmann, M.G., and Estes, M.K. (1999). X-ray crystallographic structure of the Norwalk virus capsid. Science 286, 287–290.

Prasad, B.V., Matson, D.O., and Smith, A.W. (1994). Three-dimensional structure of calicivirus. J. Mol. Biol. 240, 256–264.

Prota, A.E., Campbell, J.A., Schelling, P., Forrest, J.C., Watson, M.J., Peters, T.R., urrand-Lions, M., Imhof, B.A., Dermody, T.S., and Stehle, T. (2003). Crystal structure of human junctional adhesion molecule 1: implications for reovirus binding. Proc. Natl. Acad. Sci. U.S.A. 100, 5366–5371.

Radford, A.D., Bennett, M., McArdle, F., Dawson, S., Turner, P.C., Glenn, M.A., and Gaskell, R.M. (1997). The use of sequence analysis of a feline calicivirus (FCV) hypervariable region in the epidemiological investigation of FCV related disease and vaccine failures. Vaccine 15, 1451–1458.

Radford, A.D., Bennett, M., McArdle, F., Dawson, S., Turner, P.C., Williams, R.A., Glenn, M.A., and Gaskell, R.M. (1999a). Quasispecies evolution of a hypervariable region of the feline calicivirus capsid gene in cell culture and persistently infected cats. Vet. Microbiol. 69, 67–68.

Radford, A.D., Coyne, K.P., Dawson, S., Porter, C.J., and Gaskell, R.M. (2007). Feline calicivirus. Vet. Res. 38, 319–335.

Radford, A.D., Dawson, S., Ryvar, R., Coyne, K., Johnson, D.R., Cox, M.B., Acke, E.F., Addie, D.D., and Gaskell, R.M. (2003). High genetic diversity of the immunodominant region of the feline calicivirus capsid gene in endemically infected cat colonies. Virus Genes 27, 145–155.

Radford, A.D., Willoughby, K., Dawson, S., McCracken, C., and Gaskell, R.M. (1999b). The capsid gene of feline calicivirus contains linear B-cell epitopes in both variable and conserved regions. J. Virol. 73, 8496–8502.

Robel, I., Gebhardt, J., Mesters, J.R., Gorbalenya, A., Coutard, B., Canard, B., Hilgenfeld, R., and Rohayem, J. (2008). Functional characterization of the cleavage specificity of the sapovirus chymotrypsin-like protease. J. Virol. 82, 8085–8093.

Roberts, L.O., Al-Molawi, N., Carter, M.J., and Kass, G.E. (2003). Apoptosis in cultured cells infected with feline calicivirus. Ann. N. Y. Acad. Sci. 1010, 587–590.

Rohayem, J., Robel, I., Jager, K., Scheffler, U., and Rudolph, W. (2006). Protein-primed and de novo initiation of RNA synthesis by norovirus 3Dpol. J. Virol. 80, 7060–7069.

Ryabova, L.A., Pooggin, M.M., and Hohn, T. (2002). Viral strategies of translation initiation: ribosomal shunt and reinitiation. Prog. Nucleic Acid Res. Mol. Biol. 72, 1–39.

Ryabova, L.A., Pooggin, M.M., and Hohn, T. (2006). Translation reinitiation and leaky scanning in plant viruses. Virus Res. 119, 52–62.

Sandoval, I.V., and Carrasco, L. (1997). Poliovirus infection and expression of the poliovirus protein 2B provoke the disassembly of the Golgi complex, the or-ganelle target for the antipoliovirus drug Ro-090179. J. Virol. 71, 4679–4693.

Sarnow, P., Cevallos, R.C., and Jan, E. (2005). Takeover of host ribosomes by divergent IRES elements. Biochem. Soc. Trans. 33, 1479–1482.

Sato, Y., Ohe, K., Murakami, M., Fukuyama, M., Furuhata, K., Kishikawa, S., Suzuki, Y., Kiuchi, A., Hara, M., Ishikawa, Y., and Taneno, A. (2002). Phylogenetic analysis of field isolates of feline calicivirus (FCV) in Japan by sequencing part of its capsid gene. Vet. Res. Commun. 26, 205–219.

Schaffer, F.L., Bachrach, H.L., Brown, F., Gillespie, J.H., Burroughs, J.N., Madin, S.H., Madeley, C.R., Povey, R.C., Scott, F., Smith, A.W., and Studdert, M.J. (1980a). Caliciviridae. InterVirology 14, 1–6.

Schaffer, F.L., Ehresmann, D.W., Fretz, M.K., and Soergel, M.I. (1980b). A protein, VPg, covalently linked to 36S calicivirus RNA. J. Gen. Virol. 47, 215–220.

Schneider, R.J., and Mohr, I. (2003). Translation initiation and viral tricks. Trends Biochem. Sci. 28, 130–136.

Seal, B.S. (1994). Analysis of capsid protein gene variation among divergent isolates of feline calicivirus. Virus Res. 33, 39–53.

Seal, B.S., Neill, J.D., and Ridpath, J.F. (1994). Predicted stem–loop structures and variation in nucleotide sequence of 3′ noncoding regions among animal calicivirus genomes. Virus Genes 8, 243–247.

Seal, B.S., Ridpath, J.F., and Mengeling, W.L. (1993). Analysis of feline calicivirus capsid protein genes: identification of variable antigenic determinant regions of the protein. J. Gen. Virol. 74, 2519–2524.

Semler, B.L., and Wimmer, E. (2002). Molecular biology of picornaviruses (Washington, DC: ASM Press).

Sibilia, M., Boniotti, M.B., Angoscini, P., Capucci, L., and Rossi, C. (1995). Two independent pathways of expression lead to self-assembly of the rabbit hemorrhagic disease virus capsid protein. J. Virol. 69, 5812–5815.

Simmonds, P., Karakasiliotis, I., Bailey, D., Chaudhry, Y., Evans, D.J., and Goodfellow, I.G. (2008). Bioinformatic and functional analysis of RNA secondary structure elements among different genera of human and animal caliciviruses. Nucleic Acids Res. 36, 2530–2546.

Sosnovtsev, S., and Green, K.Y. (1995). RNA transcripts derived from a cloned full-length copy of the feline calicivirus genome do not require VpG for infectivity. Virology 210, 383–390.

Sosnovtsev, S.V., Belliot, G., Chang, K.O., Onwudiwe, O., and Green, K.Y. (2005). Feline calicivirus VP2 is essential for the production of infectious virions. J. Virol. 79, 4012–4024.

Sosnovtsev, S.V., Garfield, M., and Green, K.Y. (2002). Processing map and essential cleavage sites of the nonstructural polyprotein encoded by ORF1 of the feline calicivirus genome. J. Virol. 76, 7060–7072.

Sosnovtsev, S.V., and Green, K.Y. (2000). Identification and genomic mapping of the ORF3 and VPg proteins in feline calicivirus virions. Virology 277, 193–203.

Sosnovtsev, S.V., Prikhod'ko, E.A., Belliot, G., Cohen, J.I., and Green, K.Y. (2003). Feline calicivirus replication induces apoptosis in cultured cells. Virus Res. *94*, 1–10.

Sosnovtsev, S.V., Sosnovtseva, S.A., and Green, K.Y. (1998). Cleavage of the feline calicivirus capsid precursor is mediated by a virus-encoded proteinase. J. Virol. *72*, 3051–3059.

Sosnovtseva, S.A., Sosnovtsev, S.V., and Green, K.Y. (1999). Mapping of the feline calicivirus proteinase responsible for autocatalytic processing of the non-structural polyprotein and identification of a stable proteinase-polymerase precursor protein. J. Virol. *73*, 6626–6633.

Stacey, S.N., Jordan, D., Williamson, A.J., Brown, M., Coote, J.H., and Arrand, J.R. (2000). Leaky scanning is the predominant mechanism for translation of human papillomavirus type 16 E7 oncoprotein from E6/E7 bicistronic mRNA. J. Virol. *74*, 7284–7297.

Stuart, A.D., and Brown, T.D. (2006). Entry of feline calicivirus is dependent on clathrin-mediated endocytosis and acidification in endosomes. J. Virol. *80*, 7500–7509.

Stuart, A.D., and Brown, T.D. (2007). Alpha2, 6-linked sialic acid acts as a receptor for Feline calicivirus. J. Gen. Virol. *88*, 177–186.

Studdert, M.J. (1978). Caliciviruses. Brief review. Arch. Virol. *58*, 157–191.

Teterina, N.L., Bienz, K., Egger, D., Gorbalenya, A.E., and Ehrenfeld, E. (1997). Induction of intracellular membrane rearrangements by HAV proteins 2C and 2BC. Virology *237*, 66–77.

Thackray, L.B., Wobus, C.E., Chachu, K.A., Liu, B., Alegre, E.R., Henderson, K.S., Kelley, S.T., and Virgin, H.W.T. (2007). Murine noroviruses comprising a single genogroup exhibit biological diversity despite limited sequence divergence. J. Virol. *81*, 10460–10473.

Thumfart, J.O., and Meyers, G. (2002a). Feline calicivirus: recovery of wild-type and recombinant viruses after transfection of cRNA or cDNA constructs. J. Virol. *76*, 6398–6407.

Thumfart, J.O., and Meyers, G. (2002b). Rabbit hemorrhagic disease virus: identification of a cleavage site in the viral polyprotein that is not processed by the known calicivirus protease. Virology *304*, 352–363.

Tohya, Y., Masuoka, K., Takahashi, E., and Mikami, T. (1991). Neutralizing epitopes of feline calicivirus. Arch. Virol. *117*, 173–181.

Tohya, Y., Yokoyama, N., Maeda, K., Kawaguchi, Y., and Mikami, T. (1997). Mapping of antigenic sites involved in neutralization on the capsid protein of feline calicivirus. J. Gen. Virol. *78*, 303–305.

Vazquez, A.L., Martin Alonso, J.M., Casais, R., Boga, J.A., and Parra, F. (1998). Expression of enzymatically active rabbit hemorrhagic disease virus RNA-dependent RNA polymerase in *Escherichia coli*. J. Virol. *72*, 2999–3004.

Wai, H.H., Vu, L., Oakes, M., and Nomura, M. (2000). Complete deletion of yeast chromosomal rDNA repeats and integration of a new rDNA repeat: use of rDNA deletion strains for functional analysis of rDNA promoter elements in vivo. Nucleic Acids Res. *28*, 3524–3534.

Wardley, R.C. (1976). Feline calicivirus carrier state. A study of the host/virus relationship. Arch. Virol. *52*, 243–249.

Wardley, R.C., Gaskell, R.M., and Povey, R.C. (1974). Feline respiratory viruses – their prevalence in clinically healthy cats. J. Small Anim Pract. *15*, 579–586.

Wardley, R.C., and Povey, R.C. (1977). The clinical disease and patterns of excretion associated with three different strains of feline caliciviruses. Res. Vet. Sci. *23*, 7–14.

Willcocks, M.M., Carter, M.J., and Roberts, L.O. (2004). Cleavage of eukaryotic initiation factor eIF4G and inhibition of host-cell protein synthesis during feline calicivirus infection. J. Gen. Virol. *85*, 1125–1130.

Williams, J. C., Liu, B. L., Lambden, P. R., and Clarke, I. N. Expression of SRSV ORFs 2 and 3: assembly of virus-like particles is independent of ORF3 activity. In: Chasey, D., Gaskell, C.J., and Clarke, I.N. (1997). Proceedings of the 1st International Symposium on Caliciviruses (Reading: European Society for Veterinary Virology), pp. 51–58.

Wirblich, C., Thiel, H.J., and Meyers, G. (1996). Genetic map of the calicivirus rabbit hemorrhagic disease virus as deduced from in vitro translation studies. J. Virol. *70*, 7974–7983.

Xi, Q., Cuesta, R., and Schneider, R.J. (2005). Regulation of translation by ribosome shunting through phospho-tyrosine-dependent coupling of adenovirus protein 100k to viral mRNAs. J. Virol. *79*, 5676–5683.

Xu, Z.J., and Chen, W.X. (1989). Viral haemorrhagic disease in rabbits: a review. Vet. Res. Commun. *13*, 205–212.

Zamyatkin, D.F., Parra, F., Alonso, J.M., Harki, D.A., Peterson, B.R., Grochulski, P., and Ng, K.K. (2008). Structural insights into mechanisms of catalysis and inhibition in Norwalk virus polymerase. J. Biol. Chem. *283*, 7705–7712.

Zhou, L., Yu, Q., and Luo, M. (1994). Characterization of two density populations of feline calicivirus particles. Virology *205*, 530–533.

Caliciviruses in Swine

Yunjeong Kim and Kyeong-Ok Chang

Abstract

Viruses in three of the four established genera of the family *Caliciviridae* have been detected in pigs (*Sapovirus*, *Norovirus* and *Vesivirus*), making this animal species of particular interest in the study of calicivirus pathogenesis and host range. The Cowden strain of porcine enteric calicivirus (PEC), a sapovirus, was discovered in a diarrhoeic pig faecal sample in the US in 1980. Since then, sapoviruses have become recognized as a predominant calicivirus detected in pigs. The Cowden PEC strain grows efficiently in a unique cell culture system, and a reverse genetics system has been developed for elucidation of the mechanisms of replication and pathogenesis at the molecular level. Porcine noroviruses share genetic relatedness with those from humans, and recent studies have shown that pigs are susceptible to infection and mild diarrhoeal disease when experimentally challenged with related human norovirus strains. Research on porcine caliciviruses has yielded new insights into the mechanisms of pathogenesis, replication, and evolution of the family *Caliciviridae*.

Introduction

Caliciviruses (family *Caliciviridae*) are small, non-enveloped viruses with a diameter of 27–35 nm. They possess a single-stranded, positive-sense genomic RNA of 7–8 kb, which encodes for non-structural proteins, a major structural capsid protein (VP1) and a small basic protein (VP2) (Green *et al.*, 2001). Caliciviruses cause a wide variety of diseases in humans and animals that include gastroenteritis. At present, there are four recognized genera in the family *Caliciviridae*: *Norovirus*, *Sapovirus*, *Vesivirus* and *Lagovirus* (Green *et al.*, 2000). In addition, two new genera, which cause disease in bovine and monkey, have been proposed (Farkas *et al.*, 2008; Oliver *et al.*, 2006; Smiley *et al.*, 2002).

Despite extensive attempts to grow human noroviruses (*Norovirus*), these viruses have remained uncultivable in routine cell cultures (Duizer *et al.*, 2004). Based on their capsid sequences there are at least five genogroups (GI–GV) of noroviruses and these can be further subdivided into at least 29 genotypes (Katayama *et al.*, 2002; Wang *et al.*, 2007). GI and GII noroviruses infect humans and pigs, whereas GIII, GIV, and GV noroviruses mostly infect pigs, cattle and mice (Green, 2007; Green *et al.*, 2001; Mayo, 2002). Sapoviruses (*Sapovirus*) have been classified into at least five genogroups and at least nine genotypes based on their complete capsid sequences (Farkas *et al.*, 2004). GI, GII, GIV and GV sapoviruses infect humans whereas GIII sapoviruses infect pigs. Vesicular exanthema swine virus (VESV) (*Vesivirus*) was the first known calicivirus, initially recognized in 1932 (Studdert, 1978). Vesiviruses also infect primate, cetacean, bovine, skunk and feline species (Matson *et al.*, 1996; Neill *et al.*, 1995). Rabbit haemorrhagic disease virus (RHDV) and European brown hare syndrome virus (EBHSV) belong to the genus *Lagovirus*. RHD was first reported in 1984 in China; currently it is endemic in East Asia, Europe and Oceania (Ohlinger *et al.*, 1993). Recently, RHD outbreaks were also recorded in the US (Campagnolo *et al.*, 2003).

As noted above, pigs serve as hosts for caliciviruses in three of the four recognized genera (*Sapovirus, Norovirus* and *Vesivirus*). This chapter highlights recent findings in porcine calicivirus research, with an emphasis on the porcine sapovirus Cowden strain.

Sapoviruses in swine

The Cowden strain of porcine enteric calicivirus (PEC) was first detected in a diarrhoeic pig faecal sample by electron microscopy (EM) in the US in 1980 (Saif *et al.*, 1980). Since then, porcine sapoviruses have become recognized as the predominant caliciviruses detected in pigs (Barry *et al.*, 2008; Jeong *et al.*, 2007; Martella *et al.*, 2008; Martinez *et al.*, 2006; Mauroy *et al.*, 2008; Reuter *et al.*, 2007; Schuffenecker *et al.*, 2001; Wang *et al.*, 2006a,b; Yin *et al.*, 2006).

Prevalence of porcine sapoviruses
Porcine sapoviruses infect pigs of all age groups. Porcine sapoviruses have been reported in the US, Venezuela, South Korea, Hungary, Belgium, Brazil, Italy and Japan (Barry *et al.*, 2008; Jeong *et al.*, 2007; Martella *et al.*, 2008; Martinez *et al.*, 2006; Mauroy *et al.*, 2008; Reuter *et al.*, 2007; Schuffenecker *et al.*, 2001; Wang *et al.*, 2006b; Yin *et al.*, 2006). Based on capsid or polymerase sequences most porcine sapoviruses belong to GIII (Cowden-like), however three new unclassified porcine sapovirus genogroups have been detected in diarrhoeic or non-diarrhoeic field samples: JJ681-like, LL26/K7-like and QW19-like (Wang *et al.*, 2007; Wang *et al.*, 2006b). A prevalence study of porcine sapoviruses conducted in US pigs showed that the Cowden-like sapoviruses (GIII) were the most prevalent, whereas JJ681-like and QW19-like sapoviruses were detected infrequently (Wang *et al.*, 2006b). Recently, Martinez *et al.* detected PEC in 18% of faecal samples from pigs (0–9 weeks of age with or without diarrhoea) in Venezuela (Martinez *et al.*, 2006) and Kim *et al.* detected Cowden-like strains in 5% of diarrhoeic faecal samples from nursing and weaned pigs in South Korea (Kim *et al.*, 2006).

Genetic characterization of Cowden PEC
Wild-type Cowden PEC was initially propagated in pig kidney cells (Flynn and Saif, 1988;

Parwani *et al.*, 1991). A tissue culture adapted virus was subsequently isolated from primary pig kidney and pig kidney cells (Flynn and Saif, 1988; Parwani *et al.*, 1991). The tissue culture system required the addition of intestinal content (IC) fluid from uninfected gnotobiotic pigs. The availability of the tissue culture system facilitated genetic analysis and the full-length genome sequence of Cowden PEC was determined in 1999. The Cowden PEC genome was 7320 nucleotides (nt) in length, polyadenylated at the 3′-end and organized into two main open reading frames (ORFs) (Fig. 9.1). ORF1 encodes a polyprotein of 2254 amino acids (aa), whereas ORF2 encodes a small basic protein of 164 aa (VP2) (Guo *et al.*, 1999). The virus-encoded protease cleaves the ORF1 polyprotein into mature non-structural proteins and a major capsid protein (VP1). The Cowden PEC ORF1 polyprotein cleavage pattern of non-structural proteins was found to be: NS1 (N-terminal protein), NS2, NS3 (NTPase), NS4 (3A-like protein), NS5 (VPg) and NS6–NS7 precursor (Proteinase-RNA polymerase) (Guo *et al.*, 1999). The dipeptide cleavage sites are considered conserved among other sapoviruses.

Phylogenetic analysis showed that Cowden PEC was more closely related to human sapoviruses than to human noroviruses (Liu *et al.*, 1995), confirming the classification of Cowden PEC as a member of the genus *Sapovirus* (Guo *et al.*, 1999). Comparison with other caliciviruses showed that the Cowden PEC N-terminal region (200 aa) of the ORF1 polyprotein was highly divergent from other caliciviruses, having only 15% aa sequence identity with human sapoviruses. However, there was high aa sequence identity between Cowden PEC and a human sapovirus (Manchester strain) in the NTPase region (50%), the proteinase region (44%) and the RNA polymerase region (66%). The Cowden PEC capsid shared 39% aa sequence identity with human sapoviruses, which was higher than vesiviruses (17.1–21.7%), lagoviruses (19%) and noroviruses (15–17.1%).

The production of Cowden PEC virus-like particles (VLPs) was accomplished by expressing the complete Cowden PEC VP1 in a baculovirus expression system (Guo *et al.*, 2001b). The Cowden PEC VLPs were morphologically similar to the native virus and reacted against pig

Sapovirus: Cowden porcine enteric virus

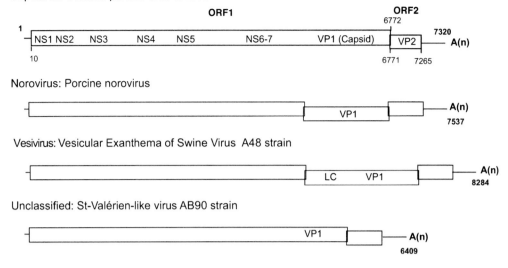

Figure 9.1 Genomic organization of representative porcine caliciviruses. Cowden porcine enteric calicivirus (sapovirus), a porcine norovirus, VESV (A48 strain) and St-Valérien-like virus (AB90 strain). While Cowden porcine enteric virus and St-Valérien-like virus have two open reading frames (ORF), porcine noroviruses and VESV have three ORF. VESV has a unique protein in ORF2 called LC (leader of the capsid).

hyperimmune and convalescent-phase sera specific for Cowden PEC (Guo *et al.*, 2001b).

Comparison of the Cowden PEC wild-type and tissue culture adapted RNA genomes showed 100% nucleotide sequence identity in the 5'-end of the genome, the NTPase, the entire ORF2 and the 3'-NTR (Guo *et al.*, 1999). However, there were a number of mutations in proteinase, polymerase and capsid genes. One silent mutation was in the proteinase, the polymerase and the capsid genes. Amino acid substitutions were in the polymerase (Y1252G and R1379K) and capsid genes (C178S, Y289H, N291D and K295R). These aa substitutions suggested that they were essential for the adaptation of Cowden PEC in tissue culture (Guo *et al.*, 1999). Three of the four aa substitutions in the capsid protein occurred in a short region (aa 289–295), which has been shown to be a receptor-binding site (P2 domain) (Prasad *et al.*, 1999; Prasad *et al.*, 1994; White *et al.*, 1996). These aa changes led to a localized higher hydrophilicity (Guo *et al.*, 1999). Interestingly, the P2 domain was also the site of aa changes during cell culture passage of murine norovirus (MNV) (Wobus *et al.*, 2004). A mixed population of viruses was detected in the early passages of MNV, in which there was either

a lysine or glutamic acid at residue 296 in the P2 domain. At the third passage, MNV became avirulent in mice, and the lysine at residue 296 emerged as the predominant residue (Wobus *et al.*, 2004). Recent studies with reverse genetics demonstrated that the substitution, lysine-to-glutamic acid, restored virulence *in vivo*, suggesting a virus-encoded molecular determinant of norovirus virulence in the P2 domain (Bailey *et al.*, 2008).

Reverse genetics for PEC

Reverse genetics systems provide an important tool for study of the molecular basis of replication and pathogenesis for caliciviruses (Chang *et al.*, 2005; Chaudhry *et al.*, 2007; Liu *et al.*, 2006, 2008; Sosnovtsev and Green, 1995; Thumfart and Meyers, 2002; Ward *et al.*, 2007; Wei *et al.*, 2008). A full-length cDNA copy of the tissue culture-adapted Cowden PEC genome was cloned and expressed in an *in vitro* transcription and translation reaction (Chang *et al.*, 2005). Compared with the wild-type sequence, the cloned genome contained ten mutations, which included four aa substitutions (two in the predicted N-terminal protein and two in the predicted NTPase protein). A radioimmunoprecipitation assay

(RIPA) demonstrated that the *in vitro* translated protein profiles were similar to those of the tissue culture-adapted infected cell lysates, indicating that the *in vitro* translated ORF1 underwent authentic proteolytic processing. Also, the *in vitro* RNA transcript was approximately 7.4 kb in size (Chang *et al.*, 2005). The RNA transcript was transfected into LLC-PK cells with a cap analogue and incubated in the absence or presence of IC. The RNA transcripts were only infectious in the presence of IC as measured by the appearance of cytopathic effects and capsid antigen expression in cells by IFA staining (Fig. 9.2) (Chang *et al.*, 2005). Virus recovery with IC was demonstrated and virus titres characteristically reached ~10^6 $TCID_{50}$/ml after transfection. Non-capped RNA transcripts were not infectious or there was no evidence of virus replication with or without IC. The pathogenicity of gnotobiotic pigs that were oral inoculated with the recovered viruses showed an attenuated phenotype similar to that of tissue culture Cowden PEC: limited virus shedding and mild or no diarrhoea in the pigs (Table 9.1).

Pathogenicity of Cowden PEC

Experimental studies of gnotobiotic pigs exposed to Cowden PEC indicated that infection resulted in profuse diarrhoea, intestinal lesions (villous atrophy) and anorexia (Flynn *et al.*, 1988). Four-day-old gnotobiotic pigs received oral challenge with Cowden PEC and developed mild or severe diarrhoea by postinoculation day (PID) 3 that persisted for 3–7 days. Virus shedding in faeces started as early as 1 PID and lasted up to 7 PID. The presence of the viral antigen in the small intestines was confirmed by a fluorescein-conjugated antibody specific to the strain and most immunofluorescence was observed in villous epithelial cells (primarily in the upper small intestines such as duodenum or jejunum)

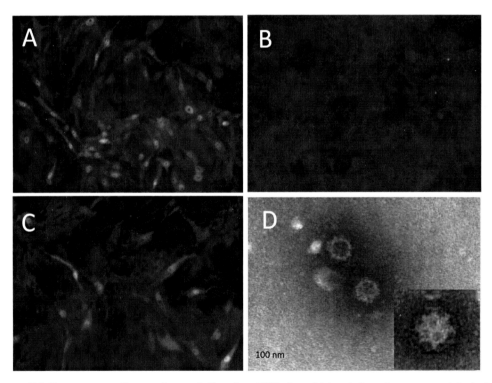

Figure 9.2 Reverse genetics system of Cowden PEC in which viral antigen was detected by immunofluorescence assay with hyperimmune guinea pig serum against VLP of Cowden PEC and by an electron microscopy after RNA transcripts were transfected into LLC-PK cells (from Chang *et al.*, with permission). Cells were transfected with RNA transcripts from pCV4A and incubated for 72 h with: (A) IC (1%); (B) mock-medium; or (C) GCDCA (200 µM). (D). for EM observation, samples were prepared after transfection of RNA transcripts and incubation with IC for 72 h, then concentrated 100× by ultracentrifugation for a negative staining (from Chang *et al.*, with permission).

Table 9.1 Pathogenesis of wild type (gnotobiotic pig-passaged) Cowden PEC and recovered Cowden PEC from the infectious clone (pCV4A) in gnotobiotic pigs (from Chang et al., 2005, with permission).

Animal	inoculum		Diarrhoea/virus shedding at post inoculation day											
			1	2	3	4	5	6	7	8	9	10	11	12
1	wtPEC	Diarrhoea	−a	−	±	±	EUT							
		Virus shedding	−b	−	+	+								
2	wtPEC	Diarrhoea	−	−	+	+	−	−	−	EUT				
		Virus shedding	−	−	+	+	+	+	+					
3	pCV4A	Diarrhoea	−	−	−	−	−	−	−	−	−	−	−	−
		Virus shedding	−	−	−	−	−	−	−	−	+	+	−	−
4	pCV4A	Diarrhoea	−	−	−	−	−	±	±	±	EUT			
		Virus shedding	−	−	−	−	−	−	+	+				

a +, faecal consistency of ≥2; ±, questionable or mild diarrhoea (faecal consistency of 1); −, no diarrhoea (normal faeces). EUT, euthanized.

b Virus shedding determined by ELISA.

in mucosal smears of inoculated pigs. The greatest number of infected enterocytes was observed in the duodenum, fewer infected enterocytes were detected in the jejunum and the fewest infected enterocytes were seen in the ileum. As with virus shedding, the viral antigen was positive at 1 PID and up to 7 PID. Histopathological findings showed villus atrophy in the duodenum and/ or the jejunum of the inoculated pigs. Short villi and long crypts were observed in the duodenum and the jejunum of the inoculated pigs from 3 to 7 PID. However, the histological sections of the ileum, colon, liver, lungs or kidney of inoculated pigs showed no lesion. Scanning EM confirmed the histopathological findings showing villous atrophy with shortening, blunting, fusion or absence of villi in the duodenum and the jejunum.

This enteropathogenicity was also confirmed by Guo et al. (Guo et al., 2001a). They performed comparative studies of tissue culture and wild-type Cowden PEC pathogenicity in gnotobiotic pigs and showed that there were marked differences in diarrhoea and faecal virus shedding between the tissue culture and wild-type Cowden PEC. The gnotobiotic pigs inoculated with wild-type Cowden PEC developed mild to moderate diarrhoea 2–4 PID and the diarrhoea persisted for 2–5 days with high cumulative faecal scores. However, there was no apparent diarrhoea observed in the pigs orally inoculated with tissue culture Cowden PEC. Interestingly, intravenous inoculation of wild-type or tissue culture Cowden PEC was comparable in levels of virus shedding and histopathology to oral inoculation in animals. Because genetic characterization of tissue culture and wild-type Cowden PEC demonstrated aa changes in the RNA polymerase and the capsid regions, the authors proposed that these changes may be critical for viral pathogenicity and cell culture adaptation.

The Cowden PEC cell culture system and bile acids

The Cowden PEC was initially isolated in cultured primary kidney cells with the addition of IC (approximately 10% volume/volume) derived from uninfected gnotobiotic pigs in the medium (Flynn and Saif, 1988). The addition of IC was to provide Cowden PEC, known to replicate in the intestinal tract, an environment that mimics in vivo conditions. In rotaviruses, an unrelated enteric virus, proteases in IC had been reported to

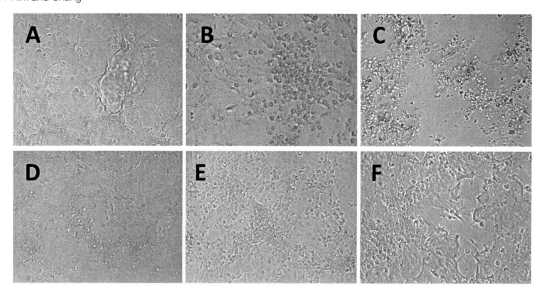

Figure 9.3 Microscopic changes (cytopathic effects) of LLC-PK cells after Cowden PEC inoculation at a MOI (multiplicity of infection) of 0.05 and incubation without (A) or with IC (B and C), or with IC and the modulators of cAMP/PKA pathway (D–F). Cells were incubated without IC for 72 h (A); with IC at 48 h (B) or 72 h (C); with IC and suramin (D), MDL12,330A (E) or NBEI (F), respectively (from Chang *et al.*, 2002, with permission).

activate virus by cleaving the outer capsid protein, VP4, allowing virus penetration and entry into the cell (Estes, 2001). Flynn and Saif prepared the PEC inoculum from the large-intestinal contents of a gnotobiotic pig infected with wild-type Cowden PEC that was passaged multiple times through gnotobiotic pigs (Flynn and Saif, 1988). These viruses were inoculated onto primary porcine kidney cells and enzymes or preparations containing molecules found in the intestinal tract were examined for their ability to support virus growth. The additives examined included IC (filtered material from the small or large intestinal contents of young uninfected gnotobiotic pigs), bile solution (filtered fluids from the gall bladders of uninfected gnotobiotic pigs), trypsin and pancreatin. Virus replication was monitored by IF assay with fluorescein isothiocyanate-conjugated-hyperimmune anti-PEC serum. After several serial passages, virus replication was evident in virus-infected cells incubated in the presence of IC (Flynn and Saif, 1988), however, there was no evidence of virus replication in the presence of trypsin or pancreatin. Increasing numbers of Cowden PEC-infected cells (up to 80%) were observed after each successive passage with IC in

the cell culture (Flynn and Saif, 1988). Limited Cowden PEC replication was observed also when virus infected cells were incubated in the presence of bile solution (Flynn and Saif, 1988). The cytopathic effects of Cowden PEC in the presence of IC included rounding of cells with eventual detachment and formation of holes in the cell monolayer (Flynn and Saif, 1988). Complete detachment of the cell monolayers usually occurred at 3–4 PID. Parwani *et al.* performed further biochemical studies of Cowden PEC grown in primary kidney cells in the presence of IC, and demonstrated the ability to amplify and purify Cowden PEC virions (Parwani *et al.*, 1990). Later, Cowden PEC was successfully adapted to serial propagation in LLC-PK cells with inclusion of IC in the medium (Parwani *et al.*, 1991). Although Cowden PEC was adapted and passaged in a continuous cell line, viral replication was continually dependent on IC. Unlike the case of rotaviruses, the mechanism did not involve enzymes of host origin or bacterial proteases (Flynn and Saif, 1988; Parwani *et al.*, 1991). Pretreatment of cells or the virus with IC did not induce virus growth unless the medium was supplemented with IC (Chang *et al.*, 2002). Likewise, transfection of

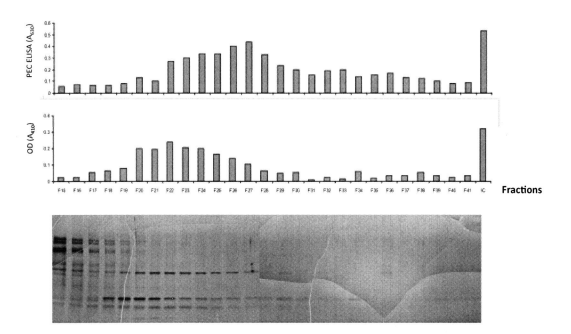

Figure 9.4 Analysis of fractions 15 to 41 obtained by size exclusion chromatography of IC. Top: the effects of each fraction on Cowden PEC growth. Virus antigen was measured by ELISA 120 h after virus infection in the presence of each fraction (1%). Virus incubated in the presence of IC (1%) was included as a control (the last lane). Middle panel, the absorbance of each fraction at 410 nm. Bottom: SDS-PAGE analysis of each fraction in a 10–20% polyacrylamide gel, and visualization of proteins by silver staining (from Chang *et al.*, with permission).

viral RNA into cells yielded virus replication but only in the presence of IC (Chang *et al.*, 2002). It was hypothesized that IC effects on virus replication might be related to IC-mediated signalling pathway. A large panel of inhibitors or promoters of cell signal transduction was examined for their effects on IC-mediated virus replication in LLC-PK cells (Chang *et al.*, 2002). The results suggested that cAMP and protein kinase A (PKA) pathway was important for IC-mediated PEC replication in cells: inhibitors of this pathway significantly reduced IC-mediated virus replication (Fig. 9.3).

Further studies were done to identify the factor(s) in IC that enabled the replication of Cowden PEC in cell culture (Chang *et al.*, 2004). Physical and chemical treatments of IC ruled out proteins and small peptides. Size fractionation studies showed that the factor(s) was small, <600 molecular weight (MW). Because the size of individual bile acid salts ranges between 300

and 500 MW and they are abundant in IC, it was hypothesized that bile acids might be the active factor in IC required for the replication of Cowden PEC in LLC-PK cells. Individual bile acids were examined for activity in Cowden PEC cell culture system, and it was discovered that several bile acids could each support Cowden PEC replication in the absence of IC. The bile acids glycochenodeoxycholic acid (GCDCA) and taurochenodeoxycholic acid (TCDCA) were most effective, and worked well at concentrations above 100 μM, comparable to the concentrations present in IC (Fig. 9.4) (Chang *et al.*, 2004). Furthermore, bile acids could be used instead of IC to recover virus from pig kidney cells following transfection of infectious RNA in the reverse genetics system (Chang *et al.*, 2005).

The identification of bile acids as the active factor in the IC fluid raised interesting questions regarding its role in the pathogenesis of Cowden PEC in the gut. Primary bile acids [cholic (CA)

A

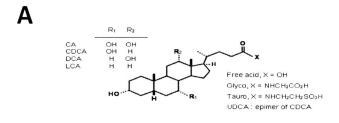

B

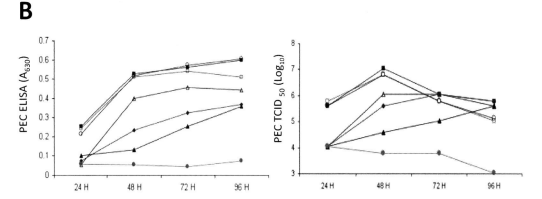

Figure 9.5 Chemical structures of bile acids and the comparison of growth kinetics as measured by ELISA or TCID$_{50}$. (A) The chemical structures of bile acids used in this study. (B) Cowden PEC was grown in the presence of mock fluid (●), IC (1%) (○), and various conjugated bile acids [GCDCA (■), TCDCA (□), GCA (▲), TLCA (△) and TDCA (◆)] at 100 mM concentration. Representative cell culture fluids assayed for viral antigen by ELISA at 24, 48, 72 and 96 h following Cowden PEC infection were tested also for the presence of infectious virus (TCID$_{50}$/ml) in order to verify that the observed increase in viral antigen was associated with an increase in virus titre (from Chang *et al.*, with permission).

and chenodeoxycholic acids (CDCA)] are synthesized from cholesterol and conjugated with glycine or taurine in the liver (Fig. 9.5). The conjugated or non-conjugated bile acids are collected and stored in the gallbladder at concentrations as high as 320 mM and their release into the duodenum is triggered primarily by hormones (cholecystokinin and secretin) in response to the presence of dietary fat in the duodenum (Johnson, 1998). While the secreted bile acids travel through the intestinal tract, they are re-absorbed into blood system (portal vein) by passive diffusion or active mechanisms with the specific bile acid receptor in the ileum (ileal bile acid transporter), which collects up to 95% of the secreted bile acids (Kullak-Ublick *et al.*, 2004). The first 2 h after a meal, the concentration of total bile acids in the upper small intestines is normally above 4 mM in humans (McLeod and Wiggins, 1968). The primary bile acids are

converted into secondary bile acids [lithocholic (LCA) and deoxycholic acids (DCA)] by normal microflora in the small intestines. The secondary bile acids enter the pool of bile acids and circulate in the system. Even though the concentration of bile acids in the systemic circulation is below 10 µM (Legrand-Defretin *et al.*, 1986), in the portal vein where the bile acids travel to the liver, the concentration of total bile acids could reach up to 80 µM in pigs (Legrand-Defretin *et al.*, 1986). This enterohepatic circulation is essential in maintaining an effective concentration of bile acids and cholesterol homeostasis. While travelling down to the intestines, most bile acids are re-absorbed, therefore the concentration of bile acids is much higher in the proximal intestinal tract (duodenum and jejunum) (Fig. 9.6) (McLeod and Wiggins, 1968). Interestingly, the viral antigen of Cowden PEC and lesions were observed in greater amounts in the proximal intestinal tract

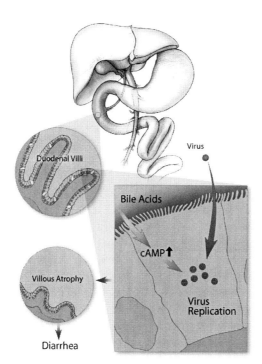

Figure 9.6 Schematic drawing for the effects of bile acids on Cowden PEC replication in the small intestines. Bile acids (~300 mM) stored in the gall bladder are released into the duodenum and actively re-absorbed in the ileum. Therefore, the concentration of bile acids is much higher in the proximal intestinal tract (duodenum and jejunum) (McLeod and Wiggins, 1968). This higher concentration of bile acids might explain the greater susceptibility of the proximal intestinal tract to Cowden PEC pathogenesis when compared with the distal intestinal tract (ileum) (Flynn *et al.*, 1988; Guo *et al.*, 2001a). Bile acids induce cAMP/PKA pathway in the susceptible cells (Chang *et al.*, 2004), and this may allow Cowden PEC to grow in the cells, which results in cell death and villus atrophy. These morphological changes disrupt the absorptive capacity of the intestinal tissue, resulting in diarrhoea.

compared with the distal intestinal tract (ileum) (Flynn *et al.*, 1988; Guo *et al.*, 2001a) which is in accordance with the concentration of bile acids. Recently it was discovered that the modulation of cholesterol pathway significantly influenced the replication of Norwalk virus in replicon-bearing cells (Chang, 2009), suggesting this model may be applicable to other enteric caliciviruses.

The mechanism by which bile acids support Cowden PEC replication in cells was examined. Because it was shown that IC-mediated growth of Cowden PEC in cells was related to cAMP/PKA pathway, this pathway was examined in the presence of various bile acids. First it was observed that bile acids significantly increased levels of intracellular cAMP in LLC-PK cells in correlation with their ability to support virus replication (Chang *et al.*, 2004). Second, inhibitors of cAMP/ PKA pathway significantly reduced the growth of

PEC in the presence of bile acids as previously reported for IC (Chang *et al.*, 2004). Several reports have shown that activation of the cAMP/ PKA pathway could cause down-regulation of STAT-1 activation which is a key molecule to IFN pathway, a major antiviral component in innate immunity (David *et al.*, 1996; Delgado, 2003; Lee and Rikihisa, 1998; Samuel, 2001; Sengupta *et al.*, 1996; Young *et al.*, 2003). Therefore, bile acid-mediated PKA was examined for STAT-1 activation in correlation with virus replication. When the activation of STAT-1 was measured in the presence of IFN in the absence or presence of bile acids, the activation of STAT-1 was significantly reduced in the presence of each bile acid. Furthermore, the inhibitors of PKA negated the effects of bile acids on STAT-1 activation. These results suggest the growth of Cowden PEC in the presence of bile acids was related to induction of PKA pathway and/or the down-regulation of innate immunity. Bile acids were also shown to promote the replication of HCV in replicon-harbouring cells at concentrations similar to those for Cowden PEC (Chang and George, 2007), and these results suggest a common mechanism is associated with virus replication in the presence of bile acids in both viruses. In the bile-rich organs such as the liver and the small intestines, it is possible that a positive factor for virus replication or a negative impact on innate immunity is provided for those viruses to exploit the condition and establish a productive infection.

Noroviruses in swine

Porcine noroviruses, represented by the Sw918 and Sw43 strains, were first detected in the caecum contents of slaughtered pigs in a meat processing facility in Japan (Sugieda et al., 1998). The striking genetic similarity with human noroviruses led to speculation that transmission between humans and pigs might occur. The ability to experimentally infect pigs with related human norovirus strains has led to increasing interest in the study of pigs as a model for both human and porcine noroviruses (Cheetham et al., 2006; Souza et al., 2007).

Prevalence of porcine noroviruses

The discovery of noroviruses in pigs by Sugieda et al. showed that 0.3% of samples (caecum contents) were positive by RT-PCR and nested PCR, suggesting a low prevalence of these viruses (Sugieda et al., 1998). However, the low prevalence in this initial study may have been due to the use of human norovirus-specific primers for screening. Recently, Wang et al. studied the prevalence of porcine noroviruses in US livestock by RT-PCR with broadly reactive primers targeting the viral polymerase region (Wang et al., 2006b). They demonstrated that porcine noroviruses genogroup II (GII) were detected exclusively in older finisher pigs (20–24 weeks of age) with an overall prevalence of 20%, while none was detected in nursing pigs and post-weaning pigs. Porcine noroviruses have been detected in pigs of the Netherlands, Belgium, Hungary, Canada, Japan and South Korea (Kim et al., 2006; Koopmans et al., 2000; Mattison et al., 2007; Mauroy et al., 2008; Park et al., 2007; Reuter et al., 2007;

Sugieda et al., 1998). A seroprevalence study of US pigs ($n = 110$) was performed with immunoassays that used recombinant norovirus VLPs produced from porcine GII strain Sw918, human GI strain Norwalk virus, or human GII strain Hawaii virus as antigen (Farkas et al., 2005). The rates of seropositive animals for each antigen were 71%, 63%, and 52%, respectively. The seroprevalence rate of antibodies specific for porcine GII norovirus (Sw918) in pigs surveyed in Japan was 36% (Farkas et al., 2005).

Genetic characterization of porcine noroviruses

The length of norovirus genome is 7.3–7.7 kb. The three ORFs encode a polyprotein (ORF1), a major capsid protein (ORF2), and a minor structural protein (ORF3), respectively. The polyprotein encoded by ORF1 is cleaved autocatalytically to produce non-structural proteins including NTPase, protease and polymerase. To date, all porcine noroviruses are grouped into GII and these could be subdivided into three genotypes (GII.11, GII.18 and GII.19). A close relatedness of porcine and human noroviruses suggested the ability of human noroviruses to replicate in gnotobiotic pigs (Cheetham et al., 2006; Souza et al., 2007). Recently, a GII.4 norovirus strain having an identical polymerase gene region to human GII.4 norovirus was found in faecal samples from pigs (Mattison et al., 2007).

Pathogenesis of porcine noroviruses

It is not clear if porcine noroviruses are significant pathogens because they have been detected in faecal samples from both healthy and diarrhoeic pigs. So far there is no report of experimental inoculation of porcine noroviruses in pigs. However, studies with human noroviruses in pigs have been conducted using the gnotobiotic pig model (Cheetham et al., 2006; Souza et al., 2007). When gnotobiotic pigs were inoculated with faecal filtrates of a norovirus GII.4 strain, 74% of pigs developed mild diarrhoea and rectal swab fluids and IC were positive for the virus for 1–4 PID. Seroconversion occurred after 21 PID in 13 of 22 pigs. Even though histopathological examination revealed only mild lesion in the proximal small intestine of one of seven pigs, immunofluorescent microscopy showed patchy infection of duodenal and jejunal enterocytes, indicating that gnotobiotic pigs could be used to study human norovirus replication (Cheetham *et al.*, 2006).

There are reports that the susceptibility of norovirus infection is related to histo-blood group antigen (HBGAs) in human (Hutson *et al.*, 2002; Lindesmith *et al.*, 2003; Tan *et al.*, 2008) (see Chapter 6). Multiple binding patterns of noroviruses to HBGAs including ABO, Lewis and secretor types of the human HBGA have been described (Huang *et al.*, 2003, 2005; Shirato *et al.*, 2008). It was reported that HBGAs in human saliva and porcine gastric mucin (PGM) bind to norovirus VLPs in a competitive manner, suggesting that human HBGA and PGM may share the same binding sites. Preincubation of VLPs with PGM completely inhibited VLP binding to Caco-2 cells (Tian *et al.*, 2005). It was found that a number of GI and GII human norovirus VLPs bound to the duodenum and buccal tissues from pigs with different A and H HBGA phenotypes (Cheetham *et al.*, 2007). When gnotobiotic pigs were challenged with a human GII norovirus strain significantly more A+ and H+ pigs shed viruses and seroconverted than non-A+ and non-H+ pigs, which was consistent with the binding profile (Cheetham *et al.*, 2007). Tian *et al.* also showed that mucin from pig stomach containing blood group A, H1 and Lewis b of HBGA binds to multiple GI and GII norovirus strains (Tian *et al.*, 2005).

Vesiviruses in swine

The first known calicivirus, VESV (genus *Vesivirus*), was initially recognized in 1932, and the disease became widespread in the US during 1950s. VESV may have arisen in pigs from feed containing marine mammals infected with San Miguel sea lion virus (SMSV) which causes vesicular lesions and reproductive failure in sea lions (Smith *et al.*, 1998). It caused vesicular lesions in pigs similar to foot-and-mouth disease, and was eradicated from pigs in the US by 1956 (Cubitt, 1994; Smith *et al.*, 1998), and it has never been reported as a natural infection in pigs in any other part of the world. The infection of VESV may cause fever up to 42°C and vesicles on the epithelium of the snout, lips, nostrils, tongue, feet and mammary glands which break within 24–48 h to form erosions (Smith *et al.*, 1980).

Newly recognized caliciviruses in swine

A novel calicivirus (St-Valérien-like viruses) was identified in the faecal samples from asymptomatic adult pigs in farms and abattoirs in Canada between 2005 and 2007 (L'Homme *et al.*, 2009). This novel calicivirus has 6409 nucleotides and two ORFs (Fig. 9.1). The ORF1 encodes a nonstructural polyprotein and a major capsid protein VP1, and ORF2 encodes a basic minor capsid protein. The genomic and phylogenetic analysis of the NTPase, polymerase and a major capsid protein places this novel virus in a unique cluster (likely a new genus) with a common root with the Tulane virus (calicivirus isolated from macaque) and the noroviruses.

Conclusion

Pigs are the host of a variety of caliciviruses that include sapovirus, norovirus and vesivirus. Epidemiological studies have demonstrated that sapoviruses are predominant in pigs and recent studies have detected closely matching human norovirus sequences in pigs. The study of porcine caliciviruses will give new insights into the mechanisms of calicivirus pathogenesis, since human noroviruses are uncultivable, and this may lead to the development of control strategies for the caliciviruses in both humans and animals.

References

Bailey, D., Thackray, L.B., and Goodfellow, I.G. (2008). A single amino acid substitution in the murine norovirus capsid protein is sufficient for attenuation in vivo. J. Virol. *82*, 7725–7728.

Barry, A.F., Alfieri, A.F., and Alfieri, A.A. (2008). High genetic diversity in RdRp gene of Brazilian porcine sapovirus strains. Vet. Microbiol. *131*, 185–191.

Campagnolo, E.R., Ernst, M.J., Berninger, M.L., Gregg, D.A., Shumaker, T.J., and Boghossian, A.M. (2003). Outbreak of rabbit hemorrhagic disease in domestic lagomorphs. J. Am. Vet. Med. Assoc. *223*, 1151–1155, 1128.

Chang, K.O. (2009). The Role of Cholesterol Pathways in Norovirus Replication. J. Virol. *83*, 8587–8595.

Chang, K.O., and George, D.W. (2007). Bile acids promote the expression of hepatitis C virus in replicon-harboring cells. J. Virol. *81*, 9633–9640.

Chang, K.O., Kim, Y., Green, K.Y., and Saif, L.J. (2002). Cell-culture propagation of porcine enteric calicivirus mediated by intestinal contents is dependent on the cyclic AMP signaling pathway. Virology *304*, 302–310.

Chang, K.O., Sosnovtsev, S.S., Belliot, G., Wang, Q., Saif, L.J., and Green, K.Y. (2005). Reverse genetics system for porcine enteric calicivirus, a prototype sapovirus in the Caliciviridae. J. Virol. 79, 1409–1416.

Chang, K.O., Sosnovtsev, S.V., Belliot, G., Kim, Y., Saif, L.J., and Green, K.Y. (2004). Bile acids are essential for porcine enteric calicivirus replication in association with down-regulation of signal transducer and activator of transcription 1. Proc. Natl. Acad. Sci. U.S.A. 101, 8733–8738.

Chaudhry, Y., Skinner, M.A., and Goodfellow, I.G. (2007). Recovery of genetically defined murine norovirus in tissue culture by using a fowlpox virus expressing T7 RNA polymerase. J. Gen. Virol. 88, 2091–2100.

Cheetham, S., Souza, M., McGregor, R., Meulia, T., Wang, Q., and Saif, L.J. (2007). Binding patterns of human norovirus-like particles to buccal and intestinal tissues of gnotobiotic pigs in relation to A/H histo-blood group antigen expression. J. Virol. 81, 3535–3544.

Cheetham, S., Souza, M., Meulia, T., Grimes, S., Han, M.G., and Saif, L.J. (2006). Pathogenesis of a genogroup II human norovirus in gnotobiotic pigs. J. Virol. 80, 10372–10381.

Cubitt, D. (1994). Caliciviruses. In Viral infections of the gastrointestinal tract, A.Z. Kapikian, ed. (New York, Marcel Dekker Inc), pp. 549–568.

David, M., Petricoin, E., 3rd, and Larner, A.C. (1996). Activation of protein kinase A inhibits interferon induction of the Jak/Stat pathway in U266 cells. J. Biol. Chem. 271, 4585–4588.

Delgado, M. (2003). Inhibition of interferon (IFN) gamma-induced Jak-STAT1 activation in microglia by vasoactive intestinal peptide: inhibitory effect on CD40, IFN-induced protein-10, and inducible nitric-oxide synthase expression. J. Biol. Chem. 278, 27620–27629.

Duizer, E., Schwab, K.J., Neill, F.H., Atmar, R.L., Koopmans, M.P., and Estes, M.K. (2004). Laboratory efforts to cultivate noroviruses. J. Gen. Virol. 85, 79–87.

Estes, M.K. (2001). Rotaviruses and Their Replication. In Fields Virology, D.M. Knipe, and P.M. Howley, eds. (Philadelphia: Lippincott Williams & Wilkins), pp. 1747–1786.

Farkas, T., Nakajima, S., Sugieda, M., Deng, X., Zhong, W., and Jiang, X. (2005). Seroprevalence of noroviruses in swine. J. Clin. Microbiol. 43, 657–661.

Farkas, T., Sestak, K., Wei, C., and Jiang, X. (2008). Characterization of a rhesus monkey calicivirus representing a new genus of Caliciviridae. J. Virol. 82, 5408–5416.

Farkas, T., Zhong, W.M., Jing, Y., Huang, P.W., Espinosa, S.M., Martinez, N., Morrow, A.L., Ruiz-Palacios, G.M., Pickering, L.K., and Jiang, X. (2004). Genetic diversity among sapoviruses. Arch. Virol. 149, 1309–1323.

Flynn, W.T., and Saif, L.J. (1988). Serial propagation of porcine enteric calicivirus-like virus in primary porcine kidney cell cultures. J. Clin. Microbiol. 26, 206–212.

Flynn, W.T., Saif, L.J., and Moorhead, P.D. (1988). Pathogenesis of porcine enteric calicivirus-like virus in four-day-old gnotobiotic pigs. Am. J. Vet. Res. 49, 819–825.

Green, K.Y. (2007). Caliciviruses: the Noroviruses, Vol 1, 5th edn (Philadelphia: Lippincott Williams & Wilkins).

Green, K.Y., Ando, T., Balayan, M.S., Berke, T., Clarke, I.N., Estes, M.K., Matson, D.O., Nakata, S., Neill, J.D., Studdert, M.J., et al. (2000). Taxonomy of the caliciviruses. J. Infect. Dis. 181 Suppl 2, S322–330.

Green, K.Y., Chanock, R.M., and Kapikian, A.Z. (2001). Human Caliciviruses. In Fields Virology, D.M. Knipe, and P.M. Howley, eds. (Philadelphia, Lippincott Williams & Wilkins), pp. 841–874.

Guo, M., Chang, K.O., Hardy, M.E., Zhang, Q., Parwani, A.V., and Saif, L.J. (1999). Molecular characterization of a porcine enteric calicivirus genetically related to Sapporo-like human caliciviruses. J. Virol. 73, 9625–9631.

Guo, M., Hayes, J., Cho, K.O., Parwani, A.V., Lucas, L.M., and Saif, L.J. (2001a). Comparative pathogenesis of tissue culture-adapted and wild-type Cowden porcine enteric calicivirus (PEC) in gnotobiotic pigs and induction of diarrhea by intravenous inoculation of wild-type PEC. J. Virol. 75, 9239–9251.

Guo, M., Qian, Y., Chang, K.O., and Saif, L.J. (2001b). Expression and self-assembly in baculovirus of porcine enteric calicivirus capsids into virus-like particles and their use in an enzyme-linked immunosorbent assay for antibody detection in swine. J. Clin. Microbiol. 39, 1487–1493.

Huang, P., Farkas, T., Marionneau, S., Zhong, W., Ruvoen-Clouet, N., Morrow, A.L., Altaye, M., Pickering, L.K., Newburg, D.S., LePendu, J., et al. (2003). Noroviruses bind to human ABO, Lewis, and secretor histo-blood group antigens: identification of 4 distinct strain-specific patterns. J. Infect. Dis. 188, 19–31.

Huang, P., Farkas, T., Zhong, W., Tan, M., Thornton, S., Morrow, A.L., and Jiang, X. (2005). Norovirus and histo-blood group antigens: demonstration of a wide spectrum of strain specificities and classification of two major binding groups among multiple binding patterns. J. Virol. 79, 6714–6722.

Hutson, A.M., Atmar, R.L., Graham, D.Y., and Estes, M.K. (2002). Norwalk virus infection and disease is associated with ABO histo-blood group type. J. Infect. Dis. 185, 1335–1337.

Jeong, C., Park, S.I., Park, S.H., Kim, H.H., Park, S.J., Jeong, J.H., Choy, H.E., Saif, L.J., Kim, S.K., Kang, M.I., et al. (2007). Genetic diversity of porcine sapoviruses. Vet. Microbiol. 122, 246–257.

Johnson, L.R. (1998). Chapter 34, Secretion. In Essential Medical Physiology, L.R. Johnson, ed. (New York, Lippincott-Raven), pp. 445–472.

Katayama, K., Shirato-Horikoshi, H., Kojima, S., Kageyama, T., Oka, T., Hoshino, F., Fukushi, S., Shinohara, M., Uchida, K., Suzuki, Y., et al. (2002). Phylogenetic analysis of the complete genome of 18 Norwalk-like viruses. Virology 299, 225–239.

Kim, H.J., Cho, H.S., Cho, K.O., and Park, N.Y. (2006). Detection and molecular characterization of porcine enteric calicivirus in Korea, genetically related to sapoviruses. J. Vet. Med. B Infect. Dis. Vet. Public Health 53, 155–159.

Koopmans, M., Vinj inverted question marke, J., de Wit, M., Leenen, I., van der Poel, W., and van Duynhoven, Y. (2000). Molecular epidemiology of human enteric caliciviruses in The Netherlands. J. Infect. Dis. *181* Suppl. 2, S262–269.

Kullak-Ublick, G.A., Stieger, B., and Meier, P.J. (2004). Enterohepatic bile salt transporters in normal physiology and liver disease. Gastroenterology *126*, 322–342.

L'Homme, Y., Sansregret, R., Plante-Fortier, E., Lamontagne, A.M., Ouardani, M., Lacroix, G., and Simard, C. (2009). Genomic characterization of swine caliciviruses representing a new genus of Caliciviridae. Virus Genes *39*, 66–75.

Lee, E.H., and Rikihisa, Y. (1998). Protein kinase A-mediated inhibition of gamma interferon-induced tyrosine phosphorylation of Janus kinases and latent cytoplasmic transcription factors in human monocytes by Ehrlichia chaffeensis. Infect. Immun. *66*, 2514–2520.

Legrand-Defretin, V., Juste, C., Corring, T., and Rerat, A. (1986). Enterohepatic circulation of bile acids in pigs: diurnal pattern and effect of a reentrant biliary fistula. Am. J. Physiol. *250*, G295–301.

Lindesmith, L., Moe, C., Marionneau, S., Ruvoen, N., Jiang, X., Lindblad, L., Stewart, P., LePendu, J., and Baric, R. (2003). Human susceptibility and resistance to Norwalk virus infection. Nat. Med. *9*, 548–553.

Liu, B.L., Clarke, I.N., Caul, E.O., and Lambden, P.R. (1995). Human enteric caliciviruses have a unique genome structure and are distinct from the Norwalk-like viruses. Arch. Virol. *140*, 1345–1356.

Liu, G., Ni, Z., Yun, T., Yu, B., Chen, L., Zhao, W., Hua, J., and Chen, J. (2008). A DNA-launched reverse genetics system for rabbit hemorrhagic disease virus reveals that the VP2 protein is not essential for virus infectivity. J. Gen. Virol. *89*, 3080–3085.

Liu, G., Zhang, Y., Ni, Z., Yun, T., Sheng, Z., Liang, H., Hua, J., Li, S., Du, Q., and Chen, J. (2006). Recovery of infectious rabbit hemorrhagic disease virus from rabbits after direct inoculation with in vitro-transcribed RNA. J. Virol. *80*, 6597–6602.

Martella, V., Banyai, K., Lorusso, E., Bellacicco, A.L., Decaro, N., Mari, V., Saif, L., Costantini, V., De Grazia, S., Pezzotti, G., et al. (2008). Genetic heterogeneity of porcine enteric caliciviruses identified from diarrhoeic piglets. Virus Genes *36*, 365–373.

Martinez, M.A., Alcala, A.C., Carruyo, G., Botero, L., Liprandi, F., and Ludert, J.E. (2006). Molecular detection of porcine enteric caliciviruses in Venezuelan farms. Vet. Microbiol. *116*, 77–84.

Matson, D.O., Berke, T., Dinulos, M.B., Poet, E., Zhong, W.M., Dai, X.M., Jiang, X., Golding, B., and Smith, A.W. (1996). Partial characterization of the genome of nine animal caliciviruses. Arch. Virol. *141*, 2443–2456.

Mattison, K., Shukla, A., Cook, A., Pollari, F., Friendship, R., Kelton, D., Bidawid, S., and Farber, J.M. (2007). Human noroviruses in swine and cattle. Emerg. Infect. Dis. *13*, 1184–1188.

Mauroy, A., Scipioni, A., Mathijs, E., Miry, C., Ziant, D., Thys, C., and Thiry, E. (2008). Noroviruses and sapoviruses in pigs in Belgium. Arch. Virol. *153*, 1927–1931.

Mayo, M.A. (2002). A summary of taxonomic changes recently approved by ICTV. Arch. Virol. *147*, 1655–1663.

McLeod, G.M., and Wiggins, H.S. (1968). Bile-salts in small intestinal contents after ileal resection and in other malabsorption syndromes. Lancet *1*, 873–876.

Neill, J.D., Meyer, R.F., and Seal, B.S. (1995). Genetic relatedness of the caliciviruses: San Miguel sea lion and vesicular exanthema of swine viruses constitute a single genotype within the Caliciviridae. J. Virol. *69*, 4484–4488.

Ohlinger, V.F., Haas, B., and Thiel, H.J. (1993). Rabbit hemorrhagic disease (RHD): characterization of the causative calicivirus. Vet. Res. *24*, 103–116.

Oliver, S.L., Asobayire, E., Dastjerdi, A.M., and Bridger, J.C. (2006). Genomic characterization of the unclassified bovine enteric virus Newbury agent-1 (Newbury1) endorses a new genus in the family Caliciviridae. Virology *350*, 240–250.

Park, S.I., Jeong, C., Kim, H.H., Park, S.H., Park, S.J., Hyun, B.H., Yang, D.K., Kim, S.K., Kang, M.I., and Cho, K.O. (2007). Molecular epidemiology of bovine noroviruses in South Korea. Vet. Microbiol. *124*, 125–133.

Parwani, A.V., Flynn, W.T., Gadfield, K.L., and Saif, L.J. (1991). Serial propagation of porcine enteric calicivirus in a continuous cell line. Effect of medium supplementation with intestinal contents or enzymes. Arch. Virol. *120*, 115–122.

Parwani, A.V., Saif, L.J., and Kang, S.Y. (1990). Biochemical characterization of porcine enteric calicivirus: analysis of structural and nonstructural viral proteins. Arch. Virol. *112*, 41–53.

Prasad, B.V., Hardy, M.E., Dokland, T., Bella, J., Rossmann, M.G., and Estes, M.K. (1999). X-ray crystallographic structure of the Norwalk virus capsid. Science *286*, 287–290.

Prasad, B.V., Rothnagel, R., Jiang, X., and Estes, M.K. (1994). Three-dimensional structure of baculovirus-expressed Norwalk virus capsids. J. Virol. *68*, 5117–5125.

Reuter, G., Biro, H., and Szucs, G. (2007). Enteric caliciviruses in domestic pigs in Hungary. Arch. Virol. *152*, 611–614.

Saif, L.J., Bohl, E.H., Theil, K.W., Cross, R.F., and House, J.A. (1980). Rotavirus-like, calicivirus-like, and 23-nm virus-like particles associated with diarrhea in young pigs. J. Clin. Microbiol. *12*, 105–111.

Samuel, C.E. (2001). Antiviral actions of interferons. Clin. Microbiol. Rev. *14*, 778–809.

Schuffenecker, I., Ando, T., Thouvenot, D., Lina, B., and Aymard, M. (2001). Genetic classification of 'Sapporo-like viruses'. Arch. Virol. *146*, 2115–2132.

Sengupta, T.K., Schmitt, E.M., and Ivashkiv, L.B. (1996). Inhibition of cytokines and JAK-STAT activation by distinct signaling pathways. Proc. Natl. Acad. Sci. U.S.A. *93*, 9499–9504.

Shirato, H., Ogawa, S., Ito, H., Sato, T., Kameyama, A., Narimatsu, H., Xiaofan, Z., Miyamura, T., Wakita, T., Ishii, K., et al. (2008). Noroviruses distinguish between type 1 and type 2 histo-blood group antigens for binding. J. Virol. *82*, 10756–10767.

Smiley, J.R., Chang, K.O., Hayes, J., Vinje, J., and Saif, L.J. (2002). Characterization of an enteropathogenic bovine calicivirus representing a potentially new calicivirus genus. J. Virol. 76, 10089–10098.

Smith, A.W., Skilling, D.E., Cherry, N., Mead, J.H., and Matson, D.O. (1998). Calicivirus emergence from ocean reservoirs: zoonotic and interspecies movements. Emerg. Infect. Dis. 4, 13–20.

Smith, A.W., Skilling, D.E., Dardiri, A.H., and Latham, A.B. (1980). Calicivirus pathogenic for swine: a new serotype isolated from opaleye *Girella nigricans*, an ocean fish. Science 209, 940–941.

Sosnovtsev, S., and Green, K.Y. (1995). RNA transcripts derived from a cloned full-length copy of the feline calicivirus genome do not require VpG for infectivity. Virology 210, 383–390.

Souza, M., Cheetham, S.M., Azevedo, M.S., Costantini, V., and Saif, L.J. (2007). Cytokine and antibody responses in gnotobiotic pigs after infection with human norovirus genogroup II.4 (HS66 strain). J. Virol. 81, 9183–9192.

Studdert, M.J. (1978). Caliciviruses. Brief review. Arch. Virol. 58, 157–191.

Sugieda, M., Nagaoka, H., Kakishima, Y., Ohshita, T., Nakamura, S., and Nakajima, S. (1998). Detection of Norwalk-like virus genes in the caecum contents of pigs. Arch. Virol. 143, 1215–1221.

Tan, M., Fang, P., Chachiyo, T., Xia, M., Huang, P., Fang, Z., Jiang, W., and Jiang, X. (2008). Noroviral P particle: structure, function and applications in virus–host interaction. Virology 382, 115–123.

Thumfart, J.O., and Meyers, G. (2002). Feline calicivirus: recovery of wild-type and recombinant viruses after transfection of cRNA or cDNA constructs. J. Virol. 76, 6398–6407.

Tian, P., Brandl, M., and Mandrell, R. (2005). Porcine gastric mucin binds to recombinant norovirus particles and competitively inhibits their binding to histo-blood group antigens and Caco-2 cells. Lett. Appl. Microbiol. 41, 315–320.

Wang, Q.H., Chang, K.O., Han, M.G., Sreevatsan, S., and Saif, L.J. (2006a). Development of a new microwell hybridization assay and an internal control RNA for the detection of porcine noroviruses and sapoviruses by reverse transcription-PCR. J. Virol. Methods 132, 135–145.

Wang, Q.H., Costantini, V., and Saif, L.J. (2007). Porcine enteric caliciviruses: genetic and antigenic relatedness to human caliciviruses, diagnosis and epidemiology. Vaccine 25, 5453–5466.

Wang, Q.H., Souza, M., Funk, J.A., Zhang, W., and Saif, L.J. (2006b). Prevalence of noroviruses and sapoviruses in swine of various ages determined by reverse transcription-PCR and microwell hybridization assays. J. Clin. Microbiol. 44, 2057–2062.

Ward, V.K., McCormick, C.J., Clarke, I.N., Salim, O., Wobus, C.E., Thackray, L.B., Virgin, H.W.t., and Lambden, P.R. (2007). Recovery of infectious murine norovirus using pol II-driven expression of full-length cDNA. Proc. Natl. Acad. Sci. U.S.A. 104, 11050–11055.

Wei, C., Farkas, T., Sestak, K., and Jiang, X. (2008). Recovery of infectious virus by transfection of in vitro-generated RNA from tulane calicivirus cDNA. J. Virol. 82, 11429–11436.

White, L.J., Ball, J.M., Hardy, M.E., Tanaka, T.N., Kitamoto, N., and Estes, M.K. (1996). Attachment and entry of recombinant Norwalk virus capsids to cultured human and animal cell lines. J. Virol. 70, 6589–6597.

Wobus, C.E., Karst, S.M., Thackray, L.B., Chang, K.O., Sosnovtsev, S.V., Belliot, G., Krug, A., Mackenzie, J.M., Green, K.Y., and Virgin, H.W. (2004). Replication of Norovirus in cell culture reveals a tropism for dendritic cells and macrophages. PLoS Biol. 2, e432.

Yin, Y., Tohya, Y., Ogawa, Y., Numazawa, D., Kato, K., and Akashi, H. (2006). Genetic analysis of calicivirus genomes detected in intestinal contents of piglets in Japan. Arch. Virol. 151, 1749–1759.

Young, D.F., Andrejeva, L., Livingstone, A., Goodbourn, S., Lamb, R.A., Collins, P.L., Elliott, R.M., and Randall, R.E. (2003). Virus replication in engineered human cells that do not respond to interferons. J. Virol. 77, 2174–2181.

Murine Norovirus Pathogenesis and Immunity

10

Stephanie M. Karst

Abstract

The first murine norovirus, murine norovirus 1 (MNV-1), was discovered in 2003. Since then, numerous murine norovirus strains have been identified and they were assigned a new genogroup in the genus *Norovirus*. Murine noroviruses share several properties with human noroviruses. Specifically, they are infectious orally, they spread between mice, and at least one strain, MNV-1, causes mild diarrhoea in wild-type hosts. Furthermore, primary MNV-1 infection fails to elicit protection from a secondary challenge with homologous virus in at least some situations, which is similar to the lack of long-term protective immunity elicited by primary human norovirus infection. Investigators have now begun to extend basic knowledge of norovirus infection and immunity using this system. In particular, studies of murine norovirus infection have provided valuable information regarding the critical nature of innate immunity in controlling infection. Mice deficient in components of the interferon signalling pathway are highly susceptible to MNV-1-induced gastroenteritis, systemic infection, and ultimately death. The precise mechanisms by which interferon protects from serious murine norovirus infection are beginning to be elucidated and will provide potential antiviral targets for combating human norovirus infections. In addition, murine norovirus infection of mice provides a useful model with which to define conditions to elicit protective immunity, potentially providing important information for human norovirus vaccine design. For example, repeated exposure to high doses of MNV-1 may provide protection from mucosal re-infection.

Introduction

Human noroviruses are estimated to be responsible for over 95% of non-bacterial epidemic gastroenteritis worldwide (Green, 2007) and to cause over 200,000 deaths in young children in developing countries annually (Patel *et al.*, 2008). Human noroviruses are thus associated with considerable morbidity and have significant economic impact. Despite the impact of human norovirus-induced disease and the potential for emergence of highly virulent strains, the pathogenic features of infection are not well understood due to the previous lack of cell culture (Duizer *et al.*, 2004) and small animal model systems. In 2003, the first murine norovirus (MNV-1) was discovered (Karst *et al.*, 2003). Since that time, a growing body of literature has defined important parameters of pathogenesis and immunity for this model system. In this chapter, both the similarities and the differences between human and murine norovirus infections will be presented. An emphasis will be placed on information gained from murine norovirus studies regarding potential antiviral strategies to treat or prevent human norovirus infections.

Pathogenesis

Human noroviruses are spread by faecal–oral transmission and by person-to-person contact, and they are highly contagious. The course of disease following human norovirus infection is rapid, with symptoms including vomiting, diarrhoea and nausea arising within 24 h and resolving 24–48 h later (Green, 2007). A recent study demonstrated the high infectivity of a human norovirus strain (Teunis *et al.*, 2008). As discussed in detail below, murine noroviruses share many of these features of infection.

Identification of the first murine norovirus

The first murine norovirus, named MNV-1, was discovered in a severely immunocompromised mouse strain that lacked intact innate and adaptive immune systems due to genetic knockouts of the STAT-1 and RAG-2 genes, respectively (RAG2/STAT1−/− mice) (Karst *et al.*, 2003). In the breeding colony of this mouse strain, sporadic deaths due to a systemic disease of unknown aetiology were observed. Serial passage of brain tissue from sick animals by intracerebral (i.c.) inoculation resulted in disease, confirming the presence of an infectious agent. Animals lacking interferon (IFN) receptors, but not wild-type mice, were found to be susceptible to disease, demonstrating that the responsible pathogen was IFN sensitive. In addition, disease was observed even after passing homogenates through a 0.22-μm filter, implicating a viral pathogen. However, standard diagnostic tests failed to detect known animal or human pathogens in the tissue inoculum used in passaging studies. Thus, a molecular technique called representational difference analysis (RDA) was employed to uncover the genetic identity of the unknown pathogen. RDA revealed numerous sequences with close homology to human noroviruses and animal caliciviruses that aligned along the length of a prototypical calicivirus genome. Despite their homology, these sequences were not identical to any known calicivirus. Using a combination of rapid amplification of cDNA ends and PCR based on sequences obtained by RDA, a consensus sequence of the 7382-nucleotide viral genome was determined. Based on phylogenetic analysis, the newly discovered virus clearly segregated in the genus *Norovirus* within the family *Caliciviridae* and was approximately 50% genetically identical to the prototype human norovirus, Norwalk virus.

These molecular results strongly indicated a previously unrecognized norovirus as the pathogen responsible for the serious disease observed in RAG2/STAT1−/− mice, thus it was named MNV-1. This conclusion was proven upon purification of virus particles from the brain tissue of a sick mouse on caesium chloride gradients – these particles had the expected buoyant density and size of norovirus particles, and they were pathogenic in RAG2/STAT1−/− mice (Karst *et al.*, 2003). While it was initially observed that i.c. inoculation of MNV-1 into RAG2/STAT1−/− mice causes CNS pathologies as well as hepatitis and pneumonia (Karst *et al.*, 2003), it has now been clearly demonstrated that peroral (p.o.) inoculation of MNV-1 into STAT1−/− mice results in severe gastroenteritis (Mumphrey *et al.*, 2007) in addition to lung, spleen and liver pathologies (Karst *et al.*, 2003). Furthermore, MNV-1 causes mild diarrhoea in wild-type mice (Mumphrey *et al.*, 2007; Liu *et al.*, 2009). Thus, despite its initial discovery in brain tissue of seriously immunocompromised mice, MNV-1 is an enteric norovirus capable of inducing gastroenteritis, demonstrating its pathogenic similarity to the human members of this virus family.

Genetic diversity

More than 100 human norovirus strains have been identified that segregate into three separate genogroups within the genus *Norovirus*; these can be further subdivided into at least 29 distinct clusters (Zheng *et al.*, 2006). Since the discovery of MNV-1, 16 additional murine norovirus strains have been identified (Fig. 10.1): three new murine norovirus strains were isolated from intestine-draining mesenteric lymph nodes (MLNs) of seropositive mice at various research colonies in the US (UM2–4) (Hsu *et al.*, 2006); three were isolated from faecal samples of mice at the Max Planck Institute and the Robert Koch Institute in Berlin, Germany (Berlin1–3) (Muller *et al.*, 2007); and ten were isolated from faecal samples of mice at Washington University School of Medicine (strains referred to as WU followed by a number) and Charles River Laboratories (strains referred to as CR followed by a number)

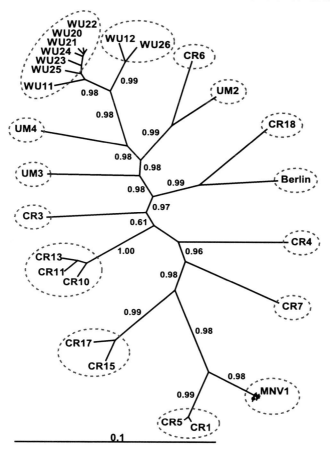

Figure 10.1 All sequenced MNV genomes comprise 17 distinct strains. A consensus Bayesian tree based on full-length MNV genomes is shown. Genetically distinct MNV strains are circled. Within circles, genetically related isolates comprising a single strain are shown. Not included in this tree are two of the German strains referred to as Berlin1–3. Adapted from Thackray et al. (2007).

in the US (Thackray et al., 2007). In the latter study, multiple virus isolates were also reported. There is suggestive evidence that recombination events occur between murine norovirus strains (Thackray et al., 2007; Muller et al., 2007). Thus, murine noroviruses appear to be extremely prevalent in colonies of research mice. Underscoring this idea, antibodies reactive to MNV-1 have been identified in 22.1% of serum samples from research colonies in both the US and Canada (Hsu et al., 2005). Moreover, 64.3% of mice tested from two research institutes in Germany were positive for murine norovirus genome in their faeces (Muller et al., 2007). Charles River Laboratories now considers murine norovirus the most prevalent endemic viruses in laboratory mice (Charles River Laboratories, 2009).

All of the newly identified murine norovirus strains share at least 87% genetic identity and segregate into a single genogroup and cluster in the genus *Norovirus* (Thackray et al., 2007). In comparison, human norovirus genomes that segregate into different genogroups and clusters can share as little as 65% genetic identity (Katayama et al., 2002). This represents a major difference between human and murine noroviruses, with human noroviruses displaying a much wider degree of genetic variation than do the murine noroviruses identified so far. There are several possible explanations for this discrepancy: (i) it may reflect the relative infancy of murine norovirus strain discovery compared to human noroviruses (17 murine norovirus strains have been identified compared to >100 human norovirus strains);

(ii) it may reflect the disparate process of their identification – human norovirus strains are identified when they cause a gastroenteritis outbreak, dictating that they are all virulent and randomly localized, while murine norovirus strains have only been identified in asymptomatic hosts and in very few specific locations; (iii) it may reflect the haplo-identical nature of inbred mouse strains in which murine noroviruses have been discovered; or (iv) it may reflect an as-yet unrecognized biological difference between norovirus infection of these two species. Continued identification of new murine norovirus strains in research colonies globally, and hopefully strains associated with gastroenteritis outbreaks in nature, will help to distinguish between these possibilities.

Cellular tropism

The precise cell tropism of human noroviruses is unknown: Researchers have been unsuccessfully attempting to identify a relevant cell culture system for many years (Duizer *et al.*, 2004). Furthermore, when intestinal biopsy sections from norovirus-infected volunteers have been analysed by electron microscopy, virus particles were not observed even though histological alterations of the intestinal cells were evident (Dolin *et al.*, 1975; Agus *et al.*, 1973). Conversely, murine noroviruses replicate in myeloid cells *in vitro* and MNV-1 has been detected in intestinal lamina propria cells *in vivo*, as discussed in detail in the following sections. It is possible that human noroviruses similarly infect intestinal myeloid cells, human noroviruses instead infect intestinal epithelial cells, or norovirus cell tropism is virus strain-dependent.

Murine noroviruses infect myeloid cells *in vitro*. Specifically, MNV-1 displays a cellular tropism for macrophages (Mφs) and dendritic cells (DCs). Primary Mφs and DCs are efficiently infected with MNV-1 *in vitro* (Fig. 10.2) (Wobus *et al.*, 2004) while other cell types including murine embryonic fibroblasts, murine hepatocyte cell lines, neuroblastoma cell lines, mouse embryonic stem cells, and embryonic stem cell-derived embryoid bodies are not permissive to MNV-1 (Karst *et al.*, 2003; Wobus *et al.*, 2004; Wobus *et al.*, 2006). Cytopathic effect (CPE) is visible in cultures of STAT1−/− Mφs and DCs (Fig. 10.2A), as well as a reduced amount in

wild-type DCs. A number of murine Mφ cell lines (RAW264.7, J774A.1, P388D1, IC-21, and the human/murine hybrid line WBC264–9C) and a murine DC line (JAWSII) also support MNV-1 replication (Wobus *et al.*, 2004; Wobus *et al.*, 2006). The RAW264.7 cell line has been used to design a typical virus plaque assay and for the growth and isolation of plaque-purified virus isolates. Such plaque-purified MNV-1 isolates display differential virulence *in vivo* (Mumphrey *et al.*, 2007; Thackray *et al.*, 2007). Moreover, passaging of purified virus in cell culture is associated with *in vivo* attenuation, as measured by lethal infection of STAT1−/− mice (Wobus *et al.*, 2004). In one specific example, this attenuation can be attributed to a single amino acid change in the hypervariable region of the capsid protein predicted to be the receptor binding domain (Bailey *et al.*, 2008). Other murine norovirus strains have also been demonstrated to infect RAW264.7 cells but some display altered growth characteristics. For example, UM2 infection does not cause CPE as rapidly as does MNV-1 (Hsu *et al.*, 2006). In addition, many of the WU and CR strains form diffuse and/or small plaques on RAW264.7 cells and are less cytopathic than MNV-1 (Thackray *et al.*, 2007). In limited studies, it appears that this decreased *in vitro* cytopathicity does not necessarily correlate with decreased replication *in vivo* (Thackray *et al.*, 2007). A more extensive comparison of murine norovirus strains is necessary to determine whether they display relevant differences in cell tropism.

Murine noroviruses may also infect myeloid cells *in vivo*. In MNV-1-infected wild-type 129SvEv mice, viral non-structural protein (indicative of intracellular replication) can be detected in lamina propria cells of rare intestinal villi at 24–48 hours post infection (hpi) (Mumphrey *et al.*, 2007). It is presently unclear if other murine noroviruses or human noroviruses infect greater numbers of intestinal cells than does MNV-1. However, it has been postulated that human noroviruses infect very few cells *in vivo* because no virions were observed when biopsy sections from norovirus-infected volunteers were analysed by EM (Dolin *et al.*, 1975; Agus *et al.*, 1973). Moreover, rotavirus infects very few intestinal cells *in vivo* but nevertheless causes significant diarrhoea (Starkey *et al.*, 1986; Offit *et al.*, 1984;

A   Cytopathic effect

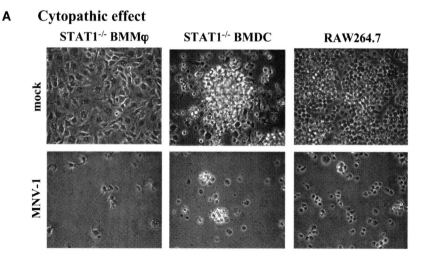

B   Multi-step growth curves

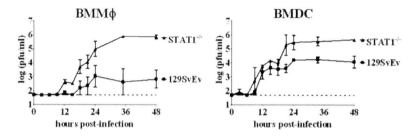

Figure 10.2 MNV-1 replicates in macrophages and dendritic cells *in vitro*. (A) Bone marrow-derived macrophages (BMMφs) and dendritic cells (BMDCs) from STAT1−/− mice, and RAW264.7 cells, were infected with MNV-1 at MOI 0.05 and monitored for cytopathic effect. Pictures were taken at 2 dpi. (B) 129SvEv and STAT1−/− BMMφs and BMDCs were infected with MNV-1 at MOI 0.05. Infected cell lysates were analysed in two to four independent experiments by plaque assay at various times post-infection to calculate standard deviations. Adapted from Wobus *et al.*, 2004.

Ramig, 2004). Secretion of a viral enterotoxin is thought to explain the discrepancy between the low number of infected cells and the severe nature of rotavirus-induced illness (Ramig, 2004). The *in vivo* infection of Mφs and DCs by MNV-1 is further supported by the immunohistochemical detection of MNV-1 antigen in cells that resemble Mφs in the liver, and Mφs and DCs in the spleen (Wobus *et al.*, 2004). It should be noted that this *in vivo* tropism has yet to be confirmed using cell-type specific markers and dual labelling techniques. It has also been reported that immunodeficient mice naturally infected with murine noroviruses contain viral antigen in the cytoplasm of inflammatory cells in the liver, splenic red and white pulp, intestinal lamina propria, intestinal lymphoid follicles, lung, pleural and peritoneal cavities, and MLNs (Ward *et al.*, 2006). This work provides suggestive evidence that some virus antigen-positive cells in the liver of these naturally infected mice express F4/80, a cell surface marker expressed by specific subsets of Mφs and DCs. Future studies examining cell tropism and pathology in specific immunodeficient mouse strains experimentally infected with murine norovirus in a controlled setting (i.e. specific virus dose and specific time points) should provide confirmation of these preliminary observations.

In the absence of an intact IFN signalling pathway, MNV-1 antigen is detected more abundantly and in additional intestinal cell types of infected mice. Specifically, viral antigen can be detected in intestinal epithelial cells of STAT1−/− mice at 12 hpi, and lamina propria and Peyer's patch cells at 24–48 hpi (Mumphrey et al., 2007). Thus, IFN may restrict the cell tropism of noroviruses, as discussed in detail below.

Infection and disease in wild-type hosts

Similar to human noroviruses, MNV-1 is infectious in wild-type hosts by the p.o. and intranasal (i.n.) routes of inoculation (Karst et al., 2003), and is spread naturally between mice (Hsu et al., 2005). Following per os (p.o.) inoculation of MNV-1, infectious virions can be detected in intestines, MLNs, spleens and livers of 129SvEv mice (Mumphrey et al., 2007; Liu et al., 2009). The course of MNV-1 infection is rapid, with peak levels of virus detected in these tissues 1–3 dpi. The infectious dose of MNV-1 in 129SvEv mice is quite low, with as few as 10 particle-forming units (pfu) initiating a low-level infection of the intestine and disseminating to at least MLNs and the spleen (Liu et al., 2009). The most extensive studies analysing MNV-1 infection of wild-type mice have been performed in the 129SvEv strain. However, MNV-1 similarly infects other wild-type mouse strains including C57Bl/6 mice (Thackray et al., 2007; Chachu et al., 2008b) and CD1 mice (Hsu et al., 2005). There have also been preliminary studies analysing infection of wild-type mice by other strains of murine norovirus. For example, WU11, CR1, CR3, CR6 and CR7 all infect inbred C57Bl/6 mice, with some achieving higher titres in the intestine and MLNs at three dpi than MNV-1 (Thackray et al., 2007). UM2, UM3, and UM4 all infect outbred CD1 mice for longer time periods than does MNV-1 (Hsu et al., 2006). Additionally, outbred Swiss Webster mice, which are commonly used as sentinels in research colonies, have been shown to contain murine norovirus-specific serum antibody (Perdue et al., 2007), supporting their permissivity to natural murine norovirus infection. In future work, it will be important to rigorously analyse the influence of both virus strain and mouse strain on murine norovirus pathogenesis. While a number of studies have failed to observe disease in murine norovirus-infected immunocompetent mice, these studies monitored infected mice only

for overt external indicators of gastroenteritis (Karst et al., 2003; Thackray et al., 2007; Hsu et al., 2005; Hsu et al., 2006). In a more rigorous examination of putative gastroenteritis induction using internal and histopathological indicators, it has now been reported that p.o. inoculation of wild-type 129SvEv mice with MNV-1 results in mild intestinal inflammation (Mumphrey et al., 2007) consistent with the moderate infiltration of inflammatory cells observed in intestines of volunteers infected with human noroviruses (Schreiber et al., 1973; Agus et al., 1973; Dolin et al., 1975; Schreiber et al., 1974). Moreover, MNV-1 infection of 129SvEv and C57Bl/6 mice induces mild diarrhoea in a majority of infected animals, as monitored by visually scoring the faecal consistency of infected and control mock-infected mice, and wet/dry ratio determination of these faecal samples (Mumphrey et al., 2007; Liu et al., 2009). Even though this phenotype is mild, it is reproducible, statistically significant upon independent blind scoring by multiple investigators, and dose-dependent with groups of mice receiving a minimum of 1000 pfu MNV-1 developing statistically significant diarrhoea compared to mock-infected control groups (Liu et al., 2009). Although this mild phenotype contrasts with the more severe diarrhoea and vomiting induced by human viruses within this genus, it is important to note that mice lack an emetic reflex and thus do not vomit. Additionally, mice are desert animals and thus display enhanced absorption of fluid from the intestinal lumen, resulting in their relative resistance to the development of diarrhoea (Buret et al., 1993). Nonetheless, the pathogenic features of MNV-1 infection recapitulate quite well those observed for human noroviruses, including oral infectivity, induction of gastroenteritis, a rapid course of infection and high infectivity.

Systemic infection

It is surprising that infectious MNV-1 particles can be detected in peripheral tissues of infected wild-type mice (Mumphrey et al., 2007; Thackray et al., 2007; Liu et al., 2009) because enteric viruses are generally considered to be limited to mucosal sites. Systemic murine norovirus infection does not appear to be limited to this virus strain: UM2, UM3, and UM4 genomes can be amplified by RT-PCR from the spleens of infected CD1 mice at 8 weeks pi (Hsu et al., 2006) and viral antigen can be detected in multiple peripheral tissues of naturally infected immunodeficient mice (Ward et al., 2006). MNV-1 not only disseminates to the spleen but also replicates quite efficiently and induces specific splenic histopathological changes, including activation of cells in the white pulp and hypertrophy of cells in the red pulp (Mumphrey et al., 2007). These changes are associated with statistically significant increases in the number of cells displaying the Mφ-specific marker F4/80, the B cell-specific marker B220, and the DC-specific marker CD11c. The physiological relevance of these splenic changes is unclear but it will be interesting to determine whether they are required for peripheral murine norovirus-specific immune responses and, if so, what role peripheral immunity plays in protecting mice from secondary murine norovirus infections. Identification of a murine norovirus strain that does not cause systemic infection would facilitate such comparative studies. Interestingly, rotaviruses have also recently been confirmed to spread to peripheral tissues and cause histological changes at these sites (Crawford et al., 2006; Blutt et al., 2006; Fenaux et al., 2006). Thus, the dogma that enteric viruses do not disseminate extraintestinally needs to be reconsidered. Supporting the possibility that human noroviruses can spread to peripheral sites, a recent study detected norovirus RNA in the serum of 15% of infected individuals (Takanashi et al., 2009). It is possible that very mild or sporadic pathologies associated with human norovirus infection of peripheral tissues have been missed due to the difficulties in their detection and the assumption that they are limited to the intestinal tract. For example, a recent case report detected norovirus RNA in the serum and cerebrospinal fluid of a child with encephalopathy (Ito et al., 2006). Furthermore, during a norovirus outbreak among military personnel in Afghanistan in May 2002, three infected patients presented with gastroenteritis, diminished alertness, headache, neck stiffness, and light sensitivity and one of these patients also displayed disseminated intravascular coagulation (Centres for Disease Control and Prevention, 2002). These case reports, along with the detection of systemic murine norovirus infection,

suggest that noroviruses should be considered as potential aetiological agents of diseases other than gastroenteritis.

The discovery that MNV-1 disseminates to MLNs that drain from the intestine (Thackray et al., 2007; Mumphrey et al., 2007; Hsu et al., 2005; Liu et al., 2009) is perhaps not surprising because virions that have access to the intestinal lamina propria can most likely diffuse in a cell-free manner to this site. As noted above, viral antigen-positive cells can be detected in the intestinal lamina propria of MNV-1-infected mice (Mumphrey et al., 2007). Moreover, MNV-1 infects DCs (Wobus et al., 2004), a cell type known to actively migrate from tissues to draining lymph nodes, so it is also possible that the virus utilizes DC infection to facilitate dissemination to MLNs. Regardless of the mechanism of transit, MLN infection is a novel finding with regard to norovirus pathogenesis. This feature of murine norovirus infection is not virus strain-specific: (i) UM2, UM3, and UM4 were originally isolated from MLNs of naturally infected mice and their viral genomes remain detectable by RT-PCR at 8 weeks pi in experimentally infected mice (Hsu et al., 2006); (ii) WU11, CR1, CR3, CR6, and CR7 all disseminate to MLNs (Thackray et al., 2007); and (iii) viral antigen was detected in MLNs of naturally infected immunodeficient mice (Ward et al., 2006). It is presently not known whether human noroviruses also disseminate to MLNs.

Persistent infection

Norovirus infections are typically described as acute; however, increasing evidence points to the potential for persistent norovirus infection. Although the symptoms caused by human norovirus infection typically resolve within several days, virus particles can be shed from asymptomatic individuals for weeks after exposure (Rockx et al., 2002; Graham et al., 1994; Patterson et al., 1993; Atmar et al., 2008). Further, symptomatic persistent infection has been documented in immunosuppressed individuals with symptoms lasting over 2 years (Gallimore et al., 2004; Morotti et al., 2004; Nilsson et al., 2003) and in immunocompetent children with symptoms lasting up to 6 weeks (Murata et al., 2007; Rockx et al., 2002). The first indication that murine noroviruses can also persistently infect their hosts came from studies of MNV-1 infection of immunodeficient RAG−/− mice in which it was observed that viral genome was retained in multiple tissues of infected mice for months following infection (Karst et al., 2003). Subsequent studies confirmed that detection of viral genome in tissues of RAG−/− mice correlates with the presence of infectious virions (Wobus et al., 2006; Chachu et al., 2008a). Importantly, these mice also persistently shed virus in their faeces.

There is also evidence that MNV-1 can persistently infect immunocompetent hosts, albeit at a drastically reduced level compared to RAG−/− mice: at 5 weeks pi, viral genome is detectable in intestines, MLNs, and spleens of CD1 mice (Hsu et al., 2005) and infectious virions are detectable in MLNs of C57Bl/6 mice (Thackray et al., 2007). Other murine norovirus strains have been reported to persistently infect immunocompetent mice for even longer periods than MNV-1. For example, UM2, UM3, and UM4 genomes can be amplified from faecal samples, intestines, MLNs, and spleens of CD1 mice for at least 8 weeks pi (Hsu et al., 2006). Similarly, CR1, CR3, CR6, and CR7 genomes can be amplified from faecal samples of C57Bl/6 mice for 5 weeks pi under conditions where MNV-1 genome cannot be detected in the faeces (Thackray et al., 2007). While some researchers have concluded that certain strains including MNV-1 fail to persistently infect wild-type mice because they are cleared more rapidly than are other strains, when interpreting such studies it is important to consider the influence of virus dose, the relative infectivity of different virus strains, and the detection methodology. Underscoring this point, the kinetics at which acute MNV-1 infection is cleared from 129SvEv mice is clearly influenced by initial inoculum dose (Liu et al., 2009). Moreover, detection of viral genome (Hsu et al., 2005) and infectious virions (Thackray et al., 2007) in certain tissues of MNV-1-infected mice at 5 weeks pi indicates that this strain is capable of establishing persistent infection under certain conditions. While it is likely that murine norovirus strains differ in their efficiency of in vivo replication during both acute and persistent infection, more rigorous studies are needed to confirm whether there is truly a distinction between strains in terms of their capacity to establish persistent infection.

Overall, human and murine noroviruses can persistently infect their hosts even in the presence of a fully functional immune system. A fundamental feature of persistent viral infections is that the host immune response is ineffective at completely clearing infectious virus (Oldstone, 2006). Typically, this is due to a combination of factors including alteration of the antigenic epitopes encoded by the virus and impairment of the normal functioning of immune cells upon infection. The mechanism(s) by which noroviruses maintain persistent infection is unclear but may play a critical role in impairing adaptive immune responses such that they fail to protect from secondary challenge. It will be interesting in future studies to examine the relationship between persistent norovirus infection and the lack of protective immunity elicited by primary infection (see below).

Immunity

Noroviruses display a rapid onset and resolution of symptoms so it is likely that components of the innate immune response are critical for restricting viral pathogenesis. However, studies of innate immune responses to human noroviruses have been greatly hindered by lack of suitable model systems. Early studies identified IFN signalling as a critical component for control of MNV-1 infection (Karst et al., 2003) and significant advances have subsequently been made in determining the mechanisms by which IFN affords this protection.

Primary human norovirus infection fails to elicit lasting protective immunity in all individuals, potentially complicating the development of effective vaccine regimes. The basis of this atypical pattern of immunity is unclear because adaptive immune responses are generated upon exposure to human noroviruses. Specifically, human norovirus infection elicits both mucosal and peripheral virus-specific antibodies in infected individuals (Green, 2007). Virus-specific serum IgG is induced and persists for months following infection, while serum IgA and IgM responses are more short-lived. Mucosal IgA is induced as well but its duration has not been determined (Agus et al., 1974). There have been no direct studies of T cell responses to human norovirus infection but surrogate cytokine studies suggest that virus-specific T cells are also induced in infected individuals. In one study (Lindesmith *et al.*, 2005), peripheral blood mononuclear cells (PBMCs) were collected from norovirus-infected volunteers at pre-challenge and post-challenge time points, incubated with norovirus virus-like particles (VLPs), and secretion of cytokines from the PBMCs was measured. The authors noted increases in the Th1 cytokines IFNγ and IL-2, as well as the Th2 cytokine IL-5. Notably though, similar increases in cytokine levels were observed when pre-challenge and post-challenge PBMCs from uninfected individuals were compared to each other, presumably indicating that controls in this study were previously exposed to natural norovirus infections. This finding highlights the difficulty of studying primary immune responses to human noroviruses where infection histories are impossible to discern.

One animal model that has been used in recent years to study human norovirus pathogenesis is inoculation of gnotobiotic pigs with a human norovirus; in this system, a majority of inoculated animals develop mild diarrhoea as assessed by visual scoring of faecal consistency (Cheetham *et al.*, 2006), similar to the disease read-out now available for MNV-1-infected wild-type mice (Mumphrey *et al.*, 2007; Liu *et al.*, 2009). This model facilitates an examination of primary immune responses to a human norovirus in a controlled setting. A recent study reported cytokine levels in human norovirus-infected pigs during primary infection (Souza *et al.*, 2007): moderate increases in several Th1 and Th2 cytokines, as well as type I IFN, were observed in the serum and intestinal contents of infected animals. While valuable information can be gained from studies in this gnotobiotic pig model, there are additional benefits of the murine model: (i) the great number of genetically engineered immunodeficient strains available to researchers; and (ii) the use of a virus in its natural host. The roles played by innate and adaptive immunity in controlling murine norovirus infection, and the factors that influence the induction of protective murine norovirus immunity, are reviewed below.

The role of innate immunity, and specifically interferon, in controlling murine norovirus infection

While MNV-1 does not cause serious disease in immunocompetent hosts, it causes severe disease in mice lacking certain components of the innate immune system. In particular, mice lacking intact IFN signalling pathways, specifically deficient in the STAT-1 molecule or both type I and type II IFN receptors (IFNαβγR−/−), are extremely sensitive to MNV-1-induced lethality following p.o., i.n., or i.c. inoculation (Karst *et al.*, 2003). Viral genomes can be detected at high numbers in the intestines, spleens, livers, lungs, brains, blood and faeces of infected STAT1−/− mice by three dpi, suggesting rapid and global virus dissemination and shedding early after infection (Karst *et al.*, 2003). This is supported by the detection of high numbers of infectious virions in intestines, serum, spleens, livers, lungs and axillary lymph nodes over a 72-h course of infection (Fig. 10.3) (Mumphrey *et al.*, 2007). The overall number of virions detected in STAT1−/− mice is between 10- and 5000-fold higher than in wild-type mice over this 72-h course of infection, demonstrating that IFN signalling pathways are absolutely essential to control MNV-1 replication *in vivo* (Mumphrey *et al.*, 2007). Comparison of the overall course of MNV-1 infection in wild-type and STAT1−/− mice suggests that IFN responses (i) directly inhibit viral replication at the site of entry, the intestine; and (ii) limit virus dissemination to, and replication in, peripheral tissues (Mumphrey *et al.*, 2007). This latter point is supported by the detection of viraemia in STAT1−/−, but not wild-type, mice (Fig. 10.3). Correlating with the protective nature of IFN in controlling murine norovirus infection, significant levels of type I IFN can be detected in intestinal homogenates and serum of MNV-1-infected wild-type, but not STAT1−/−, mice (Mumphrey *et al.*, 2007).

Interferon also protects from serious murine norovirus-induced clinical disease. Oral MNV-1 infection of STAT1−/− mice causes severe gastroenteritis characterized by rapid weight loss, dramatic gastric bloating, semi-liquid faecal contents and decreased weight of internal faeces (Fig. 10.4) (Mumphrey *et al.*, 2007). The induction of severe gastric bloating is interesting in light of the fact that human norovirus infection causes delayed gastric emptying, most likely associated with the high incidence of vomiting episodes in infected individuals (Meeroff *et al.*, 1980). While increased stomach contents leads to vomiting in the human host, mice lack an emetic reflex and

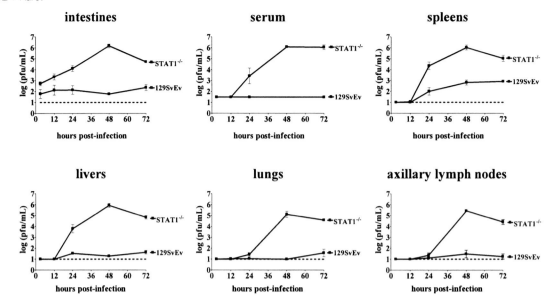

Figure 10.3 MNV-1 replicates and disseminates more efficiently in STAT1⁻ᐟ⁻ mice following oral inoculation. 129SvEv and STAT1⁻ᐟ⁻ mice were inoculated perorally with 1 × 10⁷ pfu of MNV-1.CW3, and organs were harvested at 3, 12, 24, 48, and 72 hpi. Virus titres were determined in the proximal small intestines, serum, spleens, livers, lungs, and axillary lymph nodes by plaque assay. Adapted from Mumphrey et al., 2007.

thus do not vomit. However, the observation that MNV-1 infection causes gastric bloating suggests that human and murine noroviruses both induce a similar pathophysiological outcome, supporting the argument that the study of murine norovirus pathogenesis will advance our understanding of human norovirus-induced disease. It is presently unclear why MNV-1 only appears to induce severe gastroenteritis in the absence of IFN responses while human noroviruses do so in immunocompetent hosts. One confounding factor is that almost all studies of murine norovirus pathogenesis have been performed in adult mice, which are characteristically resistant to the development of diarrhoea. While initial studies have failed to observe diarrhoea or weight loss in MNV-1-infected inbred 129SvEv (Mumphrey et al., 2007) or outbred Swiss Webster (S.M. Karst, unpublished) suckling mice, future studies analysing murine norovirus infection of suckling mice are warranted and should include comparisons of different virus and mouse strains. Pathology is not limited to the gastrointestinal tract in MNV-1-infected STAT1−/− mice- severe pneumonia and destruction of both splenic and liver tissue are also apparent during acute infection (Karst et al., 2003; Mumphrey et al., 2007).

At the histological level, MNV-1 infection of STAT1−/− mice is associated with a marked increase in the number of apoptotic cells in the intestines (Mumphrey et al., 2007). Pycnotic nuclear debris indicative of apoptosis is also present in the red and white pulp of the spleens of MNV-1-infected STAT1−/−, but not wild-type, mice (Mumphrey et al., 2007), demonstrating that IFN signalling prevents apoptotic death of both intestinal cells and splenocytes. These data highlight the absolutely critical requirement for functional IFN responses in murine norovirus protection. This information may offer insight into human norovirus infection since the course of disease suggests that innate immune components are also critical for protection.

Molecular details of interferon-mediated inhibition of murine norovirus replication are currently being investigated. The mechanism(s) of protection afforded by IFN signalling, including identification of the precise IFN-induced antiviral molecules responsible for protection and the stage of viral replication blocked upon IFN signalling, have not yet been fully elucidated. However, progress is being made on both of these fronts. With regard to the host factors responsible for IFN-mediated protection from murine norovirus

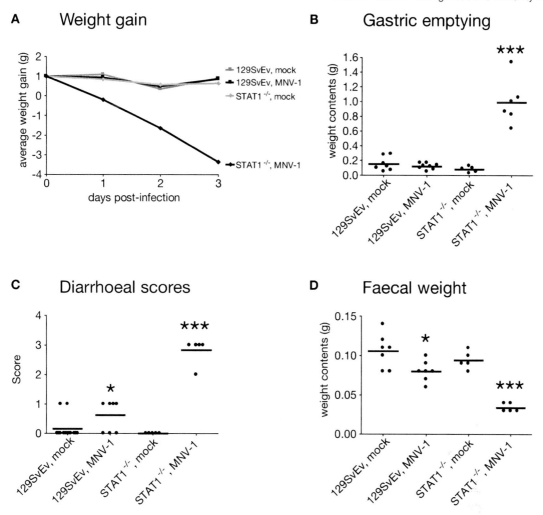

Figure 10.4 Gastroenteritis induced by MNV-1 infection is greatly reduced by STAT1-dependent responses. 129SvEv and STAT1–/– mice were inoculated perorally with 1 × 10⁷ pfu of MNV-1.CW3, or mock-infected. A) Individual mice were weighed at approximately the same time of day beginning on the day of infection (d=0), and the data is presented as the averaged weight gain of all mice in a particular group. B-D) Groups of mock-infected and MNV-1-infected mice were deprived of food for 18–20 h and then sacrificed at 72 hpi. For measuring gastric emptying (B), stomachs were ligated at both the oesophageal and the intestinal junctions, dissected, and weighed. Stomachs were then opened, rinsed, and reweighed. Faeces from each mouse was also collected, scored (C), and weighed (D). Adapted from Mumphrey et al., 2007.

infection, several studies have examined the role played by various pattern recognition receptors (PRRs) known to initiate the IFN response during viral infections. Such PRRs include (i) Toll-like receptors (TLRs), (ii) cytoplasmic RIG-I-like helicases including RIG-I itself and MDA5 and (iii) protein kinase R (PKR).

1 *Toll-like receptors* TLRs are expressed at the cell surface and on membranes of endosomal compartments. Those that would logically be suspected in norovirus recognition include TLR7 and TLR8, which recognize ssRNA, and TLR3, which recognizes dsRNA. Mice lacking TLR3 have a slight increase in virus loads in MLNs; however, there is no difference in virus loads in intestines or spleens, suggesting that TLR3 may play a minor tissue-specific role in recognition of MNV-1 (McCartney et al., 2008). Absence of TLR3 has no effect on the replication of MNV-1 in bone marrow-derived DCs (BMDCs)

in vitro, so TLR3 may play a significant role in Mφs, a DC subset that differs in its dependence on specific PRRs compared to BMDCs, or a permissive cell type that has yet to be identified. It has not been determined whether TLR7 or TLR8 recognize murine noroviruses and what contribution this recognition may play in preventing viral replication.

2 *RIG-I-like helicases* While preliminary studies suggest that RIG-I does not play a role in inhibiting the replication of a human norovirus in epithelial-like Huh-7 cells (Guix *et al.*, 2007), no studies have been performed to date analysing its potential to recognize murine noroviruses and no studies have analysed its contribution to norovirus-stimulated IFN production in permissive cell types. In contrast, a definitive role for MDA5 in the recognition of MNV-1 and subsequent activation of innate immune mediators has been reported (McCartney *et al.*, 2008). Virus loads in intestines, MLNs and spleens are significantly increased in MDA5−/− mice compared to wild-type mice at the peak of infection. Furthermore, viral replication is increased in BMDCs lacking this PRR. Interestingly, this increased replication is only observed at late times post-infection, suggesting that MDA5 only recognizes MNV-1 late in the viral replication cycle. Finally, BMDCs lacking MDA5 fail to secrete appreciable type I IFN, MCP-1, IL-6, or TNFα when infected with MNV-1. These data strongly suggest that MDA5 is the primary PRR that senses MNV-1, at least in BMDCs. MDA5 is known to be capable of recognizing dsRNA (Yoneyama *et al.*, 2005) and preliminary results suggest that this is the case for MNV-1: viral genomes that are competent for replication stimulate IFN production upon their recognition by MDA5 in BMDCs whereas UV-inactivated virions and viral genomes treated with proteinase K (to remove the covalently attached VPg that is required for replication) fail to stimulate IFN production. Thus, viral replication that is expected to generate dsRNA molecules appears to be required for MDA5 sensing of MNV-1. While MDA5 clearly recognizes

MNV-1 and plays a role *in vivo* to reduce the level of viral replication in multiple tissues, it is not necessary for the overall control of MNV-1 infection because MDA5−/− mice clear infection with similar kinetics to wild-type mice. Moreover, the increased virus loads and viral replication in BMDCs in the absence of MDA5 are not nearly as significant as the increased levels observed in STAT1−/− or IFNαβγR−/− mice (Mumphrey *et al.*, 2007) or cells (Wobus *et al.*, 2004), respectively, suggesting that additional PRRs also contribute to stimulation of innate immune effectors.

3 *Protein kinase R* PKR acts as both a PRR to recognize dsRNA and an IFN-stimulated effector molecule to inhibit viral translation initiation, thus examining its role strictly as a PRR is complicated. However, several lines of evidence suggest that PKR is not essential to controlling MNV-1 replication. First, unlike STAT1−/− and IFNαβγR−/− mice, PKR−/− mice do not succumb to lethal MNV-1 infection (Karst *et al.*, 2003). Second, MNV-1 replication is not significantly increased in PKR−/− BMMφs compared to wild-type cells (Wobus *et al.*, 2004). Finally, type I IFN signalling in BMDCs and BMMφs can inhibit the translation of MNV-1 proteins independent of PKR (Changotra *et al.*, 2009). While these studies suggest that PKR does not play a major role in controlling MNV-1 infection, they do not test the possibility that it plays a more subtle or a cell type-specific role, as MDA5 appears to do. Interestingly, type II IFN signalling in BMDCs and BMMφs inhibits the translation of MNV-1 proteins similar to type I IFN but it requires PKR for this inhibition (Changotra *et al.*, 2009).

Once PRRs have recognized viral molecular patterns and induced the expression of IFN and other pro-inflammatory cytokines, IFN is secreted from infected cells and subsequently binds to its receptor in both an autocrine and a paracrine fashion. This activates signalling cascades that ultimately induce hundreds of IFN stimulated genes (ISGs) whose protein products have potential antiviral activity. The roles of individual

ISGs in directly inhibiting murine norovirus replication have not been examined. This will be an important area of future investigation because such host factors represent potential targets for therapeutic approaches. Identification of the step in the viral replication cycle that is targeted upon IFN signalling will greatly facilitate identification of the responsible host factor(s). Recent work demonstrates that MNV-1 replication is blocked by type I and type II IFN signalling at the step of viral translation (Changotra et al., 2009). Because norovirus genomes appear to utilize a novel VPg-dependent (5′ cap- and IRES-independent) mechanism to recruit host translation initiation factors to the 5′ end of their genomes, dissecting the mechanism of IFN-mediated inhibition of norovirus translation may facilitate the discovery of a novel antiviral strategy that targets viral translation while sparing host protein synthesis.

In addition to its role in directly inhibiting viral replication in infected cells, IFN is known to restrict the cell tropism of a number of different viruses (Ryman et al., 2000; Samuel and Diamond, 2005; Wang et al., 2004). Consistent with this, there is evidence that IFN restricts the intestinal cell tropism of MNV-1: viral ProPol antigen is detectable in intestinal epithelial cells of STAT1−/−, but not wild-type mice early after infection (Mumphrey et al., 2007). While this may reflect the sensitivity of the detection method since only few viral antigen-positive cells were detected in sections from STAT1−/− mice, in vitro studies support low-level virus replication in non-myeloid cells only in the absence of intact IFN signalling pathways (Wobus et al., 2004). Moreover, Ward et al. have also detected viral antigen in intestinal epithelial cells of a naturally infected RAG1/STAT1−/− mouse (Ward et al., 2006). Viral antigen can also clearly be detected in Peyer's patch cells in STAT1−/−, but not wild-type mice (Mumphrey et al., 2007), a finding that could be due to IFN-mediated restriction of cell tropism, or could be due to IFN-mediated inhibition of viral dissemination to Peyer's patches. It will thus be enlightening to determine the precise cell types infected by MNV-1 not only in the intestinal lamina propria but also in sites of dissemination including Peyer's patches, MLNs, and the spleen.

The role of adaptive immunity in controlling murine norovirus infection

An important and surprising finding regarding the role of adaptive immunity in controlling murine norovirus infection is that RAG−/− animals lacking both B and T cells are resistant to severe MNV-1-induced disease following p.o. or i.n. inoculation (Karst et al., 2003). Although this indicates that adaptive immunity is not required to prevent lethality and no overt signs of disease are apparent in RAG−/− mice infected with MNV-1, these mice do become persistently infected with high levels of virus. Thus, B and/or T cells are required for the clearance of infection. Initial studies suggest that both types of adaptive immune cells play roles in this process, as described in the following sections.

Recent evidence points to a role for virus-specific antibody in the clearance of murine noroviruses (Chachu et al., 2008b). Specifically, murine norovirus infection induces peripheral virus-specific IgG (Karst et al., 2003; Hsu et al., 2005; Hsu et al., 2006) and IgA (Liu et al., 2009). Serum antibodies collected from MNV-1-infected mice are neutralizing (Wobus et al., 2004). An MNV-1-neutralizing monoclonal antibody (MAb) has been shown to bind to each of the 180 hypervariable protruding domains of native MNV-1 virions without inducing a major conformational change, suggesting that neutralization is instead mediated by abrogation of cellular attachment (Katpally et al., 2008). Epitope mapping studies have revealed that this MAb may bind residues suspected to be involved in carbohydrate binding by human noroviruses (Lochridge and Hardy, 2007). Thus, while human and murine noroviruses may use different receptors to attach to cells, the functional domains on the virions interacting with cells may be similar. Further supporting a role for B cells in MNV-1 clearance, μMT mice that lack B cells contain elevated virus loads in their intestines and MLNs at the peak of infection compared to wild-type mice, and they fail to clear infection from MLNs as rapidly as do wild-type mice. However, these mice do clear intestinal infection with normal kinetics, pointing to a potential important role for T cells at the mucosal site of entry. Importantly, adoptive transfer of immune splenocytes from wild-type mice into persistently infected RAG−/− mice results in nearly complete clearance of intestinal virus loads within 6 days, whereas transfer of immune splenocytes isolated from μMT mice fails to clear intestinal infection. Finally, passive transfer of murine norovirus-specific antibody, either polyclonal immune serum or neutralizing IgG MAbs directed against the capsid protein, reduces intestinal and splenic virus loads in persistently infected RAG−/− mice. Cumulatively, these data demonstrate that virus-specific antibody plays a role in clearing murine norovirus infection. The fact that virus loads are increased in μMT mice at 3 and 5 dpi compared to wild-type mice suggests that, in addition to virus-specific antibody, natural antibody may play a role in controlling acute murine norovirus infection.

The role of T cells in clearing murine norovirus infection has begun to be analysed as well (Chachu et al., 2008a). Specifically, mice lacking CD8 T cell responses have increased virus loads in both intestines and MLNs at 7 dpi, but not at earlier times in acute infection or 21 dpi. These data suggest that CD8 T cells can enhance MNV-1 clearance but that they are not essential. Mice deficient in CD4 T cells contain increased amounts of virus in their intestines at 3 and 5 dpi but they clear virus with similar kinetics compared to wild-type mice. Thus, CD4 T cells play a surprising role in reducing acute viral replication at very early times after infection in a tissue-specific manner, but they are not required for overall viral clearance. Finally, the T cell effector molecule perforin, but not IFNγ, was shown to play a role in clearing previously established persistent MNV-1 infection in the intestine (Chachu et al., 2008a).

Protective immunity

While B and T cell responses are elicited by norovirus infection and they function in clearing acute infection, it is not clear what role they play in affording protection from a secondary norovirus challenge. The primary insights into immunity to human noroviruses have been elucidated through human volunteer studies. These studies have definitively shown that not all humans develop lasting immunity upon primary exposure to a norovirus. For example, Parrino et al. found that short-term (6 to 14 weeks) immunity to a norovirus is acquired following primary

infection but long-term (27 to 42 months) immunity is not (Parrino et al., 1977), and Johnson et al. demonstrated that short-term resistance (up to 6 months) to homologous challenge is acquired but longer-term resistance is not (Johnson et al., 1990b). It is presently unclear why protection from re-infection is not sustained since virus-specific adaptive immune responses are generated. Results from many human volunteer challenge studies and data gathered from natural norovirus outbreaks demonstrate that the presence of virus-specific antibodies does not correlate with protection (Parrino et al., 1977; Graham et al., 1994; Johnson et al., 1990b; Lindesmith et al., 2003; Baron et al., 1984; Cukor et al., 1982; Blacklow et al., 1979; Johnson et al., 1990a). It should be noted that one recent study does report a correlation between rapid salivary IgA elevation and protection from human norovirus infection (Lindesmith et al., 2003). However, previous exposure histories of infected individuals were not available in any of these studies, greatly complicating their interpretation. Overall, the available information on secondary human norovirus infections supports an atypical pattern of immunity in that previously exposed individuals are susceptible to repeat infections even in the face of virus-specific memory immune responses. Understanding the basis of this lack of protection is critical not only to designing efficacious norovirus vaccines but also to our basic understanding of mucosal immune responses to viral infections.

Because mice are genetically manipulable and exposure histories can be carefully controlled, recapitulation of this atypical pattern of norovirus immunity in the mouse model would be invaluable in dissecting the basis for lack of protection to secondary norovirus infections. Indeed, a recent study by Liu et al. demonstrates that primary high-dose MNV-1 infection fails to protect from a secondary challenge with homologous virus even though these mice contain virus-specific serum IgG and IgA (Liu et al., 2009). Specifically, the incidence of diarrhoea in mice receiving primary and secondary MNV-1 infections is similar (Fig. 10.5A). The lack of protection from disease in re-challenged mice correlates with an inability of memory immune responses to provide sterilizing mucosal immunity: infectious virions can access the intestine and are found at fairly high levels in MLNs of re-challenged mice while dissemination to peripheral tissues such as the spleen is completely prevented (Fig. 10.5B). This prevention of virus spread is not surprising in light of the finding that serum antibodies directed against the virus are neutralizing (Wobus *et al.*, 2004). While enteric infection and disease are not prevented in previously exposed mice, virus is cleared more rapidly from their mucosal tissues, demonstrating that a memory immune response is generated upon primary MNV-1 exposure (Fig. 10.5B). It will be interesting to determine whether mucosal murine norovirus-specific antibody is elicited, whether it is neutralizing and, if so, to examine why it does not prevent intestinal infection. Surprisingly, there is an inverse correlation between MNV-1 inoculum dose and the magnitude of protective immunity (Liu *et al.*, 2009). It is presently unclear why lower doses of virus elicit stronger mucosal memory immune responses.

Additional important information regarding MNV-1-specific immunity has been provided in a recent study by Chachu *et al.* that reported the development of protective MNV-1 immunity when using reduced virus loads in mucosal sites at 72 hpi as a read-out for protection (Chachu *et al.*, 2008a). In this study, wild-type C57Bl/6 mice were infected with live virus by the p.o. route, boosted at 21 dpi, and challenged at various times post-boost. Regardless of the amount of time between boost and challenge (2 weeks to 24 weeks), virus loads in intestines and MLNs at 72 hpi were drastically reduced in vaccinated mice compared to control non-vaccinated mice (Fig. 10.6). Thus, boosting the immune response 21 d following

Figure 10.5 Primary high-dose MNV-1 infection fails to protect from secondary challenge even though virus is cleared more rapidly. Groups of 129SvEv mice were inoculated perorally with 10⁴ pfu MNV-1.CW3 or mock-infected. Six weeks later, the mock-infected mice were re-challenged with mock inoculum (mock; panel A only) or 10⁷ pfu MNV-1.CW3 (primary infection); the MNV-1-infected mice were re-challenged with 10⁷ pfu MNV-1.CW3 (secondary infection). (A) All faeces below the caecum was collected at 72 hpi and scored for consistency. (B) At the times indicated post infection, mice were perfused, organs were harvested, and viral burden was determined by plaque assay. Adapted from Liu *et al.* (2009).

A

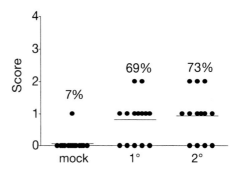

B

Virus clearance

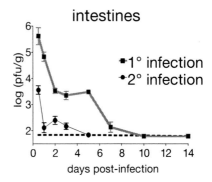

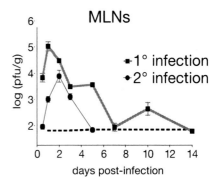

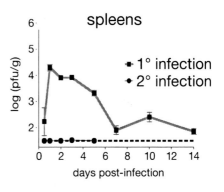

primary MNV-1 exposure may facilitate the development of protective immunity. Indeed, it has also been reported that repeated exposure to human noroviruses does elicit protection (Parrino *et al.*, 1977; Johnson *et al.*, 1990b), in contrast to a single exposure which does not. Because multiple challenge studies in human volunteers were performed with fairly short intervals between second and third exposures (six months or less), it has been postulated that short-term immunity is induced upon norovirus infection but it wanes over time. No repeat-exposure studies were performed

in the mouse model with intervals longer than six months so this possibility cannot be formally ruled out. However, it is logical to assume that six months corresponds to a long-term interval in the murine system where the life span is greatly reduced compared to a human host. Thus, these findings may instead suggest that immunity elicited by repeated norovirus exposure is maintained over time (Chachu *et al.*, 2008a), an issue that is of critical importance to vaccination regimes.

Previous efforts to design human norovirus vaccines have focused on expression of the capsid

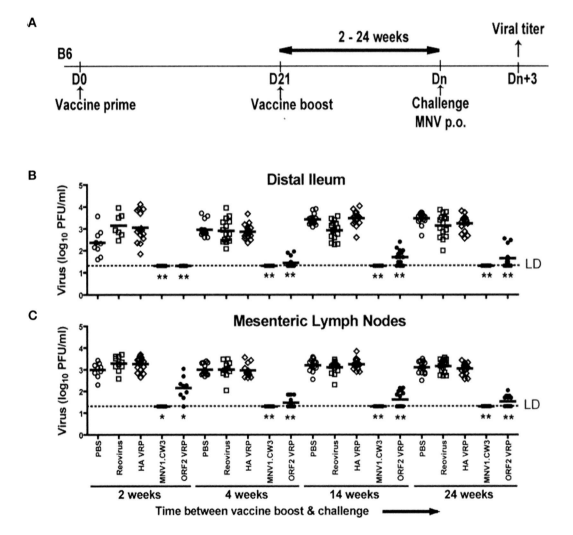

Figure 10.6 Repeated MNV-1 exposure may elicit protective immunity. (A) Vaccination regime. Virus loads in intestines (B) and mesenteric lymph nodes (C) of C57Bl/6 mice vaccinated with the indicated vaccines and challenged with 3×10^7 pfu MNV-1.CW3. VRP=Venezuelan Equine Encephalitis replicon particles; HA=haemagglutinin from influenza A virus. LD indicates the limit of detection. Adapted from Chachu *et al.*, 2008.

protein in baculovirus vectors (Jiang et al., 1992), plants (Mason et al., 1996), or Venezuelan equine encephalitis virus replicon particles (VRPs) (Baric et al., 2002). In all cases, capsid self-assembles into VLPs that are antigenically similar to native norovirus virions. VLPs produced in insect cells and in transgenic plants are immunogenic in mice (Mason et al., 1996; Ball et al., 1998; Guerrero et al., 2001) and in humans (Tacket et al., 2000; Ball et al., 1999). Similarly, inoculation of mice with VLPs produced in yeast (Xia et al., 2007) or with VEE particles expressing norovirus VLPs elicits humoral immune responses (Harrington et al., 2002; LoBue et al., 2006). Such vaccination regimes elicit both serum IgG and mucosal IgA responses but the duration of these responses and their protective potential have not been determined. While challenge studies following exposure to human norovirus VLPs have not been reported, i.n. inoculation of calves with bovine norovirus VLPs and adjuvant induces both mucosal and systemic humoral immune responses but protection from subsequent challenge is only partial and viral shedding is not prevented (Han et al., 2006). Similar studies in the mouse model may offer critical insight into the efficacy of norovirus vaccines prior to testing in humans. Initial studies examining whether exposure to a single norovirus protein can induce protection (Chachu et al., 2008a) are promising. For these studies, mice were inoculated via the footpad with VRPs expressing one or more MNV-1 proteins, boosted with the same construct, and challenged with live virus at various times post-boost. While vaccinating with the capsid protein alone was not as effective as vaccinating with replicating virus in most experiments, capsid alone was sufficient to significantly reduce virus loads in both intestines and MLNs at 72 hpi. Surprisingly, vaccinating with the ORF1 polyprotein encoding non-structural proteins also afforded some reduction in intestinal, but not MLN, virus loads. Mice vaccinated with VRPs expressing the ORF3 protein had similar mucosal virus loads compared to control non-vaccinated animals. Overall, these studies suggest that systemic exposure to a norovirus capsid protein can significantly reduce the amount of viral replication occurring in mucosal sites upon challenge with live virus. Because reductions in virus loads do not necessarily

correlate with protection from disease, it will be essential in the future to test whether repeated systemic exposure to a VRP expressing a murine norovirus capsid protein protects mice from developing murine norovirus-induced diarrhoea. It will also be informative to examine earlier time points in the course of infection in vaccinated animals because replication preceding 72 hpi may be sufficient to induce disease.

Using mice genetically deficient in specific components of the adaptive immune response, Chachu et al. determined that B cells, CD4 T cells, and CD8 T cells all play a role in reducing viral replication in the context of repeat-exposure vaccination with either live virus or capsid-expressing VRPs (Chachu et al., 2008a). Interestingly, there are tissue-specific differences as well as vaccine-specific differences in the contribution of specific cell types to virus load reductions. For example, upon live virus vaccination, B cells seem to play the most important role in reducing viral replication in MLNs but they do not appear to be critical in reducing viral replication in the intestine. In contrast, upon VRP vaccination, B cells play only a minor role in reducing viral replication in the MLNs but they play a significant role in reducing viral replication in the intestines. CD4 T cells play a role in reducing MLN and intestinal virus loads upon live virus vaccination and intestinal virus loads upon VRP vaccination, but they do not contribute to reducing viral replication in MLNs in the context of VRP vaccination. CD8 T cells modestly contribute to reducing viral replication in MLNs, but not intestines, upon live virus vaccination while they are essential to reducing both intestinal and MLN virus loads upon capsid-expressing VRP vaccination. The implications of these disparate results are presently unclear. Studies examining a complete course of infection under different conditions and parallel studies using a disease readout will greatly facilitate interpretation of these results.

Overall, several important conclusions can be drawn from these vaccination studies: (1) efficacious norovirus vaccine regimes may require multiple antigen exposures; (2) norovirus vaccines may not require oral administration; and (3) norovirus vaccines should optimally stimulate all aspects of the adaptive immune response.

Future directions

There are interesting differences between human and murine norovirus infection as well. First, human noroviruses cause more severe disease in immunocompetent hosts than do murine noroviruses. Continued murine norovirus strain identification may uncover more virulent strains. However, it is also possible that murine noroviruses are in general more adapted to their host species such that they cause only mild infections while retaining the ability to efficiently disseminate within a host population. Second, human noroviruses display a much wider antigenic diversity than do currently sequenced murine noroviruses. It will be important to determine whether this represents a true biological difference or a difference in the methods of strain discovery. Third, murine noroviruses disseminate to mesenteric lymph nodes and to peripheral tissues. It has long been assumed that human noroviruses remain confined to the intestine but recent evidence suggests that this assumption be revisited; it thus remains unclear whether disseminated infection represents an actual difference between human and murine noroviruses.

Interferon has been identified as a critical innate immune mediator of protection from serious murine norovirus-induced disease. A specific role for interferon in preventing the translation of murine norovirus proteins has been uncovered. Future studies will focus on identifying the precise host factors involved in this inhibition, with the potential to translate this knowledge into an effective anti-norovirus therapy. Because viruses routinely co-opt cellular processes in order to replicate, a common barrier to new antiviral therapies is their detrimental bystander effects on the host cell. Noroviruses utilize a virus-specific process of translation initiation involving VPg-mediated recruitment of host translation initiation factors; thus, targeting the process of norovirus translation as an antiviral strategy has the potential to inhibit viral protein synthesis while leaving host translation unaffected.

Another important parallel between human and murine norovirus infections is that they do not elicit protective immunity upon a primary exposure, at least under certain conditions. This feature of infection has important implications to the design of candidate norovirus vaccines. For example, using the mouse model of infection, it has now been demonstrated that inoculum dose inversely correlates with the magnitude of mucosal memory immune responses. Initial studies in the murine system also suggest that repeated exposure to a murine norovirus does elicit protection, in contrast to a single exposure. Similarly, repeated exposure to the murine norovirus capsid protein in the absence of virus replication also results in a reduction in virus loads. Both humoral and cell-mediated arms of the adaptive immune system contribute to this protection. It will be critical to extend these studies using disease as a read-out of protection. Ultimately, this type of investigation will be invaluable to our understanding of human norovirus vaccine failure or success.

Conclusion

Many of the pathogenic features of human norovirus infection have been recapitulated in the murine system, including faecal-oral spread, a low infectious dose, a rapid course of infection, induction of gastroenteritis, and persistent infection that extends well beyond the period of symptomatic infection. These findings solidify the utility of murine norovirus infection as a model system to study critical aspects of human norovirus pathogenesis and immunity, with the ultimate goal being the identification of antiviral strategies to combat norovirus infections in the human population.

References

Agus, S.G., Dolin, R., Wyatt, R.G., Tousimis, A.J., and Northrup, R.S. (1973). Acute infectious nonbacterial gastroenteritis: intestinal histopathology. Histologic and enzymatic alterations during illness produced by the Norwalk agent in man. Ann. Intern. Med. 79, 18–25.

Agus, S.G., Falchuk, Z.M., Sessoms, C.S., Wyatt, R.G., and Dolin, R. (1974). Increased jejunal IgA synthesis in vitro during acute infectious nonbacterial gastroenteritis. Am. J. Dig. Dis. 19, 127–131.

Atmar, R.L., Opekun, A.R., Gilger, M.A., Estes, M.K., Crawford, S.E., Neill, F.H., and Graham, D.Y. (2008). Norwalk virus shedding after experimental human infection. Emerg. Infect. Dis. 14, 1553–1557.

Bailey, D., Thackray, L.B., and Goodfellow, I.G. (2008). A single amino acid substitution in the murine norovirus capsid protein is sufficient for attenuation in vivo. J. Virol. 82, 7725–7728.

Ball, J.M., Graham, D.Y., Opekun, A.R., Gilger, M.A., Guerrero, R.A., and Estes, M.K. (1999). Recombinant

Norwalk virus-like particles given orally to volunteers: phase I study. Gastroenterology *117*, 40–48.

Ball, J.M., Hardy, M.E., Atmar, R.L., Conner, M.E., and Estes, M.K. (1998). Oral immunization with recombinant Norwalk virus-like particles induces a systemic and mucosal immune response in mice. J. Virol. *72*, 1345–1353.

Baric, R.S., Yount, B., Lindesmith, L., Harrington, P.R., Greene, S.R., Tseng, F.C., Davis, N., Johnston, R.E., Klapper, D.G., and Moe, C.L. (2002). Expression and self-assembly of norwalk virus capsid protein from venezuelan equine encephalitis virus replicons. J. Virol. *76*, 3023–3030.

Baron, R.C., Greenberg, H.B., Cukor, G., and Blacklow, N.R. (1984). Serological responses among teenagers after natural exposure to Norwalk virus. J. Infect. Dis. *150*, 531–534.

Blacklow, N.R., Cukor, G., Bedigian, M.K., Echeverria, P., Greenberg, H.B., Schreiber, D.S., and Trier, J.S. (1979). Immune response and prevalence of antibody to Norwalk enteritis virus as determined by radioimmunoassay. J. Clin. Microbiol. *10*, 903–909.

Blutt, S.E., Fenaux, M., Warfield, K.L., Greenberg, H.B., and Conner, M.E. (2006). Active viremia in rotavirus-infected mice. J. Virol. *80*, 6702–6705.

Buret, A., Hardin, J., Olson, M.E., and Gall, D.G. (1993). Adaptation of the small intestine in desert-dwelling animals: morphology, ultrastructure and electrolyte transport in the jejunum of rabbits, rats, gerbils and sand rats. Comp. Biochem. Physiol. Comp. Physiol. *105*, 157–163.

Centers for Disease Control and Prevention (2002). Outbreak of acute gastroenteritis associated with Norwalk-like viruses among British military personnel-Afghanistan, May 2002. MMWR Morb. Mortal. Wkly. Rep. *51*, 477–479.

Chachu, K.A., LoBue, A.D., Strong, D.W., Baric, R.S., and Virgin, H.W. (2008a). Immune mechanisms responsible for vaccination against and clearance of mucosal and lymphatic norovirus infection. PLoS. Pathog. *4*, e1000236.

Chachu, K.A., Strong, D.W., LoBue, A.D., Wobus, C.E., Baric, R.S., and Virgin, H.W. (2008b). Antibody is critical for the clearance of murine norovirus infection. J. Virol. *82*, 6610–6617.

Changotra, H., Jia, Y., Moore, T.N., Liu, G., Kahan, S.M., Sosnovtsev, S.V., and Karst, S.M. (2009). Type I and type II interferons inhibit the translation of murine norovirus proteins. J. Virol. *83*, 5683–5692.

Charles River Laboratories. (2009). Murine norovirus. http://www.criver.com/SiteCollectionDocuments/rm_ld_r_Murine_Norovirus.pdf

Cheetham, S., Souza, M., Meulia, T., Grimes, S., Han, M.G., and Saif, L.J. (2006). Pathogenesis of a genogroup II human norovirus in gnotobiotic pigs. J. Virol. *80*, 10372–10381.

Crawford, S.E., Patel, D.G., Cheng, E., Berkova, Z., Hyser, J.M., Ciarlet, M., Finegold, M.J., Conner, M.E., and Estes, M.K. (2006). Rotavirus viremia and extraintestinal viral infection in the neonatal rat model. J. Virol. *80*, 4820–4832.

Cukor, G., Nowak, N.A., and Blacklow, N.R. (1982). Immunoglobulin M responses to the Norwalk virus of gastroenteritis. Infect. Immun. *37*, 463–468.

Dolin, R., Levy, A.G., Wyatt, R.G., Thornhill, T.S., and Gardner, J.D. (1975). Viral gastroenteritis induced by the Hawaii agent. Jejunal histopathology and serologic response. Am. J. Med. *59*, 761–768.

Duizer, E., Schwab, K.J., Neill, F.H., Atmar, R.L., Koopmans, M.P., and Estes, M.K. (2004). Laboratory efforts to cultivate noroviruses. J. Gen. Virol. *85*, 79–87.

Fenaux, M., Cuadras, M.A., Feng, N., Jaimes, M., and Greenberg, H.B. (2006). Extraintestinal spread and replication of a homologous EC rotavirus strain and a heterologous rhesus rotavirus in BALB/c mice. J. Virol. *80*, 5219–5232.

Gallimore, C.I., Lewis, D., Taylor, C., Cant, A., Gennery, A., and Gray, J.J. (2004). Chronic excretion of a norovirus in a child with cartilage hair hypoplasia (CHH). J. Clin. Virol. *30*, 196–204.

Graham, D.Y., Jiang, X., Tanaka, T., Opekun, A.R., Madore, H.P., and Estes, M.K. (1994). Norwalk virus infection of volunteers: new insights based on improved assays. J. Infect. Dis. *170*, 34–43.

Green, K.Y. (2007). Fields Virology: Caliciviridae: The Noroviruses Chapter 28. (Philadelphia: Lippincott, Williams & Wilkins).

Guerrero, R.A., Ball, J.M., Krater, S.S., Pacheco, S.E., Clements, J.D., and Estes, M.K. (2001). Recombinant Norwalk virus-like particles administered intranasally to mice induce systemic and mucosal (fecal and vaginal) immune responses. J. Virol. *75*, 9713–9722.

Guix, S., Asanaka, M., Katayama, K., Crawford, S.E., Neill, F.H., Atmar, R.L., and Estes, M.K. (2007). Norwalk Virus RNA is infectious in mammalian cells. J. Virol. *81*, 12238–12248

Han, M.G., Cheetham, S., Azevedo, M., Thomas, C., and Saif, L.J. (2006). Immune responses to bovine norovirus-like particles with various adjuvants and analysis of protection in gnotobiotic calves. Vaccine *24*, 317–326.

Harrington, P.R., Yount, B., Johnston, R.E., Davis, N., Moe, C., and Baric, R.S. (2002). Systemic, mucosal, and heterotypic immune induction in mice inoculated with Venezuelan equine encephalitis replicons expressing Norwalk virus-like particles. J. Virol. *76*, 730–742.

Hsu, C.C., Riley, L.K., Wills, H.M., and Livingston, R.S. (2006). Persistent infection with and serologic cross-reactivity of three novel murine noroviruses. Comp. Med. *56*, 247–251.

Hsu, C.C., Wobus, C.E., Steffen, E.K., Riley, L.K., and Livingston, R.S. (2005). Development of a microsphere-based serologic multiplexed fluorescent immunoassay and a reverse transcriptase PCR assay to detect murine norovirus 1 infection in mice. Clin. Diagn. Lab. Immunol. *12*, 1145–1151.

Ito, S., Takeshita, S., Nezu, A., Aihara, Y., Usuku, S., Noguchi, Y., and Yokota, S. (2006). Norovirus-associated encephalopathy. Pediatr. Infect. Dis. J. *25*, 651–652.

Jiang, X., Wang, M., Graham, D.Y., and Estes, M.K. (1992). Expression, self-assembly, and antigenicity of the Norwalk virus capsid protein. J. Virol. 66, 6527–6532.

Johnson, P.C., Hoy, J., Mathewson, J.J., Ericsson, C.D., and DuPont, H.L. (1990a). Occurrence of Norwalk virus infections among adults in Mexico. J. Infect. Dis. 162, 389–393.

Johnson, P.C., Mathewson, J.J., DuPont, H.L., and Greenberg, H.B. (1990b). Multiple-challenge study of host susceptibility to Norwalk gastroenteritis in US adults. J. Infect. Dis. 161, 18–21.

Karst, S.M., Wobus, C.E., Lay, M., Davidson, J., and Virgin, H.W. (2003). STAT1-dependent innate immunity to a Norwalk-like virus. Science 299, 1575–1578.

Katayama, K., Shirato-Horikoshi, H., Kojima, S., Kageyama, T., Oka, T., Hoshino, F., Fukushi, S., Shinohara, M., Uchida, K., Suzuki, Y., Gojobori, T., and Takeda, N. (2002). Phylogenetic analysis of the complete genome of 18 Norwalk-like viruses. Virology 299, 225–239.

Katpally, U., Wobus, C.E., Dryden, K., Virgin, H.W., and Smith, T.J. (2008). Structure of antibody-neutralized murine norovirus and unexpected differences from viruslike particles. J. Virol. 82, 2079–2088.

Lindesmith, L., Moe, C., Lependu, J., Frelinger, J.A., Treanor, J., and Baric, R.S. (2005). Cellular and humoral immunity following Snow Mountain virus challenge. J. Virol. 79, 2900–2909.

Lindesmith, L., Moe, C., Marionneau, S., Ruvoen, N., Jiang, X., Lindblad, L., Stewart, P., Lependu, J., and Baric, R. (2003). Human susceptibility and resistance to Norwalk virus infection. Nat. Med. 9, 548–553.

Liu, G., Kahan, S.M., Jia, Y., and Karst, S.M. (2009). Primary high-dose murine norovirus 1 infection fails to protect from secondary challenge with homologous virus. J. Virol. 83, 6963–6968.

LoBue, A.D., Lindesmith, L., Yount, B., Harrington, P.R., Thompson, J.M., Johnston, R.E., Moe, C.L., and Baric, R.S. (2006). Multivalent norovirus vaccines induce strong mucosal and systemic blocking antibodies against multiple strains. Vaccine 24, 5220–5234.

Lochridge, V.P., and Hardy, M.E. (2007). A single-amino-acid substitution in the P2 domain of VP1 of murine norovirus is sufficient for escape from antibody neutralization. J. Virol. 81, 12316–12322.

Mason, H.S., Ball, J.M., Shi, J.J., Jiang, X., Estes, M.K., and Arntzen, C.J. (1996). Expression of Norwalk virus capsid protein in transgenic tobacco and potato and its oral immunogenicity in mice. Proc. Natl. Acad. Sci. U.S.A. 93, 5335–5340.

McCartney, S.A., Thackray, L.B., Gitlin, L., Gilfillan, S., Virgin, H.W., and Colonna, M. (2008). MDA-5 recognition of a murine norovirus. PLoS. Pathog. 4, e1000108.

Meeroff, J.C., Schreiber, D.S., Trier, J.S., and Blacklow, N.R. (1980). Abnormal gastric motor function in viral gastroenteritis. Ann. Intern. Med. 92, 370–373.

Morotti, R.A., Kaufman, S.S., Fishbein, T.M., Chatterjee, N.K., Fuschino, M.E., Morse, D.L., and Magid, M.S. (2004). Calicivirus infection in pediatric small intestine transplant recipients: pathological considerations. Hum. Pathol. 35, 1236–1240.

Muller, B., Klemm, U., Mas, M.A., and Schreier, E. (2007). Genetic diversity and recombination of murine noroviruses in immunocompromised mice. Arch. Virol. 152, 1709–1719.

Mumphrey, S.M., Changotra, H., Moore, T.N., Heimann-Nichols, E.R., Wobus, C.E., Reilly, M.J., Moghadamfalahi, M., Shukla, D., and Karst, S.M. (2007). Murine norovirus 1 infection is associated with histopathological changes in immunocompetent hosts but clinical disease is prevented by STAT1-dependent interferon responses. J. Virol. 81, 3251–3263.

Murata, T., Katsushima, N., Mizuta, K., Muraki, Y., Hongo, S., and Matsuzaki, Y. (2007). Prolonged norovirus shedding in infants <or=6 months of age with gastroenteritis. Pediatr. Infect. Dis. J. 26, 46–49.

Nilsson, M., Hedlund, K.O., Thorhagen, M., Larson, G., Johansen, K., Ekspong, A., and Svensson, L. (2003). Evolution of human calicivirus RNA in vivo: accumulation of mutations in the protruding P2 domain of the capsid leads to structural changes and possibly a new phenotype. J. Virol. 77, 13117–13124.

Offit, P.A., Clark, H.F., Kornstein, M.J., and Plotkin, S.A. (1984). A murine model for oral infection with a primate rotavirus (simian SA11). J. Virol. 51, 233–236.

Oldstone, M.B. (2006). Viral persistence: parameters, mechanisms and future predictions. Virology 344, 111–118.

Parrino, T.A., Schreiber, D.S., Trier, J.S., Kapikian, A.Z., and Blacklow, N.R. (1977). Clinical immunity in acute gastroenteritis caused by Norwalk agent. N. Engl. J. Med. 297, 86–89.

Patel, M.M., Widdowson, M.A., Glass, R.I., Akazawa, K., Vinje, J., and Parashar, U.D. (2008). Systematic literature review of role of noroviruses in sporadic gastroenteritis. Emerg. Infect. Dis. 14, 1224–1231.

Patterson, T., Hutchings, P., and Palmer, S. (1993). Outbreak of SRSV gastroenteritis at an international conference traced to food handled by a post-symptomatic caterer. Epidemiol. Infect. 111, 157–162.

Perdue, K.A., Green, K.Y., Copeland, M., Barron, E., Mandel, M., Faucette, L.J., Williams, E.M., Sosnovtsev, S.V., Elkins, W.R., and Ward, J.M. (2007). Naturally occurring murine norovirus infection in a large research institution. J. Am. Assoc. Lab Anim Sci. 46, 39–45.

Ramig, R.F. (2004). Pathogenesis of intestinal and systemic rotavirus infection. J. Virol. 78, 10213–10220.

Rockx, B., De, W.M., Vennema, H., Vinje, J., De, B.E., Van, D.Y., and Koopmans, M. (2002). Natural history of human calicivirus infection: a prospective cohort study. Clin. Infect. Dis. 35, 246–253.

Ryman, K.D., Klimstra, W.B., Nguyen, K.B., Biron, C.A., and Johnston, R.E. (2000). Alpha/beta interferon protects adult mice from fatal Sindbis virus infection and is an important determinant of cell and tissue tropism. J. Virol. 74, 3366–3378.

Samuel, M.A., and Diamond, M.S. (2005). Alpha/beta interferon protects against lethal West Nile virus infection by restricting cellular tropism and enhancing neuronal survival. J. Virol. *79*, 13350–13361.

Schreiber, D.S., Blacklow, N.R., and Trier, J.S. (1973). The mucosal lesion of the proximal small intestine in acute infectious nonbacterial gastroenteritis. N. Engl. J. Med. *288*, 1318–1323.

Schreiber, D.S., Blacklow, N.R., and Trier, J.S. (1974). The small intestinal lesion induced by Hawaii agent acute infectious nonbacterial gastroenteritis. J. Infect. Dis. *129*, 705–708.

Souza, M., Cheetham, S.M., Azevedo, M.S., Costantini, V., and Saif, L.J. (2007). Cytokine and antibody responses in gnotobiotic pigs after infection with human norovirus genogroup II.4 (HS66 strain). J. Virol. *81*, 9183–9192.

Starkey, W.G., Collins, J., Wallis, T.S., Clarke, G.J., Spencer, A.J., Haddon, S.J., Osborne, M.P., Candy, D.C., and Stephen, J. (1986). Kinetics, tissue specificity and pathological changes in murine rotavirus infection of mice. J. Gen. Virol. *67*, 2625–2634.

Tacket, C.O., Mason, H.S., Losonsky, G., Estes, M.K., Levine, M.M., and Arntzen, C.J. (2000). Human immune responses to a novel norwalk virus vaccine delivered in transgenic potatoes. J. Infect. Dis. *182*, 302–305.

Takanashi, S., Hashira, S., Matsunaga, T., Yoshida, A., Shiota, T., Tung, P.G., Khamrin, P., Okitsu, S., Mizuguchi, M., Igarashi, T., and Ushijima, H. (2009). Detection, genetic characterization, and quantification of norovirus RNA from sera of children with gastroenteritis. J. Clin. Virol. *44*, 161–163.

Teunis, P.F., Moe, C.L., Liu, P., Miller, S.E., Lindesmith, L., Baric, R.S., Le, P.J., and Calderon, R.L. (2008). Norwalk virus: how infectious is it? J. Med. Virol. *80*, 1468–1476.

Thackray, L.B., Wobus, C.E., Chachu, K.A., Liu, B., Alegre, E.R., Henderson, K.S., Kelley, S.T., and Virgin, H.W. (2007). Murine noroviruses comprising a single genogroup exhibit biological diversity despite limited sequence divergence. J. Virol. *81*, 10460–10473.

Wang, F., Ma, Y., Barrett, J.W., Gao, X., Loh, J., Barton, E., Virgin, H.W., and McFadden, G. (2004). Disruption of Erk-dependent type I interferon induction breaks the myxoma virus species barrier. Nat. Immunol. *5*, 1266–1274.

Ward, J.M., Wobus, C.E., Thackray, L.B., Erexson, C.R., Faucette, L.J., Belliot, G., Barron, E.L., Sosnovtsev, S.V., and Green, K.Y. (2006). Pathology of immunodeficient mice with naturally-occurring murine norovirus infection. Toxicol. Pathol. *34*, 708–715.

Wobus, C.E., Karst, S.M., Thackray, L.B., Chang, K.O., Sosnovtsev, S.V., Belliot, G., Krug, A., Mackenzie, J.M., Green, K.Y., and Virgin, H.W. (2004). Replication of Norovirus in cell culture reveals a tropism for dendritic cells and macrophages. PLoS. Biol. *2*, e432.

Wobus, C.E., Thackray, L.B., and Virgin, H.W. (2006). Murine norovirus: a model system to study norovirus biology and pathogenesis. J. Virol. *80*, 5104–5112.

Xia, M., Farkas, T., and Jiang, X. (2007). Norovirus capsid protein expressed in yeast forms virus-like particles and stimulates systemic and mucosal immunity in mice following an oral administration of raw yeast extracts. J. Med. Virol. *79*, 74–83.

Yoneyama, M., Kikuchi, M., Matsumoto, K., Imaizumi, T., Miyagishi, M., Taira, K., Foy, E., Loo, Y.M., Gale, M., Jr., Akira, S., Yonehara, S., Kato, A., and Fujita, T. (2005). Shared and unique functions of the DExD/H-box helicases RIG-I, MDA5, and LGP2 in antiviral innate immunity. J. Immunol. *175*, 2851–2858.

Zheng, D.P., Ando, T., Fankhauser, R.L., Beard, R.S., Glass, R.I., and Monroe, S.S. (2006). Norovirus classification and proposed strain nomenclature. Virology *346*, 312–323.

Murine Norovirus Translation, Replication and Reverse Genetics

11

Akos Putics, Surender Vashist, Dalan Bailey and Ian Goodfellow

Abstract

Murine norovirus, currently the only norovirus that replicates efficiently in tissue culture, has offered scientists the first chance to study the entire norovirus life cycle in the laboratory. In addition, the development of reverse genetics for murine norovirus has provided the ideal opportunity for researchers to determine how variation at the genetic level affects pathogenicity in the natural host. Despite differences in the diseases caused by human and murine noroviruses, they possess a significant amount of genetic similarity; hence the general mechanisms of viral genome translation and replication are likely to be highly conserved. Here we aim to summarize our current understanding of the mechanisms of norovirus translation and replication, highlighting the important role of murine norovirus as a model system in the study of norovirus biology.

Introduction

Significant advances have been made in the study of human norovirus replication in tissue culture, namely the generation of stable replicon containing cell lines (Chang et al., 2006), the observation that norovirus replication and packaging can be driven in tissue culture (Asanaka et al., 2005; Katayama et al., 2006) and the demonstration that norovirus RNA purified from stool samples is infectious (Guix et al., 2007). However, despite epic efforts (Duizer et al., 2004), a reproducible system to allow the study of the complete human norovirus life cycle remains elusive. Preliminary results have suggested that a highly differentiated tissue culture system may allow human norovirus propagation (Straub et al., 2007), however further studies are required to validate these observations. The identification of murine norovirus (MNV) in 2003 heralded a new era for the study of the basic mechanisms of norovirus translation and replication, as for the first time it provided researchers with a norovirus capable of a full infectious cycle in tissue culture (Karst et al., 2003; Wobus et al., 2004). Until the 'missing-link' which will allow human noroviruses to grow efficiently in tissue culture is identified, which recent studies suggest may be at the level of virus entry (Guix et al., 2007), MNV offers the best readily available and easily manipulated norovirus experimental system. In this chapter we aim to summarize how the MNV model has been used to further our understanding of norovirus translation and replication.

Murine norovirus genome structure

MNV possesses the typical calicivirus genome layout as highlighted in Fig. 11.1, with an open reading frame (ORF) encoding a large polyprotein at the 5' end of the genome (ORF1), which is processed into the non-structural proteins (discussed further below). ORF1 is followed immediately by ORF2 encoding the major capsid protein (VP1) and ORF3 encoding a small basic protein which is thought to be a minor component of the virion (VP2) (Sosnovtsev and Green, 2000). In addition however, analysis of the sequence of MNV 1 published in 2003 revealed the presence of an additional fourth ORF (ORF4), not described in the initial publication (Karst et al., 2003). Further

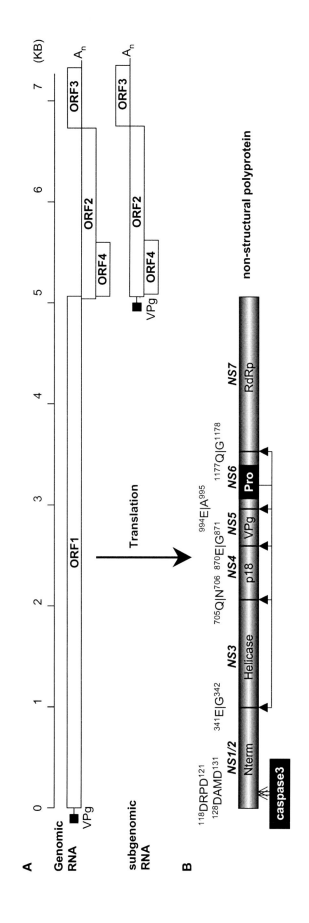

Figure 11.1 Genome organization and proteolytic processing of the non-structural polyprotein of MNV. (A) The structure of the 3' polyadenylated genomic RNA and sgRNA is shown. ORFs are depicted as white boxes while VPg is represented as a black square at the 5' end of the genome. (B) The domain organization and the proteolytic processing of the MNV non-structural polyprotein encoded by ORF1. The grey boxes represent the non-structural proteins (*NS1–2–NS7*) released by the viral 3CL-protease. Proteases involved in the processing of the polyprotein are indicated in black. The cleavage sites, represented as arrows, along with the P1 and P1' residues are also given. Abbreviations: Nterm, N-terminal protein; Helicase, Helicase/NTPase; p18, a protein with a molecular weight of 18 kDa, also termed 3A-like protein; VPg, viral protein, genome-linked; Pro, viral 3C-like protease; RdRp, RNA dependent RNA polymerase.

large scale sequence analysis of several MNV isolates demonstrated that ORF4 is present in all MNV isolates identified to date and shows only a limited degree of variation (Thackray *et al.*, 2007). This degree of conservation indicates a potential role in the virus life cycle. Our preliminary studies suggest that ORF4 protein expression may be associated with virulence in STAT1 knockout mice, but it is not required for growth of the virus in cell culture (our unpublished observations). Further analysis will be needed to identify a function for this protein.

Murine norovirus protein synthesis

VPg-dependent translation initiation

Viral genome translation occurs once the viral RNA has been released into the cytoplasm of the target cell. Characterization of the mechanisms that caliciviruses use for protein synthesis has lagged behind that of many related positive strand RNA viruses, with the study of norovirus translation being particularly hampered due to the lack of a suitable source of authentic VPg-linked viral RNA. The discovery of MNV has provided researchers with the first consistent source of VPg-linked template RNA for the study of norovirus translation initiation. Early insights into calicivirus translation initiation came from the laboratory of the late Fred Brown at Pirbright in the 1970s. Using the animal calicivirus vesicular exanthema of swine virus (VESV), they demonstrated that calicivirus RNA was covalently linked to a low molecular weight protein and that treatment with proteinase K reduced the infectivity of viral RNA (Burroughs and Brown, 1978). In 1997, work with feline calicivirus (FCV) further demonstrated that this was a common feature of many caliciviruses and that proteinase K removal of this virus encoded protein (VPg) resulted in a reduction in viral protein synthesis *in vitro* using the well established rabbit reticulocyte lysate system (RRL) (Herbert *et al.*, 1997). This work also demonstrated that at least for FCV, translation *in vitro* was insensitive to cap analogue and high concentrations of potassium chloride, both of which readily inhibit normal cap-dependent mRNA translation initiation. These data indicated that VPg was required for efficient protein

synthesis but further studies on the mechanism did not appear for another 6 years. In 2003, work using recombinant Norwalk virus (NV) VPg protein and a yeast two hybrid screen identified a component of the eIF3 complex as a direct binding partner (Daughenbaugh *et al.*, 2003). eIF3 is a large complex of 11 proteins and functions by stabilizing the interaction of the Met-tRNAiMet-eIF2–GTP ternary complex with the 40S ribosomal subunit, leading to the formation of the 43S pre-initiation complexes (Hershey and Merrick, 2000). The eIF3d subunit was identified as the direct binding partner of the NV VPg protein using pull down of *in vitro* translated eIF3d and yeast two hybrid analysis; however the entire eIF3 complex could be isolated from cells using GST-NV VPg, suggesting that the NV VPg protein in fact recruited the eIF3 complex to the 5′ end of the viral RNA genome. This is an attractive hypothesis given that it would lead directly to ribosome assembly on the 5′ end of viral RNA. Owing to the unavailability of a source of norovirus VPg-linked viral RNA, the authors were unable to conduct a detailed investigation of the role of the VPg–eIF3 interaction in viral translation initiation. Interestingly, the addition of GST-tagged forms of VPg to RRL lead to a significant inhibition of protein synthesis from a variety of mRNA templates, including capped and IRES containing RNA. Translation from the cricket paralysis virus intergenic region IRES, which does not require eIF3 and can assemble the 80S ribosomes without the requirement for any initiation factors (Wilson *et al.*, 2000), was also inhibited by the norovirus VPg. The authors suggested that their observed (unpublished observations as stated in Daughenbaugh *et al.* 2003) affinity of VPg for the 40S ribosomal subunit may have contributed to this observed inhibition. This conclusion was later questioned by Cotton *et al.* given their observation that purified NV GST-VPg fusion proteins, but not GST, possess RNase activity (Cotton *et al.*, 2006). This appears somewhat counterintuitive given that the virus has an RNA genome and it may simply reflect the observation that NV VPg, and indeed many calicivirus VPg proteins appear to possess RNA binding activity (our unpublished observations). Although not examined in any great detail, the cluster of basic amino acids at the N-terminus of calicivirus VPg proteins may be responsible for such binding. RNA binding activity can often pose problems if robust measures to ensure the removal of bacterial RNA from protein preparations are not followed, as it can lead to the co-purification of bacterial RNA binding proteins, including those with RNase activity. This RNA binding ability may offer another possible explanation for the reported RNase activity as well as the ability of norovirus GST-VPg fusion proteins to inhibit the translation of any reporter RNA examined, although further studies are warranted. Our unpublished observations using a native non-tagged human norovirus (Lordsdale virus strain) and MNV VPg proteins, purified as described in Chaudhry *et al.* 2007 and Goodfellow *et al.* 2005, indicate that neither of these proteins possess RNase activity. Studies of potyvirus VPg have also questioned the presence of an RNase activity for both potyvirus and norovirus VPg proteins (Miyoshi *et al.*, 2008).

These early studies highlighted that the calicivirus family as a whole, may share some similarities with the potyvirus family of plant viruses which also appears to use a virus encoded protein to recruit translation initiation factors to the 5′ end of the viral RNA genome (Wittmann *et al.*, 1997). Independent work in several laboratories has now confirmed that a number of cellular translation initiation factors interact with the calicivirus VPg during virus infection. Studies from our laboratory have demonstrated that FCV, human norovirus (Lordsdale virus strain) and MNV VPg proteins bind to the cap binding protein eIF4E (Chaudhry *et al.*, 2007; Goodfellow *et al.*, 2005). This interaction was found also to occur during virus replication as we showed that VPg, as well as VPg containing precursors, could be co-purified with eIF4E from infected cells using M7-GTP Sepharose (Chaudhry *et al.*, 2007; Goodfellow *et al.*, 2005). *In vitro* pull down analysis from the Hardy laboratory provided direct evidence also that eIF4E is a component of the MNV VPg-translation initiation complex (Daughenbaugh *et al.*, 2006).

The eIF4F complex is a key component of the translation initiation machinery and is composed of the cap binding protein (eIF4E), a scaffold protein (eIF4G) which links the eIF4F complex to eIF3, and an RNA helicase (eIF4A)

(Fig. 11.2A). Further detailed analysis of the role of the individual eIF4F components highlighted that significant differences exists in the translation mechanisms within the calicivirus family (Chaudhry *et al.*, 2006). Despite the observed MNV VPg–eIF4E interaction, depletion of eIF4E from *in vitro* translation reactions, performed using RRLs, led to a significant reduction in FCV VPg-dependent translation, yet had little or no effect on MNV translation (Chaudhry *et al.*, 2006). This may simply reflect that the RRL system, which contains very high concentrations of many of the initiation factors, may not represent the most authentic system for the study of norovirus translation initiation as other factors present at high concentrations may complement for the absence of eIF4E. Further studies to address the role of eIF4E in MNV translation initiation are currently under way, and detailed proteomic analysis of the MNV translation initiation complex, purified from cells expressing a tagged form of the MNV VPg protein, indicates that eIF4E is present in this complex (our unpublished data).

Further analysis of the remaining components of the eIF4F complex has been performed using a variety of approaches (Chaudhry *et al.*, 2006). *In vitro* cleavage of eIF4G using recombinant FMDV L-protease, which separates the eIF4A and eIF4E binding sites, also resulted in differential effects on FCV and MNV, with FCV being inhibited and MNV translation being stimulated (Chaudhry *et al.*, 2006). The apparent stimulation of MNV translation is likely to simply reflect decrease in competition from cellular mRNAs in the viral RNA preparation or release of eIF4G from the eIF4E–eIF4G complex. Hence further studies are required to determine which domains of eIF4G are required for MNV translation initiation.

Studies on the role of the RNA helicase component of eIF4F, namely eIF4A, indicate that both FCV and MNV require this factor for efficient translation *in vitro* and in tissue culture although their relative dependence differs (Chaudhry *et al.*, 2006). Dominant negative forms of eIF4A and a small molecule inhibitor of eIF4A have been used to highlight that MNV has a greater requirement for eIF4A than FCV, an observation which may be as a direct result of the increased amount of RNA secondary structure

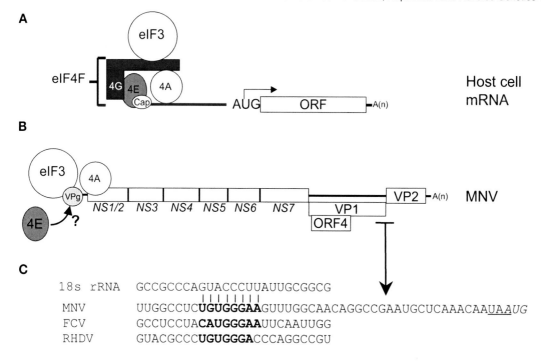

Figure 11.2 Schematic representation of the translation initiation complexes and RNA sequences involved in MNV translation initiation. (A) Components of the eukaryotic translation initiation apparatus responsible for translation initiation on 5′ capped host mRNA molecules. For simplicity only the eIF4F components and eIF3 are shown. (B) The MNV genome (MNV) highlighting the translation initiation factors thought to be responsible for translation initiation. Note that the role of eIF4E–VPg interaction in MNV translation is still unknown as highlighted with a question mark. (C) Sequence of the putative MNV TURBS sequence. The sequence of the core motif of the previously reported TURBS sequences responsible termination–reinitiation of ribosomes on the VP1–VP2 stop–start sequence is highlighted in bold. Note that, of the sequences shown, only feline calicivirus (FCV) and rabbit haemorrhagic disease virus (RHDV) have been experimentally confirmed. Sequences with complementary to nucleotides 1101–1125 of the 18S rRNA sequence are highlighted in bold. The MNV VP1 stop codon is underlined and the VP2 start codon highlighted in italics.

present in the MNV genome (Simmonds et al., 2008). The natural small molecule eIF4A inhibitor hippuristanol, purified from the sea fan coral *Isis hippuris* and identified via a high throughput screen for translation inhibitors (Bordeleau et al., 2006), was found to inhibit MNV translation (and subsequent virus production) very effectively in tissue culture (Chaudhry et al., 2006). Our initial interpretation was that this observed effect was simply due to the inability of the MNV RNA to be efficiently translated in the infected cell. However, recent data have demonstrated that hippuristanol treatment of cells rapidly leads to the redistribution of several translation initiation factors and the formation of cytoplasmic stress granules (Mazroui et al., 2006). Hence, treatment of cells with hippuristanol not only prevents

the binding of eIF4A to mRNA, but also results in the sequestration of several host cell factors, including eIF4A, in defined complexes within the cytoplasm of the infected cell, which in the case of MNV, may lead to an enhancement of the observed effect.

Further studies into this novel paradigm of protein-directed translation initiation, illustrated in Fig.11. 2B, will undoubtedly result in the identification of additional canonical and non-canonical translation initiation factors required for protein synthesis. Indeed, a full scale proteomics analysis of the MNV translation initiation complex has recently been completed and indicated that numerous non-canonical translation initiation factors can be found associated with VPg (our unpublished observations).

Synthesis of the minor capsid protein VP2

The synthesis of VP2 from the MNV subgenomic RNA (sgRNA) has yet to be characterized in any great detail, however one might assume that a similar termination-reinitiation of translation is likely to occur as documented for FCV and RHDV (Luttermann and Meyers, 2007; Meyers, 2003, 2007; Poyry *et al.*, 2007). This mechanism, whereby ribosomes terminating on the stop codon of VP1 subsequently reinitiate translation on the start codon of VP2, relies on a bipartite RNA sequence upstream of the VP1 termination codon-VP2 initiation codon overlap. This motif has been termed the termination upstream binding site (TURBS). There are currently two proposed models, which are not mutually exclusive, on the function of this motif, the first of which builds on the observation that motif 1 appears to have complementarity to 18S rRNA (Fig. 11.2C). This model supposes that complementarity with 18S rRNA prevents ribosome disassembly during termination thereby allowing ribosomes to subsequently reinitiate on the start codon of VP2. In agreement with this model, mutations which disrupt the predicted interaction with 18S rRNA are deleterious to VP2 expression (Luttermann and Meyers, 2007). A second model suggests that eIF3 binding to the TURBS motif may prevent ribosome disassembly and allow reinitiation to occur (Poyry *et al.*, 2007). This is strengthened by the observation that addition of purified eIF3 to *in vitro* translation reactions stimulates the reinitiation event (Poyry *et al.*, 2007). Analogous studies for MNV have yet to be performed but the presence of potential TURBS-like motifs with complementarity to 18S rRNA (Fig. 11.2C) would suggest a similar mechanism exists.

Proteolytic processing of the MNV non-structural polyprotein

Translation of ORF1 of the viral genomic RNA produces a large, 187.5-kDa protein; the so called non-structural polyprotein (Fig. 11.1B). Comparative sequence analysis of 15 MNV strains revealed that the ORF1 encoded polyprotein is highly conserved with up to 19% divergence at the nucleotide level and no more than 11% at the protein level (Thackray *et al.*,

2007). Further sequence comparison analysis of the MNV genome with those of other RNA viruses showed that the MNV ORF1 encodes amino acid motifs which are homologous to those of caliciviruses and picornaviruses (Karst *et al.*, 2003). The MNV non-structural polyprotein was predicted to contain six non-structural protein (NS) domains which are released by the viral 3CL-protease (Fig. 11.1B). On the basis of the sequence homology to other calicivirus and picornavirus proteins, the following domains were identified: N-terminal protein (NS1–2), helicase/NTPase (NS3 or 2CL protein), a protein with a molecular weight of 18 kDa (NS4, p18 or 3A-like protein), VPg (NS5), viral 3C-like protease (NS6 or 3CL-protease) and the RNA dependent RNA polymerase (RdRp, NS7 or 3DL polymerase). Sosnovtsev *et al.* demonstrated the existence of these six cleavage products (NS1–2–NS7) both *in vitro* as well as in MNV infected cells (Sosnovtsev *et al.*, 2006) and mapped the dipeptide cleavage sites recognized by the protease (Fig. 11.1B).

In MNV, similarly to other noroviruses, the proteolytic processing is mediated by the viral 3CL-protease (NS6) which is autocatalytically released from the polyprotein precursor. The MNV 3CL-protease shares high amino acid sequence similarity with those of other well-characterized noroviruses, like NV, Southampton virus (SHV), Chiba virus (CV) and MD145 virus (Belliot *et al.*, 2003; Blakeney *et al.*, 2003; Liu *et al.*, 1996; Someya *et al.*, 2005; Sosnovtsev *et al.*, 2006). It has been already shown that the norovirus protease is a member of the chymotrypsin-like serine protease superfamily. The crystal structure of the Chiba virus 3CL-protease revealed that the protease has a chymotrypsin-like fold, consisting of an N-terminal and a C-terminal domain. The active site was shown to be located in a hydrophobic cleft between the N- and C-terminal domains. It includes a catalytic dyad consisting of Cys139 and His30 (Nakamura *et al.*, 2005; Someya *et al.*, 2005). The presence of a catalytic dyad (Cys/His) is not an unique feature among viral 3CL-proteases since the active site of the 3CL-proteases of another family of positive single-stranded RNA viruses, the coronaviruses, were also shown to consist of a dyad of Cys and His (Anand *et al.*,

2002). Although the mechanism of proteolysis remains unclear, recently published results of a site-saturation mutagenesis study of the Chiba virus 3CL-protease raised an interesting possibility that the norovirus 3CL-protease uses a papain-like proteolytic mechanism (Someya *et al.*, 2008). Owing to the lack of a crystal structure or extensive mutagenesis analysis, we have very limited information on the structure and the active site of the MNV 3CL-protease. However, the putative nucleophilic active site Cys1133 of the MNV 3CL-protease was shown to be essential for the proteolytic activity, strongly supporting its presumed catalytic role (Sosnovtsev *et al.*, 2006).

Despite the variability of the ORF1 polyprotein sequences at the amino acid level, both the number and the location of the cleavage sites and the scissile bonds cleaved between the P1 and P1′ amino acids were found to be well-conserved among noroviruses (Sosnovtsev *et al.*, 2006). Similarly to other noroviruses, five 3CL-protease cleavage sites were identified in the MNV-1 non-structural polyprotein: ^{341}E|G^{342}, ^{705}Q|N^{706}, ^{870}E|G^{871}, ^{994}E|A^{995}, ^{1177}Q|G^{1178} (Sosnovtsev *et al.*, 2006) (Fig. 11.1B). Regarding the P1 residues, these data correlate with those obtained from the analysis of the 3CL-protease cleavage sites of other noroviruses, like SHV, MD145 or NV, indicating that the norovirus 3CL-protease has a preference for glutamic acid or glutamine at the P1 position (Belliot *et al.*, 2003; Blakeney *et al.*, 2003; Hardy *et al.*, 2002; Liu *et al.*, 1999). Comparison of the genome sequences of 26 MNV strains showed that the cleavage sites were highly conserved at the P1 position (Thackray *et al.*, 2007). In contrast, some minor amino acid exchanges were found at the P1′ position such as the substitution of N^{760} to S in the NS3/NS4 cleavage site in MNV strains CR3, CR6, CR7 and UM2, and the substitution of G^{871} to S in the NS4/NS5 cleavage site of strains CR3, CR4, CR10, CR11, CR13, UM3 (Thackray *et al.*, 2007). Similar to other noroviruses, the MNV 1 3CL-protease recognizes either glycine or alanine at the P1′ position, with the exception of the NS3/NS4 cleavage site which has a polar P1′ asparagine (Sosnovtsev *et al.*, 2006). The same glutamine-asparagine scissile bound was also found to be cleaved in FCV between NS3 and NS4 (Sosnovtsev *et al.*, 2002).

Because proteolytic processing of calicivirus proteins by the viral 3CL-protease occurs in the cytoplasm of the infected host cell, certain cellular proteases may also recognize and cleave viral protein targets. The first evidence of the involvement of cellular proteases in the proteolytic processing of MNV was shown by Sosnovtsev *et al.* who identified two proteolytic cleavage sites in the MNV non-structural polyprotein recognized by a cellular enzyme, caspase 3 (Sosnovtsev *et al.*, 2006). These sites were mapped *in vitro* to the non-structural polyprotein sequences ^{118}DRPD^{121} and ^{128}DAMD^{131} which are located in NS1–2 (Fig. 11.1B). Caspase 3 is a key cellular cysteine protease that cleaves a range of protein substrates within the cell to trigger apoptosis (Chowdhury *et al.*, 2008). Of note, some features characteristic of apoptosis (e.g. the activation of caspase 3, 9 and 9) can be observed in MNV infected cells, and proteolytic processing of NS1–2 by caspase 3 was proposed as responsible for the observed cleavage of this protein over time in cells (Sosnovtsev *et al.*, 2006). The activation of caspase 3, together with additional components of the mitochondrial pathway of apoptosis activation, was found also in FCV infected cells (Natoni *et al.*, 2006). Furthermore, an *in vitro* cleavage assay revealed that caspase-2, as well as caspase-6, cleaved the FCV capsid protein to generate a 40-kDa fragment suggesting a possible role of capsid cleavage in the life cycle of FCV infection (Al-Molawi *et al.*, 2003). It will be important to verify whether caspase or other cell protease-mediated cleavages are required for replication of the virus, or merely a by-product of apoptosis or necrotic cell death.

Another interesting question in murine norovirus biology is whether the MNV 3CL-protease itself has cellular targets. Previous studies with the human norovirus MD145-12 and the FCV 3CL-proteases have already demonstrated their ability to cleave poly(A)-binding protein (PABP), thereby inhibiting cellular translation (Kuyumcu-Martinez *et al.*, 2004). Cellular substrates of the MNV 3CL-protease are likely to exist but have yet to be identified.

The kinetics of the release of the non-structural proteins might be a crucial regulatory point of the viral life cycle because this process may determine at which time point each NS protein gains its functionally active form during the course of virus infection. Mutagenesis data showed that the cleavage between the 3CL-protease and the RdRp is essential for the recovery of infectious MNV (Ward *et al.*, 2007). In contrast, the FCV protease–polymerase precursor seems to be active because the mutation of the putative cleavage site between the protease and polymerase is not lethal (Sosnovtsev *et al.*, 2002). An *in vitro* study characterizing the catalytic activity and substrate specificity of the human norovirus 3CL-protease suggested that the 3CL-protease fused to RdRp has a different cleavage pattern than the 3CL-protease alone (Scheffler *et al.*, 2007). In MNV infected cells, the predominant form of the 3CL-protein was shown to be the mature version of the protease; however fusions of the protease to p18-VPg, VPg or RdRp could be also observed (Sosnovtsev *et al.*, 2006). The kinetics of the release of the MNV non-structural proteins remains to be investigated but they undoubtedly play a key role in the regulation of the various aspects of the MNV life cycle.

Replication

Most of our understanding of the mechanisms of norovirus genome replication is drawn by analogy or from parallel studies on related animal caliciviruses and/or other positive strand RNA viruses. Here we will discuss the current state of knowledge on the replication of MNV in comparison with other related positive strand RNA viruses, highlighting specific data on MNV wherever it is available. The replication of positive strand RNA viruses, including caliciviruses is believed to occur in or on the surface of replication complexes (RCs) formed by virus induced host intracellular membranous structures (Green *et al.*, 2002; Love and Sabine, 1975; Studdert and O'Shea, 1975). These membranous structures are induced by the expression of the viral non-structural proteins, usually as a result of virus mediated modulation of the host cell secretory pathway (Denison, 2008). These replication complexes are thought to act as a surface on which the replication factors, both viral and cellular, are localized and concentrated for assembly. Along with the viral non-structural proteins, these RCs also contain host proteins that function in viral replication, as well as viral double stranded and single stranded

RNA molecule intermediates in replication. Enzymatically active forms of these replication complexes have been isolated for various positive strand RNA viruses including FCV (Green *et al.*, 2002) and MNV (Wobus *et al.*, 2004). The rearrangement of intracellular membranes has been observed during MNV 1 infection of RAW 264.7 cells (Wobus *et al.*, 2004). After infection, a striking change in overall morphology of cells and intracellular organization of organelles was observed as compared to the mock-infected cells. Single or double membraned vesicles appear 12 hours post infection, which seem to increase in size and number over time. Virus particles can also be detected within or close to these areas. These vesicles appear to be the result of rearrangement of endoplasmic reticulum and loss of intact Golgi apparatus. However, they are often surrounded by mitochondria which, given that mitochondria are generally associated with autophagocytic vesicles (McCartney *et al.*, 2008), may lead to the hypothesis that the mechanism of autophagy may be involved in their formation. Clearly, further studies in this area are warranted but it is possible that MNV may use the natural response of cells to infection (i.e. induction of autophagy) for replication complex formation. This has yet to be examined in any detail but has been suggested for other viruses as well (Taylor and Kirkegaard, 2008; Wileman, 2006).

For all positive strand RNA viruses to replicate, the genome must act both as mRNA for the production of viral proteins required for genome replication and as a template for the synthesis of negative-strand RNA. This negative strand RNA then acts as a template to produce large amounts of positive strand genomic RNA (Fig. 11.3B). The RdRp encoded by the viral genome along with other viral and cellular factors binds at specific promoter sequences present at the 3' end of both positive and negative strand RNAs in order to drive and regulate RNA synthesis (Fig. 11.3B). MNV, as other caliciviruses, is also known to produce a sgRNA corresponding to the 3' end of the full length genomic RNA (Wobus *et al.*, 2004). A detailed review of sgRNA synthesis by various positive strand RNA viruses has been published previously (Miller and Koev, 2000), however, how the MNV sgRNA is produced is still unclear. By analogy with other RNA viruses there are two

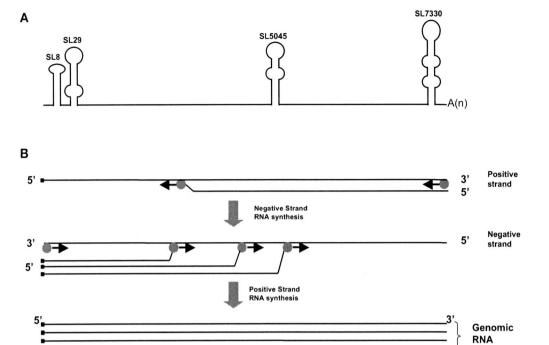

Figure 11.3 Murine norovirus genome replication. (A) The MNV genome depicting various predicted RNA secondary structures that were shown to affect MNV replication (see text for further details). Stem loop (SL) 8 and SL29 are at the 5′end of the positive strand genomic RNA within the region coding for the NS1–2 protein, whereas SL7330 is present in the non-coding region at the 3′end. SL5045 is at the NS/S junction and can be predicted to form in both the positive and negative strand. The nomenclature used highlights the first base of the predicted structure i.e. SL8 forms a stable stem–loop structure starting at nucleotide 8 of the MNV genome. (B) Schematic overview of murine norovirus genome replication. Positive sense genomic RNA is uncoated after viral entry into the cell and subsequently translated to produce the viral proteins, including the RNA dependent RNA polymerase (RdRp), responsible for genome replication. The RdRp then binds to the 3′- end of the incoming positive strand genomic RNA and generates a negative strand RNA copy. This negative strand RNA is then used as a template by the MNV RdRp to produce large numbers of positive strand genomic and sgRNAs. The possible mechanisms of sgRNA synthesis are described in the text and Fig. 11.4.

most widely accepted possible mechanisms for sgRNA production:

1   *Internal initiation (Fig. 11.4A)* The viral RdRp NS7 binds at a promoter sequence/ structure present internally in the negative strand RNA, which in the case of MNV would be predicted to lie upstream of the start codon of ORF2 in the genomic strand, and subsequently initiates positive strand sgRNA synthesis (Miller *et al.*, 1985). Such promoters have been identified for various positive RNA viruses including caliciviruses,

but in the latter case only *in vitro* data is available, warranting further studies as discussed below (French and Ahlquist, 1988; Levis *et al.*, 1990; Morales *et al.*, 2004).

2   *Premature termination (Fig. 11.4B)* This model predicts that during the synthesis of negative strand RNA using the positive strand genomic RNA as a template, the NS7 RdRp terminates at a region upstream of start codon of ORF2 producing subgenomic length negative strand (Sit *et al.*, 1998). This negative strand then acts as a template for (end to end) positive-strand sgRNA synthesis.

A

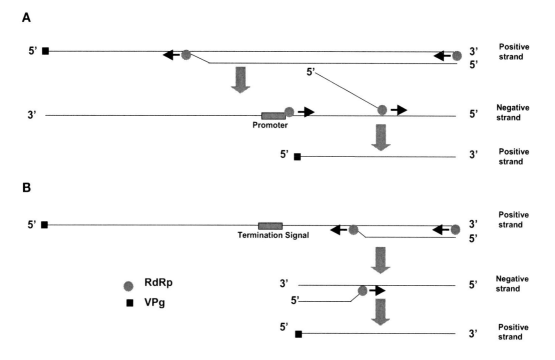

B

Figure 11.4 Two models for murine norovirus sgRNA RNA synthesis. For simplicity, the genomes, which are highly structured, are shown as straight lines and promoters/termination signal are indicated by grey rectangles. (A) Internal Initiation: The viral RNA dependent RNA polymerase (RdRp) binds to the sgRNA promoter located downstream of the sgRNA start site on the negative strand. sgRNA synthesis then occurs to produce VPg-linked positive strand sgRNA. (B) Premature Termination: The RdRp binds the 3′ end of the positive strand genomic RNA and initiates negative strand synthesis. During elongation the RdRp terminates at a specific sequence located at the 3′ end of the negative strand sgRNA. This results in the production of negative strand sgRNA, which then acts as a template to produce positive strand sgRNA.

Various studies have indicated the possible involvement of the internal initiation mechanism for sgRNA synthesis, and the promoter sequence required for sgRNA synthesis, at least *in vitro*, has been mapped for RHDV (Morales *et al.*, 2004). In this study, using an *in vitro* assay, a nested set of negative strand RNA templates with a common 5′ end (complementary to the 3′ end of genomic RNA) were used that extend to different lengths after the predicted sgRNA initiation site. It was found that at least 50 nucleotides beyond the +1 sgRNA start were required to produce a sgRNA of the expected size. Such promoters have been mapped in other viruses as well and at least a portion of each promoter is located immediately upstream of the 5′ end of the sgRNA (French and Ahlquist, 1988; Levis *et al.*, 1990). It has been proposed that the presence of a negative strand sgRNA would support the use of the premature

termination mechanism of sgRNA synthesis, but it has been shown that negative strand sgRNA can be extracted from BMV infected cells, though it clearly employs an internal initiation mechanism (Ishikawa *et al.*, 1997; Miller *et al.*, 1985). It is possible that after internal initiation has occurred to produce newly synthesized VPg-linked sgRNA, this RNA may then be picked up by the viral replication machinery and replicated in a manner similar to the genomic RNA, effectively producing a negative strand sgRNA molecule. More focussed studies would be required to decipher the production of sgRNA synthesis using a more authentic system as the minimal promoter requirements *in vitro* are generally insufficient *in vivo* (French and Ahlquist, 1988). Our published data have demonstrated that all caliciviruses appear to retain a stem loop structure 6 nucleotides upstream in the positive strand and downstream in the negative strand, of the sgRNA initiation site (Simmonds et al., 2008) (SL 5045 in Fig. 11.3A). Thermodynamic analysis would indicate that, generally speaking, this structure is more stable in the negative strand than in the positive strand and is referred to as Sla 5045 in Simmonds et al. (2008). Mutational analysis of this stem loop using the MNV reverse genetics system, described in more detail below, has demonstrated that the stability of this RNA structure is critical for MNV replication (Simmonds et al., 2008). Ongoing analysis in our laboratory focuses on the further characterization of the role of this RNA structure in MNV sgRNA synthesis.

The genome of MNV is covalently linked to the viral protein VPg, which is known to be important in translation of the genome (Goodfellow et al., 2005). However it is not known whether the linkage of this viral protein to the 5′ end of the genome occurs prior to genome replication in order to serve as a protein primer as has been demonstrated for picornaviruses (Paul et al., 1998) or after replication. Recent reports suggested that VPg of a human norovirus could be nucleotidylated by the precursor form of the viral RdRp in a template independent manner at tyrosine 27 (Belliot et al., 2008). It has been suggested in an in vitro study that replication of norovirus genomic RNA is VPg primed as norovirus RdRp could initiate replication of viral genome only after elongation of VPg, where as replication of negative strand RNA occurs de novo (Rohayem et al., 2006). This study provides in vitro evidence of the involvement of VPg in replication by acting as a protein primer but as discussed below, the in vitro requirements of viral RdRp might not be sufficient in vivo.

The roles of cis-acting RNA and trans-acting proteins in the regulation of positive strand RNA virus replication are widely accepted (Ortin and Parra, 2006). Conserved RNA structures or sequences, usually at the extremities of the positive or negative strands of viral RNA are known to play a role in replication, e.g. by providing promoter/binding sites for viral and host proteins, by forming inter- or intra–molecular interactions with other RNA structures, or by acting as promoter/termination signals. In the majority of positive strand RNA viruses, these elements are usually present in the non-coding regions

(NCRs) of viral genomes (Foy et al., 2003; Mirmomeni et al., 1997; Pelletier and Sonenberg, 1988; Rohll et al., 1995; Tsukiyama Kohara et al., 1992). In MNV or for all caliciviruses in general, the 5′ NCRs are very small and it is believed that the conserved structures/sequences present at terminal regions or non-structural/structural protein ORF (NS/S) junction (ORF1 and ORF2) play similar roles. Studies in our laboratory have revealed the presence of conserved RNA secondary structures at the 5′ and 3′ ends, as well as at the NS/S junctions as predicted by various bioinformatics approaches such as suppression of synonymous site variability (SSSV), high resolution thermodynamic scanning, Alifold and Pfold (Simmonds et al., 2008). These structures, highlighted in Fig. 11.3A as SL8, SL29, SL5045, and SL7330 respectively, were also analysed functionally using the reverse genetics system developed in our laboratory and described in more detail below (Chaudhry et al., 2007). Mutations were introduced into RNA stem–loop structures as predicted for the 5′ end of the genome (SL8 and SL29), NS/S junction (SL5045) and for the 3′ end region of the genome (SL7330). These mutations are predicted to disrupt the structures in both the positive and negative strand. It is important to note that SL5045 can be predicted in both the positive and negative strand of the viral RNA and this structure is referred to as Sla5045 in Simmonds et al. (2008), due the increased stability of the structure in the negative strand. The replication of the mutants was analysed using a reverse genetics system which allows genome replication/packaging and release of virions, but not subsequent rounds of infection, hence virus obtained after 24 h represents a single round of replication. All the introduced mutations affected viral titre, although to varying degrees, ranging from a 15-fold reduction to a complete abrogation of virion production. Mutations were also introduced into the stem–loop upstream of the NS/S junction (SL5045, Fig. 11.3A), restoring the ability of this RNA structure to form a stable stem–loop, the effect of which was to restore virus replication. This study confirmed the requirement for conserved secondary structures in the MNV genome for efficient genome replication.

As discussed above for sgRNA production, host factors are involved in positive strand RNA virus genome replication. These host factors are also thought to contribute to host and tissue specificity, as well as the pathogenicity of infection. The role of these host factors has been predicted to vary and can range from acting as RNA chaperones to bridging the 5'- and 3'-ends during replication. Characterization of the interactions of these host proteins with viral RNA and/or proteins is one of the key aspects of understanding the mechanism of replication as disruption of these interactions might serve as a potential therapeutic target. For other positive strand RNA viruses, the majority of the host proteins which bind to the viral genome have been shown to interact with 5'- and 3'- extremities of viral RNAs (Ahlquist *et al.*, 2003; Lai, 1998). In fact a number of host proteins have been reported to interact with the 5'- and 3'- ends of the NV genome (Gutierrez-Escolano *et al.*, 2003), though their role in replication is yet to be described. Unpublished data from our lab has shown that many of the host proteins such as La, PTB, and PCBP also interact with MNV genomic RNA and sgRNA extremities. Data from our lab showed that the interaction of polypyrimidine tract binding protein (PTB) with the FCV genome was essential for virus replication in a temperature dependent manner (Karakasiliotis *et al.*, 2006) and ongoing studies focus on the role of other host factors in viral replication.

Reverse genetics

The synthesis of viable virus from cloned DNA and the subsequent analysis of specific genotypic changes on phenotypic traits (reverse genetics) is a powerful technique in virus research. Prior to 2007 the only reverse genetics systems available for calicivirus research were that of FCV (Sosnovtsev and Green, 1995) and porcine enteric calicivirus (PEC) (Chang *et al.*, 2005). The lack of a norovirus reverse genetics system, combined with the non-cultivatable nature of the human viruses in tissue culture has long hampered research into this genus. Recently two independent systems have been reported for MNV, both of which utilize the CW1 strain derived from the original MNV isolate (Karst *et al.*, 2003; Wobus *et al.*, 2004) and build on the successful growth of MNV in immune derived cell-lines (e.g. RAW 264.7 cells) (Wobus *et al.*, 2004).

Pol II-driven recovery of MNV

The system described by Ward *et al.* presents various separate mechanisms for generating infectious MNV derived from cDNA (Ward *et al.*, 2007) (Fig. 11.5A–C). The first is a two-component baculovirus expression system that is used to transduce mammalian cells (Fig. 11.5A). Transfection of the bacmids used to generate the recombinant baculoviruses may also be used to generate infectious MNV (Fig. 11.5B). A further variation using a single plasmid system whereby RNA expression is under the control of a minimal CMV promoter element was also reported (Fig. 11.5C).

In the baculovirus system, two separate bacmid constructs are used to generate baculovirus stocks capable of driving expression of the MNV genome in cells. One virus (BACTETMNV) contains the MNV genome under the control of a minimal CMV promoter. This promoter is linked to the tetracycline (tet) operator/repressor (tetO/tetR) system and has been used previously for hepatitis C virus (McCormick *et al.*, 2002). DNA dependent RNA-polymerase Pol II-mediated expression of a tetR/VP16 transactivator from a second baculovirus (BACTETtTA) activates the tetO/minCMV leading to production of genomic sense MNV-like viral RNA. To briefly outline the mechanism: the mRNA produced from BACTETtTA must be exported from the nucleus and translated into the tetR/VP16 transactivator protein. This protein then returns to the nucleus and activates the minCMV promoter of BACTETMNV driving expression of viral genome- like RNA in the nucleus. This large RNA must then be capped to increase its stability and exported from the nucleus while simultaneously avoiding any mRNA processing pathways.

Based on data from other viruses it was assumed that the 5' end of the MNV genomic RNA was critical for viral replication. The integrity of this terminus was maintained by its correct positioning in relation to the CMV transcriptional start. A correct 3' end, also assumed to be crucial for replication, was generated using the hepatitis delta ribozyme sequence. The baculoviruses are initially generated in the insect cell line Sf9 (by transfection of the bacmids), the recovered viruses amplified by subsequent passage in the same cell line and highly concentrated virus stocks

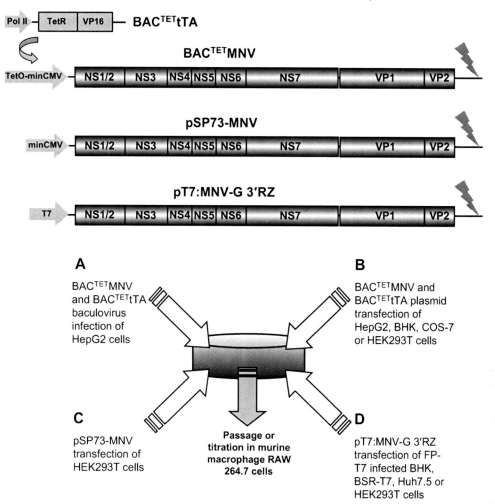

Figure 11.5 Diagrammatic representation of the available murine norovirus reverse genetics systems. Top: Schematics of the various constructs used in reverse genetics rescue of MNV. The following abbreviations are used: minCMV, minimal cytomegalovirus promoter; T7, T7 promoter element; TetO/TetR, tetracycline operator/repressor; VP16, VP16 trans-activator. Bottom: Simplified flow-diagram of the available mechanisms for rescuing MNV from cDNA namely (A) the dual-baculovirus TetO/TetR transduction system, (B) the dual-plasmid TetO/TetR, (C) the single plasmid CMV promoter driven system and (D) the recombinant T7 expressing pox driven cytoplasmic replication system. All systems require the subsequent passage of viruses in the murine macrophage RAW 264.7 cell line in order to obtain high-titre virus stock. The lightning bolt symbols highlight the location of self-cleaving ribozyme sequence. See main text for further details.

generated by centrifugation. Co-infection of human hepatocellular liver carcinoma (HepG2) cells with BACTETMNV and BACTETtTA results in the appearance of significant cytopathic effect in HepG2 cells, in contrast to singular infection with either baculovirus. The observed cytopathic effect is probably due to the primary production of viral proteins within transduced cells, some of which may be toxic due to the ability to modify or inhibit host cell processes, rather than secondary rounds of infection of neighbouring un-transduced HepG2 cells by newly formed infectious virus, since HepG2 cells themselves are not naturally permissive to MNV infection. Virus is isolated by freeze thawing and either titred, or passaged/amplified in the permissive cell line, RAW 264.7. Typical primary yields from the baculovirus system are reported to be 2×10^2 pfu/ml. Of note, direct baculovirus mediated transduction of RAW 264.7 cells is inefficient and does not yield recoverable virus, presumably due to low protein expression.

An alternative system also presented by Ward *et al.* describes transfection of eukaryotic cells with the cDNA constructs used to generate the baculoviruses described above. Virus was recovered from HepG2, BHK-21, COS-7 and HEK293T cells directly transfected with the BACTETMNV and BACTETtTA constructs. Interestingly, further research showed the HEK293T cells produced virus after transfection with BACTETMNV alone. A construct similar to BACTETMNV but containing the MNV genome under the control of a minimal CMV promoter (Fig. 11.5C), also produced recoverable virus after transfection of HEK293T cells, with this system typically yielding $5x10^3$ pfu/ml. Recently, a similar Pol II-driven expression system was used to recover RHDV (Liu *et al.*, 2008).

T7 RNA polymerase driven recovery of MNV

An additional reverse genetics system for MNV is based on T7 RNA polymerase promoter-driven expression of cDNA clones (Chaudhry *et al.*, 2007) (Fig. 11.5D). When a cDNA copy of the MNV genome, under the control of a bacteriophage T7 RNA polymerase promoter, was transfected directly into eukaryotic cells expressing T7 polymerase, genome-like viral RNA was generated. Conventionally, in this sort of system, the T7 polymerase used to drive expression is provided by pre-infection of permissive cells with the vaccinia recombinant MVA-T7. However, no recoverable MNV was detected after MVA-T7 infection. By transfection of permissive cells with VPg-linked viral RNA and analysis of protein and virus yields, we were able to demonstrate that MVA-T7 infection of cells resulted in a 3–4 $\log_{10}$ reduction in MNV yield. In contrast, a fowlpox virus recombinant expressing T7 (FP-T7), which had been successfully used for the recovery of other viruses (Das *et al.*, 2000) had no deleterious effect on MNV replication. Pre-infection of cells with FP-T7 followed by transfection of a plasmid containing the MNV genome under the control of a T7 RNA polymerase promoter generated infectious MNV at yields of $2x10^3$ TCID-50/35 mm dishes, as determined by titration in the permissive murine macrophage cell line RAW 264.7. In addition, there was no requirement for concurrent expression of the sgRNA in this rescue system. The inhibitory effect of MVA on murine norovirus infection may be simply the result of a more efficient shut-off of host cell-translation than FP.

This simple system was further refined by the addition of a hepatitis-delta ribozyme sequence at the 3′ end of the RNA genome. This MNV clone (pT7:MNV-G 3′RZ) was found to be more efficient at generating infectious virus, yielding on average 3.56×10^4 TCID-50/35mm dish (Fig. 11.5D). The system was also shown to work in other cell lines including Huh 7.5, 293T and BSR-T7s (BHKs endogenously expressing T7). However in the latter case, the endogenously expressed T7 RNA polymerase in BSR-T7 cells was insufficient to rescue MNV and these cells also required pre-infection with FP-T7. This may in part be due to the ability of the helper virus to supply the enzymatic activities required for cyto-plasmic capping of the RNA as has been observed for other poxviruses (Fuerst and Moss, 1989). Attempts to rescue MNV directly in RAW 264.7 cells were unsuccessful, possibly because of the typically low transfection rate of these cells.

Similarly to the baculovirus derived system the authors demonstrated the recombinant viruses were genetically identical to the cloned progenitor MNV1.CW1.P3 and had not undergone any additional changes. The presence of viral proteins and genomic RNA were confirmed by western blot and RT-PCR and the stringency of the rescue system was proved through rescue of a genetically defined (marked) mutant.

Applications of MNV reverse genetics

Although it has been only a few years since their generation, the reverse genetics systems described above have been utilized to dissect fundamental aspects of MNV replication. These include:

Pol II system
- The successful cleavage of MNV protease-polymerase precursor form of the RdRP into separate units has been shown to be essential for replication, in contrast to FCV where this is not essential (Ward *et al.*, 2007).

T7 RNA polymerase system

- The correct sequence of the last nucleotide of the MNV genome, immediately upstream of the polyA tail, was shown to be essential for recovery of infectious virus in a reverse genetics-based assay (Chaudhry *et al.*, 2007).
- A single amino acid substitution in the VP1 protein of MNV 1 was proven to attenuate the virus in the STAT −/− virulent mouse model (Bailey *et al.*, 2008).
- Small RNA structures (stem–loops) in the MNV genome were shown to be essential for virus replication, and in some instances, when mutated, were found be under selection pressure to revert (Simmonds *et al.*, 2008 and unpublished data).
- The MNV genome encodes a fourth open reading frame (internal ORF in ORF2/ VP1), producing a protein of expected size in infected cells. This protein is not produced by a recombinant virus engineered to lack an intact version of the open reading frame (our unpublished data).
- Identification of amino acids within the MNV VPg protein critical for the interaction with translation initiation factors and virus replication (our unpublished data).

Whilst a direct side-by-side comparison of the efficiency of the various systems has not been performed, published data would indicate that the yield of infectious virus from the T7 RNA polymerase system is typically >10 fold higher than that obtained using pol II. The obvious drawback of the T7-driven system is the requirement to use a helper virus, preparation of which requires primary chick embryo fibroblasts. However, an unexpected benefit of the use of T7 RNA polymerase, and one which relies on the inherent error rate of T7 (Huang *et al.*, 2000), is the ability to readily generate revertant and second site suppressors of debilitating mutations. Our unpublished work has indicated that in many cases where the introduced mutations are severely debilitating to virus replication, 'blind passage' of tissue culture supernatants, often leads to the recovery of viruses which have regained the ability to replicate efficiently in tissue culture. Such viruses often contain revertant or second

site suppressor mutations which overcome the apparent defect. This type of analysis can lead to the identification of protein-protein binding sites and/or functional sites within key proteins involved in virus replication and can aid our understanding of the various interactions required for virus replication.

Future applications of MNV reverse genetics

The advantages of a reverse genetics system lie in the ability to focus on specific aspects of the virus life cycle by targeting certain genes and untranslated regions for mutation. By analysing the resultant effect on the virus phenotype (*in vivo* and *in vitro*) researchers can elucidate the mechanisms of virus replication. Future research with the MNV reverse genetics systems described above will undoubtedly focus on molecular determinants of pathogenicity, immune responses to infection, virus replication and translation. In addition, through construction of human/murine norovirus chimeras the MNV systems may aid in unlocking the problems surrounding human norovirus cultivation in tissue culture.

Conclusion

In conclusion, the past few years have seen a re-invigoration of the study of norovirus molecular biology which (at least partially) has been the result of the identification of MNV. Using this system, researchers have begun to unravel the intricacies of norovirus translation and replication, work which can only benefit one of the key goals of many in the field, i.e. the establishment of a fully infectious system for the study of human norovirus biology. In its own right MNV is a significant and prevalent pathogen of laboratory mice. Understanding this virus in its natural host and applying that knowledge to other caliciviruses should facilitate and expedite vaccine and anti-viral development in the future.

References

Ahlquist, P., Noueiry, A.O., Lee, W.M., Kushner, D.B., and Dye, B.T. (2003). Host factors in positive-strand RNA virus genome replication. J. Virol. 77, 8181–8186.

Al-Molawi, N., Beardmore, V.A., Carter, M.J., Kass, G.E., and Roberts, L.O. (2003). Caspase-mediated cleavage of the feline calicivirus capsid protein. J. Gen. Virol. 84, 1237–1244.

Anand, K., Palm, G.J., Mesters, J.R., Siddell, S.G., Ziebuhr, J., and Hilgenfeld, R. (2002). Structure of coronavirus main proteinase reveals combination of a chymotrypsin fold with an extra alpha-helical domain. EMBO J. *21*, 3213–3224.

Asanaka, M., Atmar, R.L., Ruvolo, V., Crawford, S.E., Neill, F.H., and Estes, M.K. (2005). Replication and packaging of Norwalk virus RNA in cultured mammalian cells. Proc. Natl. Acad. Sci. U.S.A. *102*, 10327–10332.

Bailey, D., Thackray, L.B., and Goodfellow, I.G. (2008). A single amino acid substitution in the murine norovirus capsid protein is sufficient for attenuation in vivo. J. Virol. *82*, 7725–7728.

Belliot, G., Sosnovtsev, S.V., Chang, K.O., McPhie, P., and Green, K.Y. (2008). Nucleotidylylation of the VPg protein of a human norovirus by its proteinase-polymerase precursor protein. Virology *374*, 33–49.

Belliot, G., Sosnovtsev, S.V., Mitra, T., Hammer, C., Garfield, M., and Green, K.Y. (2003). In vitro proteolytic processing of the MD145 norovirus ORF1 nonstructural polyprotein yields stable precursors and products similar to those detected in calicivirus-infected cells. J. Virol. *77*, 10957–10974.

Blakeney, S.J., Cahill, A., and Reilly, P.A. (2003). Processing of Norwalk virus nonstructural proteins by a 3C-like cysteine proteinase. Virology *308*, 216–224.

Bordeleau, M., Mori, A., Oberer, M., Lindqvist, L., Chard, L.S., Higa, T., Belsham, G.J., Wagner, G., Tanaka, J., and Pelletier, J. (2006). Functional characterization of IRESes by an inhibitor of the RNA helicase eIF4A. Nature Chem. Biol. *2*, 213–220.

Burroughs, J.N., and Brown, F. (1978). Presence of a covalently linked protein on calicivirus RNA. J. Gen. Virol. *41*, 443–446.

Chang, K.-O., Sosnovtsev, S.S., Belliot, G., Wang, Q., Saif, L.J., and Green, K.Y. (2005). Reverse Genetics System for Porcine Enteric Calicivirus, a Prototype Sapovirus in the Caliciviridae. J. Virol. *79*, 1409–1416.

Chang, K.-O., Sosnovtsev, S.V., Belliot, G., King, A.D., and Green, K.Y. (2006). Stable expression of a Norwalk virus RNA replicon in a human hepatoma cell line. Virology *353*, 463–473.

Chaudhry, Y., Nayak, A., Bordeleau, M.-E., Tanaka, J., Pelletier, J., Belsham, G.J., Roberts, L.O., and Goodfellow, I.G. (2006). Caliciviruses differ in their functional requirements for eIF4F components. J. Biol. Chem. *281*, 25315–25325.

Chaudhry, Y., Skinner, M.A., and Goodfellow, I.G. (2007). Recovery of genetically defined murine norovirus in tissue culture by using a fowlpox virus expressing T7 RNA polymerase. J. Gen. Virol. *88*, 2091–2100.

Chowdhury, I., Tharakan, B., and Bhat, G.K. (2008). Caspases – an update. Comp. Biochem. Physiol. B Biochem. Mol. Biol. *151*, 10–27.

Cotton, S., Dufresne, P.J., Thivierge, K., Ide, C., and Fortin, M.G. (2006). The VPgPro protein of Turnip mosaic virus: in vitro inhibition of translation from a ribonuclease activity. Virology *351*, 92–100.

Das, S.C., Baron, M.D., Skinner, M.A., and Barrett, T. (2000). Improved technique for transient expression and negative strand virus rescue using fowlpox

T7 recombinant virus in mammalian cells. J. Virol. Methods *89*, 119–127.

Daughenbaugh, K.F., Fraser, C.S., Hershey, J.W., and Hardy, M.E. (2003). The genome-linked protein VPg of the Norwalk virus binds eIF3, suggesting its role in translation initiation complex recruitment. EMBO J. *22*, 2852–2859.

Daughenbaugh, K.F., Wobus, C.E., and Hardy, M.E. (2006). VPg of murine norovirus binds translation initiation factors in infected cells. Virology J *3*, 33.

Denison, M.R. (2008). Seeking membranes: positive-strand RNA virus replication complexes. PLoS Biol. *6*, e270.

Duizer, E., Schwab, K.J., Neill, F.H., Atmar, R.L., Koopmans, M.P.G., and Estes, M.K. (2004). Laboratory efforts to cultivate noroviruses. J. Gen. Virol. *85*, 79–87.

Foy, E., Li, K., Wang, C., Sumpter, R., Jr., Ikeda, M., Lemon, S.M., and Gale, M., Jr. (2003). Regulation of interferon regulatory factor-3 by the hepatitis C virus serine protease. Science *300*, 1145–1148.

French, R., and Ahlquist, P. (1988). Characterization and engineering of sequences controlling in vivo synthesis of brome mosaic virus subgenomic RNA. J. Virol. *62*, 2411–2420.

Fuerst, T.R., and Moss, B. (1989). Structure and stability of mRNA synthesized by vaccinia virus- encoded bacteriophage T7 RNA polymerase in mammalian cells. Importance of the 5′ untranslated leader. J. Mol. Biol. *206*, 333–348.

Goodfellow, I., Chaudhry, Y., Gioldasi, I., Gerondopoulos, A., Natoni, A., Labrie, L., Lailiberte, J., and Roberts, L. (2005). Calicivirus translation initiation requires an interaction between VPg and eIF4E. EMBO Reports *6*, 968–972.

Green, K.Y., Mory, A., Fogg, M.H., Weisberg, A., Belliot, G., Wagner, M., Mitra, T., Ehrenfeld, E., Cameron, C.E., and Sosnovtsev, S.V. (2002). Isolation of enzymatically active replication complexes from feline calicivirus-infected cells. J. Virol. *76*, 8582–8595.

Guix, S., Asanaka, M., Katayama, K., Crawford, S.E., Neill, F.H., Atmar, R.L., and Estes, M.K. (2007). Norwalk virus RNA is infectious in mammalian cells. J. Virol. *81*, 12238–12248.

Gutierrez-Escolano, A.L., Vazquez-Ochoa, M., Escobar-Herrera, J., and Hernandez-Acosta, J. (2003). La, PTB, and PAB proteins bind to the 3(′) untranslated region of Norwalk virus genomic RNA. Biochem. Biophys. Res. Commun. *311*, 759–766.

Hardy, M.E., Crone, T.J., Brower, J.E., and Ettayebi, K. (2002). Substrate specificity of the Norwalk virus 3C-like proteinase. Virus Res. *89*, 29–39.

Herbert, T.P., Brierley, I., and Brown, T.D. (1997). Identification of a protein linked to the genomic and subgenomic mRNAs of feline calicivirus and its role in translation. J. Gen. Virol. *78*, 1033–1040.

Hershey, J., and Merrick, W. (2000). Pathway and mechanism of initiation of protein synthesis, 2nd edn (Cold Spring Harbor, NY: Cold Spring Harbor Laboratory.).

Huang, J., Brieba, L.G., and Sousa, R. (2000). Misincorporation by wild-type and mutant T7 RNA polymerases: identification of interactions that reduce misincorporation rates by stabilizing the catalytically incompetent open conformation. Biochemistry (Mosc) 39, 11571–11580.

Ishikawa, M., Janda, M., Krol, M.A., and Ahlquist, P. (1997). In vivo DNA expression of functional brome mosaic virus RNA replicons in *Saccharomyces cerevisiae*. J. Virol. *71*, 7781–7790.

Karakasiliotis, I., Chaudhry, Y., Roberts, L.O., and Goodfellow, I.G. (2006). Feline calicivirus replication: requirement for polypyrimidine tract-binding protein is temperature dependent. J. Gen. Virol. *87*, 3339–3347.

Karst, S.M., Wobus, C.E., Lay, M., Davidson, J., and Virgin, H.W. t. (2003). STAT1-dependent innate immunity to a Norwalk-like virus. Science *299*, 1575–1578.

Katayama, K., Hansman, G.S., Oka, T., Ogawa, S., and Takeda, N. (2006). Investigation of norovirus replication in a human cell line. Arch. Virol. *151*, 1291–1308.

Kuyumcu-Martinez, M., Belliot, G., Sosnovtsev, S.V., Chang, K.O., Green, K.Y., and Lloyd, R.E. (2004). Calicivirus 3C-like proteinase inhibits cellular translation by cleavage of poly(A)-binding protein. J. Virol. *78*, 8172–8182.

Lai, M.M. (1998). Cellular factors in the transcription and replication of viral RNA genomes: a parallel to DNA-dependent RNA transcription. Virology *244*, 1–12.

Levis, R., Schlesinger, S., and Huang, H.V. (1990). Promoter for Sindbis virus RNA-dependent subgenomic RNA transcription. J. Virol. *64*, 1726–1733.

Liu, B., Clarke, I.N., and Lambden, P.R. (1996). Polyprotein processing in Southampton virus: identification of 3C-like protease cleavage sites by in vitro mutagenesis. J. Virol. *70*, 2605–2610.

Liu, B.L., Viljoen, G.J., Clarke, I.N., and Lambden, P.R. (1999). Identification of further proteolytic cleavage sites in the Southampton calicivirus polyprotein by expression of the viral protease in *E. coli*. J. Gen. Virol. *80*, 291–296.

Liu, G., Ni, Z., Yun, T., Yu, B., Chen, L., Zhao, W., Hua, J., and Chen, J. (2008). A DNA-launched reverse genetics system for rabbit hemorrhagic disease virus reveals that the VP2 protein is not essential for virus infectivity. J. Gen. Virol. *89*, 3080–3085.

Love, D.N., and Sabine, M. (1975). Electron microscopic observation of feline kidney cells infected with a feline calicivirus. Arch. Virol. *48*, 213–228.

Luttermann, C., and Meyers, G. (2007). A bipartite sequence motif induces translation reinitiation in feline calicivirus RNA. J. Biol. Chem. *282*, 7056–7065.

Mazroui, R., Sukarieh, R., Bordeleau, M.E., Kaufman, R.J., Northcote, P., Tanaka, J., Gallouzi, I., and Pelletier, J. (2006). Inhibition of ribosome recruitment induces stress granule formation independently of eukaryotic initiation factor 2alpha phosphorylation. Mol. Biol. Cell *17*, 4212–4219.

McCartney, S.A., Thackray, L.B., Gitlin, L., Gilfillan, S., Virgin, H.W., and Colonna, M. (2008). MDA-5 recognition of a murine norovirus. PLoS Pathog. *4*, e1000108.

McCormick, C.J., Rowlands, D.J., and Harris, M. (2002). Efficient delivery and regulable expression of hepatitis C virus full-length and minigenome constructs in hepatocyte-derived cell lines using baculovirus vectors. J. Gen. Virol. *83*, 383–394.

Meyers, G. (2003). Translation of the minor capsid protein of a calicivirus is initiated by a novel termination-dependent reinitiation mechanism. J. Biol. Chem. *278*, 34051–34060.

Meyers, G. (2007). Characterization of c in RNA of the calicivirus rabbit hemorrhagic disease virus. J. Virol. *81*, 9623–9632.

Miller, W.A., Dreher, T.W., and Hall, T.C. (1985). Synthesis of brome mosaic virus subgenomic RNA in vitro by internal initiation on (-)(–)sense genomic RNA. Nature *313*, 68–70.

Miller, W.A., and Koev, G. (2000). Synthesis of subgenomic RNAs by positive-strand RNA viruses. Virology *273*, 1–8.

Mirmomeni, M.H., Hughes, P.J., and Stanway, G. (1997). An RNA tertiary structure in the 3' untranslated region of enteroviruses is necessary for efficient replication. J. Virol. *71*, 2363–2370.

Miyoshi, H., Okade, H., Muto, S., Suehiro, N., Nakashima, H., Tomoo, K., and Natsuaki, T. (2008). Turnip mosaic virus VPg interacts with *Arabidopsis thaliana* eIF(iso)4E and inhibits in vitro translation. Biochimie *90*, 1427–1434.

Morales, M., Barcena, J., Ramirez, M.A., Boga, J.A., Parra, F., and Torres, J.M. (2004). Synthesis in vitro of rabbit hemorrhagic disease virus subgenomic RNA by internal initiation on (-)sense genomic RNA: mapping of a subgenomic promoter. J. Biol. Chem. *279*, 17013–17018.

Nakamura, K., Someya, Y., Kumasaka, T., Ueno, G., Yamamoto, M., Sato, T., Takeda, N., Miyamura, T., and Tanaka, N. (2005). A norovirus protease structure provides insights into active and substrate binding site integrity. J. Virol. *79*, 13685–13693.

Natoni, A., Kass, G.E., Carter, M.J., and Roberts, L.O. (2006). The mitochondrial pathway of apoptosis is triggered during feline calicivirus infection. J. Gen. Virol. *87*, 357–361.

Ortin, J., and Parra, F. (2006). Structure and function of RNA replication. Annu. Rev. Microbiol. *60*, 305–326.

Paul, A.V., vanBoom, J.H., Filippov, D., and Wimmer, E. (1998). Protein-primed RNA synthesis by purified poliovirus RNA polymerase. Nature *393*, 280–284.

Pelletier, J., and Sonenberg, N. (1988). Internal initiation of translation of eukaryotic mRNA directed by a sequence derived from poliovirus RNA. Nature *334*, 320–325.

Poyry, T.A., Kaminski, A., Connell, E.J., Fraser, C.S., and Jackson, R.J. (2007). The mechanism of an exceptional case of reinitiation after translation of a long ORF reveals why such events do not generally occur in mammalian mRNA translation. Genes Dev. *21*, 3149–3162.

Rohayem, J., Robel, I., Jager, K., Scheffler, U., and Rudolph, W. (2006). Protein-Primed and De Novo Initiation of RNA Synthesis by Norovirus 3Dpol. J. Virol. *80*, 7060–7069.

Rohll, J.B., Moon, D.H., Evans, D.J., and Almond, J.W. (1995). The 3'-untranslated region of picornavirus

RNA – features required for efficient genome replication. J. Virol. *69*, 7835–7844.

Scheffler, U., Rudolph, W., Gebhardt, J., and Rohayem, J. (2007). Differential cleavage of the norovirus polyprotein precursor by two active forms of the viral protease. J. Gen. Virol. *88*, 2013–2018.

Simmonds, P., Karakasiliotis, I., Bailey, D., Chaudhry, Y., Evans, D.J., and Goodfellow, I.G. (2008). Bioinformatic and functional analysis of RNA secondary structure elements among different genera of human and animal caliciviruses. Nucleic Acids Res. *36*, 2530–2546.

Sit, T.L., Vaewhongs, A.A., and Lommel, S.A. (1998). RNA-mediated trans-activation of transcription from a viral RNA. Science *281*, 829–832.

Someya, Y., Takeda, N., and Miyamura, T. (2005). Characterization of the norovirus 3C-like protease. Virus Res. *110*, 91–97.

Someya, Y., Takeda, N., and Wakita, T. (2008). Saturation mutagenesis reveals that GLU54 of norovirus 3C-like protease is not essential for the proteolytic activity. J Biochem. *144*, 771–780.

Sosnovtsev, S., and Green, K.Y. (1995). RNA transcripts derived from a cloned full-length copy of the feline calicivirus genome do not require VPg for infectivity. Virology *210*, 383–390.

Sosnovtsev, S.V., Belliot, G., Chang, K.-O.K., Prikhodko, V.G., Thackray, L.B., Wobus, C.E., Karst, S.M., Virgin, H.W., and Green, K.Y. (2006). Cleavage map and proteolytic processing of the murine norovirus nonstructural polyprotein in infected cells. J. Virol. *80*, 7816–7831.

Sosnovtsev, S.V., Garfield, M., and Green, K.Y. (2002). Processing map and essential cleavage sites of the nonstructural polyprotein encoded by ORF1 of the feline calicivirus genome. J. Virol. *76*, 7060–7072.

Sosnovtsev, S.V., and Green, K.Y. (2000). Identification and genomic mapping of the ORF3 and VPg proteins in feline calicivirus virions. Virology *277*, 193–203.

Straub, T.M., Bentrup, K.H. z., Orosz-Coghlan, P., Dohnalkova, A., Mayer, B.K., Bartholomew, R.A.,

Valdez, O., C., Bruckner-Lea, C.J., Gerba, C.P., Abbaszadegan, M., and Nickerson, C.A. (2007). In vitro cell culture infectivity assay for human noroviruses. Emerg. Infect. Dis. *13*, 396–403.

Studdert, M.J., and O'Shea, J.D. (1975). Ultrastructural studies of the development of feline calicivirus in a feline embryo cell line. Arch. Virol. *48*, 317–325.

Taylor, M.P., and Kirkegaard, K. (2008). Potential subversion of autophagosomal pathway by picornaviruses. Autophagy *4*, 286–289.

Thackray, L.B., Wobus, C.E., Chachu, K.A., Liu, B., Alegre, E.R., Henderson, K.S., Kelley, S.T., and Virgin, H.W. t. (2007). Murine noroviruses comprising a single genogroup exhibit biological diversity despite limited sequence divergence. J. Virol. *81*, 10460–10473.

Tsukiyama Kohara, K., Iizuka, N., Kohara, M., and Nomoto, A. (1992). Internal ribosome entry site within hepatitis C virus RNA. J. Virol. *66*, 1476–1483.

Ward, V.K., McCormick, C.J., Clarke, I.N., Salim, O., Wobus, C.E., Thackray, L.B., Virgin, H.W. t., and Lambden, P.R. (2007). Recovery of infectious murine norovirus using pol II-driven expression of full-length cDNA. Proc. Natl. Acad. Sci. U.S.A. *104*, 11050–11055.

Wileman, T. (2006). Aggresomes and autophagy generate sites for virus replication. Science *312*, 875–878.

Wilson, J.E., Powell, M.J., Hoover, S.E., and Sarnow, P. (2000). Naturally occurring dicistronic cricket paralysis virus RNA is regulated by two internal ribosome entry sites. Mol. Cell Biol. *20*, 4990–4999.

Wittmann, S., Chatel, H., Fortin, M.G., and Laliberte, J.F. (1997). Interaction of the viral protein genome linked of turnip mosaic potyvirus with the translational eukaryotic initiation factor (iso) 4E of *Arabidopsis thaliana* using the yeast two-hybrid system. Virology *234*, 84–92.

Wobus, C.E., Karst, S.M., Thackray, L.B., Chang, K.O., Sosnovtsev, S.V., Belliot, G., Krug, A., Mackenzie, J.M., Green, K.Y., and Virgin, H.W. (2004). Replication of Norovirus in cell culture reveals a tropism for dendritic cells and macrophages. PLoS Biol. *2*, e432.

Rabbit Haemorrhagic Disease Virus and Other Lagoviruses

12

Vernon K. Ward, Brian D. Cooke and Tanja Strive

Abstract

Rabbit haemorrhagic disease virus (RHDV) is a pathogen of rabbits that causes major problems throughout the world where rabbits are reared for food and clothing, make a significant contribution to ecosystem ecology, and where they support valued wildlife as a food source. The high mortality caused by RHDV has driven research in protecting rabbits from infection. However, RHDV is an unusual calicivirus in that it has served also as an important model in the family *Caliciviridae* by providing a range of beneficial outcomes as diverse as the creation of virus-like particles (VLPs) for vaccine and therapeutics delivery, the elucidation of calicivirus replication and structural features at the molecular level, and the biological control of a vertebrate pest.

Introduction

Lagoviruses are spread diversely throughout Asia, Europe and Australasia (reviewed in Chasey, 1997) with incursion outbreaks occurring in the US and other countries (McIntosh *et al.*, 2007). The two recognized members of the genus *Lagovirus*, Rabbit haemorrhagic disease virus (RHDV) and European brown hare syndrome virus (EBHSV) (Koopmans *et al.*, 2005), cause pathogenic disease in lagomorphs. Until recently, there was only one RHDV genotype (Capucci *et al.*, 1995); however, a new antigenic variant termed RHDVa (Capucci *et al.*, 1998) is displacing the initial RHDV strain (Farnos *et al.*, 2007; Le Gall-Recule *et al.*, 2003; van de Bildt *et al.*, 2006) leading to the suggestion that RHDVa represents a new pandemic strain of RHDV

(McIntosh *et al.*, 2007). There are also multiple strains of these viruses as well as related non-pathogenic rabbit caliciviruses (Capucci *et al.*, 1996; Strive *et al.*, 2009). This chapter focuses primarily upon RHDV, however EBHSV and other non-pathogenic related viruses are discussed.

RHDV was first reported in Angora rabbits in Jiangzu province in the People's Republic of China in 1984 (Liu *et al.*, 1984). The virus was subsequently reported in Europe in 1989 but it rapidly became apparent that the virus had been in Europe prior to 1984 (Moussa *et al.*, 1992) with serum samples from 1978 testing positive for RHDV (Rodak *et al.*, 1990) and samples from the 1950s being positive by nested reverse transcription-PCR (RT-PCR) (Moss *et al.*, 2002). The epidemic strains of RHDV appeared to have emerged from circulating non-pathogenic strains (Forrester *et al.*, 2006). EBHSV was reported in 1980 and positive hare sera have been identified from samples collected in 1971 (Chasey and Duff, 1990) and by RT-PCR in paraffin-embedded tissue samples from the 1970s (Ros Bascunana *et al.*, 1997).

RHDV has been shown by many studies as the causative agent of rabbit haemorrhagic disease (RHD) (Chasey, 1997; Moussa *et al.*, 1992; Ohlinger *et al.*, 1989; Ohlinger *et al.*, 1993), including the identification of CsCl purified 40 nm particles that resemble caliciviruses with a 60-kDa capsid protein (VP60) and the analysis and characterization of the viral genome (Ohlinger *et al.*, 1990). RHDV has been detected only in the European rabbit *Oryctolagus cuniculus* thus far (Gregg *et al.*, 1991).

RHD is usually fatal in *O. cuniculus*, causing death within 48–72 h with obvious involvement of the lungs, liver, kidneys and spleen (Marcato *et al.*, 1991; Ohlinger *et al.*, 1989). The gross pathology of RHD (Fig. 12.1) includes severe disseminated necrotic hepatitis, multifocal petechial haemorrhages in the liver and other organs, liver discoloration and friability, and an enlarged and discoloured spleen. The lungs show alveolar oedema and infiltrating granulocytes can be observed in the lungs and at early stages of infection in the liver. Effects upon coagulation factors leads to multiple organ failure with overall disseminated intravascular coagulation (Marcato *et al.*, 1991). Clinical symptoms may include convulsions, lateral torsion, paresis, bloody nasal secretions

and fever (Xu and Chen, 1989). Mortality rates can be as high as 90%, although rabbits under 4 weeks generally remain unaffected by the virus (Mocsari *et al.*, 1991; Ohlinger *et al.*, 1993). However, when these young animals do become infected, liver damage is observed and the rabbits may become long-term carriers, possibly acting as a source for transmission of the virus (Ferreira *et al.*, 2004). Between 4 and 8 weeks old, young rabbits increase in susceptibility to disease upon first exposure possibly through a progressive increase in the levels of A and H type 2 receptors on the surface of cells (Ruvoen-Clouet *et al.*, 2000).

EBHSV causes disease in *Lepus europaeus* and *L. timidus* (Le Gall *et al.*, 1996) with a high fatality rate, although infection of *L. timidus* appears

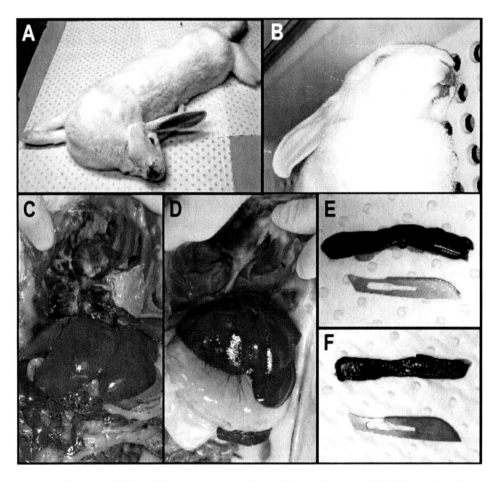

Figure 12.1 Pathology of RHD. (A) Typical posture of a rabbit that died from RHDV infection. This animal died while eating, with food still in its mouth. (B) Bleeding from the nostrils is frequently observed in rabbits that die from acute RHD. (C) Internal organs of a rabbit that died from RHD. The haemorrhagic lungs and the discoloured liver are visible. (D) The internal organs of a healthy rabbit, with normal lungs and dark glossy liver. (E) Enlarged spleen of a rabbit that died from RHD and (F) a normal spleen for comparison.

to only occur where its range overlaps that of *L. europaeus* (Gavierwiden and Morner, 1993) suggesting that the European brown hare is the true host. The epidemiology and symptoms are similar to RHDV, with marked hepatocyte degeneration and liver necrosis (Fuchs and Weissenbock, 1992). Although EBHSV and RHDV represent distinct genetic lineages in the genus *Lagovirus* with different host specificities, they share 71% overall nucleotide identity in their genomes (Le Gall *et al.*, 1996) and are antigenically and morphologically related (Capucci *et al.*, 1991; Chasey *et al.*, 1992).

One of the key issues associated with RHDV infection is the significant effects of infection upon organs such as the liver. Initial microscopic analysis suggested that liver damage is caused by widespread apoptosis of infected cells including the classic apoptosis markers of nuclear condensation and marginalisation (Alonso *et al.*, 1998) and confirmed by TUNEL analysis and DNA fragmentation patterns in infected cells (Jung *et al.*, 2000). Subsequent studies investigating RHD as a model for fulminant liver failure (Sanchez-Campos *et al.*, 2004; Tunon *et al.*, 2003) have investigated RHDV-induced apoptosis. Both the extrinsic and intrinsic pathways of apoptosis are triggered with increases in cytochrome C release from mitochondria, FasL increase, PARP activation and adjustments in the ratios of pro-apoptotic Bax to anti-apoptotic Bcl2 occurring during infection. The use of *N*-acetylcysteine to interrupt the intrinsic pathway, slowed but did not prevent rabbit death, confirming the role of both pathways in RHDV-induced cell death (San-Miguel *et al.*, 2006). Interestingly, Bok *et al.* (2009) have shown that murine norovirus downregulates survivin levels during infection, suggesting that apoptosis is an important process in calicivirus infection.

Analysis of intravascular infection by RHDV demonstrated that macrophages and monocytes are extensively infected with RHDV in a range of organs (Ramiro-Ibanez *et al.*, 1999). It was proposed that tissue tropism, in particular tropism to cells such as macrophages and internal organs, with associated apoptosis, is a major pathogenicity factor in RHDV and EBHSV and may play a role in the disseminated intravascular coagulation characteristic of RHD.

Molecular characterization

Rabbit haemorrhagic disease virus

The RHDV genome was the first complete calicivirus genome to be sequenced (Meyers *et al.*, 1991b). The positive-sense single-stranded RNA genome was 7437 nucleotides with a 3′ polyA tail. The genome was observed in infected cells and from purified particles as a full-length genomic RNA and a 2.2 kb subgenomic RNA (sgRNA) commencing at nucleotide 5296 (Meyers *et al.*, 1991a) that was co-linear with the 3′ end of the genome (Fig. 12.2) (Moussa *et al.*, 1992). Both the genomic RNA and sgRNA encode and express the VP60 (Boniotti *et al.*, 1994; Parra *et al.*, 1993) and both RNAs have a viral genome-linked protein (VPg) attached to their 5′ terminus (Meyers *et al.*, 1991a). Gradient purification of viruses suggests that the genomic RNA and sgRNA are both packaged into virus particles, indicating a packaging signal must be on both RNA species, although this signal remains unknown. The first evidence of a subgenomic promoter in caliciviruses was obtained by a demonstration of synthesizing sgRNA *in vitro* from a genomic negative-sense RNA (Morales *et al.*, 2004).

The lagovirus genome possesses two open reading frames (ORFs) (Fig. 12.2). ORF1 represents the majority (7 kb) of the genome while the non-essential ORF2 (Liu *et al.*, 2008) is only 351 nucleotides long (Wirblich *et al.*, 1996). ORF1 encodes a 257 kDa polyprotein that undergoes proteolytic processing to generate the non-structural proteins and the capsid protein through the activity of a virus encoded 3C-like cysteine protease (Meyers *et al.*, 2000; Wirblich *et al.*, 1995) that can function in both *cis* and *trans* (Boniotti *et al.*, 1994). The polyprotein undergoes cleavage by the 3C-like protease at defined cleavage sites (Martin Alonso *et al.*, 1996; Meyers *et al.*, 2000; Wirblich *et al.*, 1996) to generate the range of mature viral proteins that includes the viral helicase, VPg, protease, polymerase and VP60 proteins. This process is a step-wise cleavage that sees the initial generation of primary cleavage products P16, P60, P41, P72 and P60 (VP1, VP60; Fig. 12.2). These products are then cleaved to generate the final products with an alternative cleavage activity of an undefined non-viral protease generating a P23/2 protein from

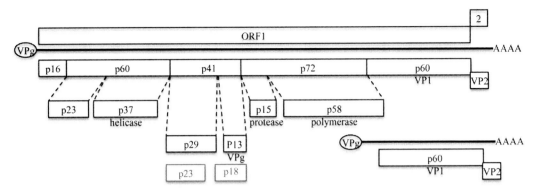

Figure 12.2 RHDV genome. The genome of RHDV possesses a VPg protein attached to the 5′ end of the gRNA and sgRNA and a 3′ polyA tail. The genome encodes two open reading frames (*orf1* and 2) while the sgRNA is collinear with the 3× end of the viral genomic RNA and encodes the VP60 capsid and VP2 protein. The ORF1 polyprotein undergoes autocatalytic processing through the activity of the 3C-like protease (p15). A range of peptides have been identified including the p16 protein, a p60 protein that is further cleaved into the p23 and p37 (helicase) proteins, a p41 protein that is either cleaved into p29 and p13(VPg) or an undefined alternative cleavage to generate p23 and p18. The p72 product is cleaved into the protease and polymerase components and the p60 capsid (VP1) protein is released. The majority of the VP1 protein is produced from the sgRNA while VP2 is expressed through a novel 'TURBS' mechanism.

the P41 protein (Thumfart and Meyers, 2002). It is considered likely that the partially processed polyproteins have variant activities to the fully mature proteins and hence provide a wider range of functionality than that seen for the fully processed mature proteins alone. The sgRNA produces the VP60 (Boga *et al.*, 1992) and VP2, the non-essential minor capsid protein (Liu *et al.*, 2008) postulated to associate with viral RNA (Wirblich *et al.*, 1996).

ORF2 is translated by a novel mechanism first identified in RHDV (Meyers, 2003) and subsequently found in bovine noroviruses (McCormick *et al.*, 2008). The expression of ORF2 involves reinitiation of protein synthesis after the termination of ORF1 translation. This requires 84 nucleotides at the end of ORF1, and the ORF1 stop codon is essential (Meyers, 2003). The motif includes a sequence complementary to 18s rRNA that was postulated to tether the ribosome to the RNA to allow binding of reinitiation factors (Meyers, 2007). This has been shown to facilitate the acquisition of the met-tRNA ternary complex (Poyry *et al.*, 2007). The sequence motif was termed TURBS for termination upstream ribosomal binding site (Meyers, 2003). How common this mechanism will be in other caliciviruses or indeed other virus families is yet to be determined.

Analysis of the role(s) of RHDV genes has been boosted by recent developments in reverse genetics for RHDV (Liu *et al.*, 2006). Initial experiments used *in vitro* transcripts of cDNA clones that were injected into rabbits. These full-length transcripts were able to reconstitute infectious RHDV. More recently, this system was adapted to cDNA transfection of RK13 (rabbit kidney) cells and used to show that VP2 was not essential, thus providing the benefit of both a reverse genetics system for lagoviruses and overcoming the lack of an *in vitro* system within which to study RHDV (Liu *et al.*, 2008).

European brown hare syndrome virus

EBHSVs are closely related to RHDV with the viral particles being morphologically identical. While EBHSV and RHDV share some antigenic epitopes, baculovirus-produced virus-like particles (VLPs) from each do not cross-protect (Laurent *et al.*, 1997) and the viruses are antigenically distinct (Capucci *et al.*, 1991). The EBHSV genome is 7442 nucleotides of positive-sense single-stranded RNA and like RHDV, contains a large ORF1 (encoding non-structural and capsid proteins) and a second small ORF2. RHDV and EBHSV share 71% identity at the nucleotide level with base substitutions scattered uniformly over the genome (Le Gall *et al.*, 1996). The 5′ UTR for each virus is almost identical while the 3' UTR vary but possess 22 closely matching nucleotides at the end of their genomes that are predicted to form a hairpin (Le Gall *et al.*, 1996). Infected cells produce both genomic RNA and sgRNA of similar size to RHDV (Wirblich *et al.*, 1994). The ORF1 polyprotein is 2334 amino acids (256 kDa) with 78% identity (87% similarity) to RHDV. The orthologous genes are present on the genome, including a 2C-like helicase, 3C-like protease and 3D-like RNA dependent RNA polymerase.

Structure of the RHDV particle

The RHDV virus particle is approximately 40 nm in diameter and displays the characteristic cup-shaped surface derived from the surface projections of the P2 domain that gives caliciviruses their name. A low-resolution 32 Å cryo-EM reconstruction model of the VLPs in association with monoclonal antibody E3 has been determined (Thouvenin *et al.*, 1997). The VLPs were empty and displayed a spherical shell with protruding capsomers. The E3 bivalent monoclonal antibody interacts with adjacent VP60 molecules on the capsid surface, and was postulated to be protective through preventing decapsidation, or by interference with binding to the cell receptor. The model showed the particles to be similar to that of the Norwalk virus, for which the X-ray diffraction structure was determined (Prasad *et al.*, 1994). In particular, the general features of VP60 with an N-terminal shell domain followed by C-terminal P domains comprised of a P1 domain into which the surface P2 domain was inserted are apparent. A subsequent 24-Å reconstruction (Barcena *et al.*, 2004) showed a typical calicivirus structure. The three quasi-equivalent conformations required for a single protein to assemble into icosahedra were observed with 90 dimers assembling to form a 38-Å-thick continuous shell from which there are 53-Å protrusions. The dimers clearly interact with their neighbouring subunits and the 12 pentameric units and the 20 hexameric depressions can be clearly seen as part of the T=3 icosahedral structure.

A low resolution cryo-EM reconstruction of RHDV VLP is shown in Fig. 12.3C, however, a high-resolution structure of the RHDV particle by cryo-EM or X-ray diffraction remains to be determined, and this will provide insights on how the particle varies from other caliciviruses, in particular the presentation of the external P2 domain. Modelling of the structure by SWISS-MODEL using the Norwalk virus VP60 structure showed the expected domains as detailed above (Barcena *et al.*, 2004).

The ability to form RHDV VLPs in a range of expression systems has allowed the modification of the VP60 capsid protein. Deletions in VP60 have shown that the N-terminus was essential for the formation of the 40 nm T=3 particles. Removal of this molecular switch leads to the formation of approximately 27–30 nm T=1 particles comprised of 60 dimers (Laurent *et al.*, 2002; Nagesha *et al.*, 1999). Different researchers have reported variable outcomes from the deletion of N-terminal amino acids, but it is clear that the N-terminal loop (Barcena *et al.*, 2004) dictates the size of the particle assembled, with the T=3, 40 nm, 90 dimer particle forming when all of the VP60 is present and a T=1, 60 dimer, 27–30 nm particle forming when approximately 30 amino acids at the N-terminus are deleted (Barcena *et al.*, 2004; Laurent *et al.*, 2002; Nagesha *et al.*, 1999). This T=1 particle appears to require an acidic pH to be stable and this may explain why there was some variation between researchers (Barcena *et al.*, 2004). It is interesting that the extreme N-terminus of VP60 appears to be unstructured and this region of disorder was important in the assembly of the virus particles, supporting the concept that regions of disorder in proteins are important in protein–protein interactions. The removal of 75 amino acids from the N-terminus (Laurent *et al.*, 2002) and deletions at the C-terminus of VP60 prevents assembly of particles, although additions are tolerated (Barcena *et al.*, 2004).

RHDV vaccines

RHDV is an important pathogen of rabbits in over 40 countries (Carman *et al.*, 1998). European rabbitries that raise animals for food and fur are estimated to have lost 100–200 million rabbits between 1988–1990 (Morisse *et al.*, 1991) and epizootics have occurred throughout wild populations (Villafuerte *et al.*, 1994). Rabbits are an integral part of the scrubland ecosystem in many countries and provide an important food source

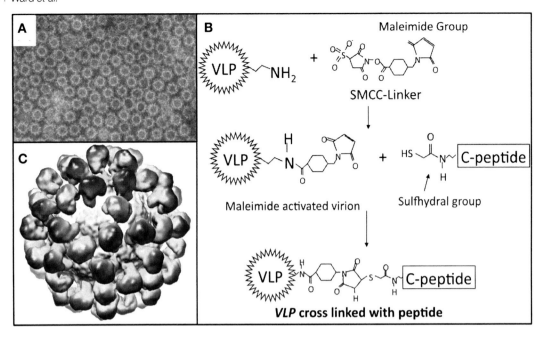

Figure 12.3 RHDV virus-like particle modification. (A) RHDV VLP (40 nm particles) expressed from recombinant AcMNPV baculovirus in *Spodoptera frugiperda* Sf9 cells. (B) Coupling of peptides to amines on the surface of the RHDV particle. The heterobifunctional linker SMCC is coupled to amines through a succinimide ester linkage to functionalise the surface with a maleimide group. Peptides with a cysteine residue containing a sulphydryl side chain form a covalent linkage with the maleimide group allowing use of the VLP as a carrier platform. (C) Cryo-electronmicroscopy reconstruction of RHDV VLP. The virus particle displays a shell with distinctive protruding P domains. Image kindly provided by Drs Tom Smith and Umesh Katpally, Donald Danforth Plant Centre, St Louis, MO, USA.

for predators (Cooke, 2002; Delibes-Mateos *et al.*, 2007). Rabbits are utilized also for laboratory research and competitive breeding. The high mortality and impact of this disease in rabbits has driven research into the development of effective and practical vaccines.

Tissue derived inactivated vaccines

Current vaccines are predominantly derived from infected liver extracts (Arguello Villares, 1991) that are inactivated, usually with formalin or beta-propiolactone. Different manufacturers use a range of adjuvants, including Freund's incomplete adjuvant, aluminium hydroxide, or oil formulations. These vaccines are effective, generating antibodies that protect adult rabbits from infection by the virus. These inactivated vaccines work well against the predominant strain of RHDV as injected vaccines, show some efficacy against the variant strain of RHDV and can be used in rabbitries and for pet rabbits.

Regular vaccination of wild populations by injection is not feasible, necessitating the development of alternative strategies for the production and delivery of vaccines. These have focussed upon the development of potential subunit vaccines produced by a range of recombinant systems including *Escherichia coli* (Boga *et al.*, 1994), baculovirus (Gromadzka *et al.*, 2006; Laurent *et al.*, 1994), poxviruses (Bertagnoli *et al.*, 1996a), plants (Castanon *et al.*, 1999) and yeasts (Farnos *et al.*, 2005). All of these alternatives have used the VP60 capsid protein with or without the 12-kDa ORF-2 protein.

Baculovirus derived vaccines

Baculoviruses are used for the production of VLPs in insect cells (Noad and Roy, 2003) (Fig. 12.3A). Insect cell expression has proved to be exceptionally good for the production of VLPs (Roy and Noad, 2008), including calicivirus particles such as the Norwalk virus (Prasad *et* al., 1999), San Miguel sea lion virus (Chen *et al.*, 2003) and primate calicivirus (Prasad *et al.*, 1994). The expression of RHDV VP60 in insect cells using the baculovirus system was shown to produce particles that are morphologically and antigenically indistinguishable from native RHDV (Laurent *et al.*, 1994; Nagesha *et al.*, 1995). VLPs yields of 10^6 particles per ml of cell culture were achieved by Nagesha *et al.* (Nagesha *et al.*, 1995), while Plana-Duran *et al.* (1996) optimized this system through the use of the strong late P10 promoter to obtain yields of 200 µg/10^6 cells. VLPs generated in this system were formulated with oil, alumina or QuilA adjuvants, all of which were effective with high doses of >10 µg providing protection against challenge with a single dose and even 0.5 µg protecting against RHDV challenge when a booster immunization was employed.

Despite these excellent yields, the cost of production in *Spodoptera frugiperda* insect cell culture remains very high, and the use of larval insect biofactories has been trialled (Perez-Filgueira *et al.*, 2007). The intrahaemocoelic injection of recombinant baculovirus particles into fourth instar *Trichoplusia ni* larvae yielded 2 mg of VLP per larva and was shown to be effective as crude lyophilised insect homogenates that were stable for years at room temperature. Single intramuscular doses of 2 µg in the presence of adjuvant were 100% protective, providing 1000 i.m. doses per insect larva at approximately 500-fold lower cost than in cell culture.

Of particular note is the observation that baculovirus-derived RHDV VLPs can be protective when delivered orally (Plana-Duran *et al.*, 1996) at a very low dose of 3 µg per rabbit. Key to this observation was the treatment of VLPs with binary ethylenimine. This virus-inactivating agent was postulated to stabilize the VLPs for survival in the gut so that high antibody titres were achieved in four of five animals vaccinated orally. These four animals survived challenge while the animal that did not respond succumbed to infection. This report of oral protection is in contrast to many other reports, however, it opens the possibility of oral delivery to wild populations of rabbits without using live recombinant organisms and with the added safety of binary ethylenimine inactivated material.

Live recombinant vaccines

The use of live recombinant viruses is advantageous for the delivery of vaccines to wildlife, although there are drawbacks that are discussed later. One example is the use of recombinant vaccinia virus expressing the G protein of rabies virus for the oral vaccination of foxes in Europe to control rabies (Blancou *et al.*, 1986; Brochier *et al.*, 1991; Kieny *et al.*, 1984; Wiktor *et al.*, 1984). The attenuated Copenhagen strain of vaccinia virus has been modified to express VP60 (Bertagnoli *et al.*, 1996b) using the p7.5 promoter and the live recombinant virus was used to inoculate rabbits intradermally and per os at doses of 10^8 and 10^9 viruses per animal respectively (Bertagnoli *et al.*, 1996b). Both routes generated detectable antibody responses and provided protection against challenge with $1000 LD_{50}$ of RHDV.

The use of a recombinant myxoma virus expressing VP60 has been proposed to provide a vaccine against myxoma and RHDV (Bertagnoli *et al.*, 1996a). The cell culture attenuated SG33 strain of myxoma has been engineered to replace either the thymidine kinase gene or parts of both the MGF and M11L genes with VP60. These genes have been shown to be involved in pathogenicity and their deletion is intended to provide less pathogenic strains of the parental virus (Buller *et al.*, 1985; Opgenorth *et al.*, 1992). Animals vaccinated with both myxoma recombinants survived RHDV challenge even where only low antibody titres were observed. However, a repeat vaccination was required to provide full immunity to myxoma virus, though this was normal for protection against myxomatosis.

Recombinant plant vaccines

A number of recombinant plant systems have been trialled. These offer the advantage of large-scale production and the potential of simple oral vaccination. The expression of VP60 in potato tubers (Martin-Alonso *et al.*, 2003) yielded 3.5 µg/mg of total soluble protein and the material could be freeze-dried in a stable form. The use of repeat oral boosts of 500 µg of VP60 led to four of five animals developing antibodies, however, only the animal with the highest antibody level survived the challenge with RHDV. Other examples of plant expression have involved the extraction of material from leaves and formulation for use by injection or intranasal vaccination. Expression in arabidopsis (Gil *et al.*, 2006) through a range of constructs and fusion partners showed that only unmodified VP60 assembled into higher structures and that formulation with adjuvants and boosts were required to generate antibodies. Castanon and colleagues (Castanon *et al.*, 1999) also looked at enhancing expression in plants through the use of transcriptional enhancers in combination with a 35S promoter. While no VLPs were observed, VP60 could be extracted at 12 µg/ml from leaf tissue and immunization followed by boosts in adjuvant was effective for subcutaneous and intramuscular injection, however, insufficient material was produced for oral vaccination.

An alternative plant system was the engineering of plum pox virus to express VP60 (Fernandez-Fernandez *et al.*, 2001). The VP60 gene was engineered between the N1b and CP genes of this potyvirus while retaining the viral protease cleavage sites so that the VP60 would be released automatically from the viral polyprotein. The clarified plant extract when formulated with adjuvant and used in a single boost regime provided 100% protection against intranasal challenge from RHDV, although the presence of intact VLPs was uncertain and yields were not determined.

Yeast systems

Both *Saccharomyces cerevisiae* (Boga *et al.*, 1997) and *Pichia pastoris* (Farnos *et al.*, 2006, 2008) have been used to express VP60. Yields of 1.5 g per litre have been reported from *P. pastoris* (Farnos *et al.*, 2005), although this was somewhat offset by the processing required to extract the protein. Despite not forming VLPs the expressed material was able to generate both humoral and cell-mediated immunity in mice as determined by IgG1, IgG2a, IgG2b and IgA production, as well as *ex vivo* stimulation of interferon gamma and IL-12 (Farnos *et al.*, 2006). Refining of the *P. pastoris* expression constructs to generate soluble VP60 led to yields of 480 mg/l in the form of aggregates that generate high haemagglutination titres and apparent cross strain protection (Farnos *et al.*, 2008).

RHDV VLPs as epitope carriers

Recombinant VLPs have been used successfully in several viral systems for the generation of vaccines against their cognate diseases in animals and humans. The use of RHDV VLPs as a vaccine against RHD is described above. Examples in human disease include the hepatitis B vaccine Engerix, the highest-revenue vaccine in the world, and the recently developed human papillomavirus vaccine that should significantly reduce the incidence of cervical cancer by providing protection against the virus. A greater challenge will be the use of VLPs as carriers for foreign epitopes where the immunogenicity of the VLPs can be used to elicit a response against the displayed protein or peptide. Recent work has shown that the RHDV VLPs may be useful for this purpose.

Nagesha *et al.* (1999) used a natural BamHI site near the N-terminus of VP60 to replace the N-terminal 30 amino acids with the bluetongue virus Btag epitope representing a six-residue epitope from the VP7 protein of this virus. Baculovirus expression led to the formation of 27 nm particles while the C-terminal addition of this epitope generated 40 nm particles (Nagesha *et al.*, 1999). The direct addition of peptides to the N-terminus rather than replacing the first 30 amino acids led to the formation of 40 nm particles. This engineering of peptide epitopes into the VLPs has been undertaken subsequently by others (Peacey *et al.*, 2007) with most using the N-terminal addition of foreign peptides. T and B cell epitopes to a range of targets have been packaged in this manner. Structural analysis indicates that these epitopes will be on the inner surface of the assembled particle, making this a suitable site for the presentation of helper and other T-cell epitopes. However, the presentation of B-cell epitopes for the generation of antibodies will probably require the display of epitopes on the viral surface.

Drawbacks to the encapsidation of immune epitopes by engineering are the limits that can be placed upon insert size, the reduction in VLP yields upon engineering of the particles and the solely internal packaging of the epitope when placed at the N-terminus of VP60. An alternative approach is to couple molecules to the surface of the RHDV particles.

Coupling molecules to the surface of RHDV VLPs

Peacey and colleagues (Peacey *et al.*, 2007) have shown that heterobifunctional linkers such as sulpho-SMCC [sulphosuccinimidyl 4-(*N*-maleimidomethyl)-cyclohexane-1-carboxylate] can be used to couple peptides and proteins to the RHDV VLPs. Sulfo-SMCC forms a succinimidyl ester with primary amines on the surface of the VLPs and contains a free sulphydryl reactive maleimide group that can be used to couple peptides and proteins through the SH group on the side chain of cysteine (Fig. 12.3B). The synthesis of peptides with an N- or C-terminal cysteine residue can be coupled directly at neutral pH to the activated particles. Alternatively, the addition of a heterobifunctional molecule such as SATA (*N*-succinimidyl-*S*-acetylthioacetate) to proteins through linkage to primary amine groups followed by deprotection to activate a sulphydryl group on SATA allows the direct coupling of whole proteins to the surface of the viral particles. This overcomes the size and surface constraints of genetically engineering the particles and removes the need for using defined peptide epitopes for immune presentation. Green fluorescent protein and ovalbumin have been coupled to purified RHDV VLPs derived from the baculovirus system (Peacey *et al.*, 2007) and been used to track the uptake of VLPs into dendritic cells and to vaccinate against an engineered mouse melanoma presenting an ovalbumin peptide (Peacey *et al.*, 2008).

The coupling of peptides can lead to the assembly of multiple peptides on each VLP through peptide–peptide interactions with up to 1800 peptides being present on a single VLP in some cases (Peacey *et al.*, 2007). This high loading provides a high dose of peptide epitope to the immune system.

Modified VLPs can elicit humoral and cellular immune responses

The dual coupling of CD4 and CD8 ovalbumin peptides to VLPs has been used successfully to vaccinate mice (with CpG adjuvant) against the B16 OVA tumour model illustrating that RHDV can be used as a delivery vehicle to present epitopes in animals other than rabbits (Peacey *et al.*, 2008). That the modified VLPs were able to generate a protective immune response when injected as an exogenous antigen illustrates that the mechanism of VLPs processing in mice involves the cross-presentation of antigen on MHCI, a critical advantage for the generation of a protective immune response against cancerous tissues. The coupling of NHS-dylight-488 reagent (Pierce Chemical Company) to the VLPs has been used to track VLPs uptake and processing in human and mouse dendritic cells (Win, unpublished). Further, the VP60 protein contains a single cysteine residue and, while VLPs do not aggregate through disulphide bond formation, the sulphydryl side chain is accessible for coupling with maleimide-based cross-linking reagents (Ward and Young, unpublished).

The particulate nature of VLPs makes them particularly attractive delivery vehicles (Nieba and Bachmann, 2000; Noad and Roy, 2003; Roy and Noad, 2008) and the flexibility of genetic modification and surface coupling chemistry makes RHDV particles very attractive. That RHDV does not infect humans (Carman *et al.*, 1998) is no apparent barrier to VLP uptake by human dendritic cells and the lack of pre-existing immunity to the carrier particle may provide an advantage over human specific VLPs where the pre-existing immune response may deplete the response to the foreign molecule being presented by the VLP carrier. For example, Da Silva *et al.* (2001) demonstrated that HPV16 VLP carrying E7 protein protected mice from E7-positive tumour challenge; however, mice that possessed neutralizing antibodies through pre-vaccination with HPV VLP were not protected from HPV16 E7-positive TC-1 tumour challenge.

RHDV VLPs are amenable to delivery through a variety of routes. Subcutaneous, intraperitoneal and intranasal delivery routes have all been used. In addition, the appropriate formulation of VLPs allows for transcutaneous delivery (Young *et al.*, 2006). The observation that binary ethylenimine treatment greatly enhances the effectiveness of RHDV VLPs as an anti-RHDV vaccine in rabbits through the oral route may facilitate this route of delivery in other animals.

Encapsidation of molecules inside the RHDV VLPs

The application of RHDV VLPs as carriers for the delivery of nucleic acids into target cells has been investigated. The addition of the DNA binding sequence from HPV16 to the N-terminus of VP60 (El Mehdaoui *et al.*, 2000) generated particles that could package DNA. The disassembly of the modified VLPs through the use of EGTA as a chelating agent combined with DTT as a reducing agent led to disassembly of the virus particles into smaller subunits. The disassembled VLPs were incubated with a GFP-encoding plasmid in a reassembly solution consisting of DMSO and increasing concentrations of $CaCl_2$. Following transduction of lagomorph cells with the reassembled VLPs and packaged plasmid DNA, GFP expression was observed in the cells. The extension of this disassembly and reassembly process to package a wide variety of molecules is particularly attractive.

In summary, the RHDV VLPs are proving to be an excellent scaffold for the development of a heterologous delivery vehicle for peptides, proteins and nucleic acids. The VLPs can be expressed in large amounts in a wide range of expression systems and these are sufficiently stable and can be purified to near homogeneity. The VLPs elicit effective immune responses and the host restriction that applies to infection does not appear to translate to a lack of VLP recognition or processing by animal immune systems.

RHDV for biological control of rabbits

The current application of RHDV as a biocontrol agent for the rabbit – while controversial to some – is driving significant advances in our understanding calicivirus ecology, wildlife ecology, calicivirus evolution and the impacts of foreign pests species in new environments. Underlying this application of RHDV are the problems caused by rabbits when they were introduced to countries such as Australia and New Zealand. There has been significant discussion on the merits and problems associated with the use of RHDV that will not be readdressed in this chapter.

History of rabbits in Australasia

Unlike many other parts of the world, rabbits are ecological and economic pests in Australasia. Rabbits were aboard the ships of the First Fleet reaching Australia and both domestic and wild rabbits were introduced on several occasions from 1799 onwards, and populations remained small and localized. This changed in 1859, when Sir Thomas Austin imported 26 wild rabbits to establish a rabbit colony for hunting purposes on his property in Victoria. A bushfire destroyed the enclosures, clearing the way for the rabbits' colonization of the continent (Myers, 1970). Rabbits were first introduced into New Zealand for sport and meat with reports of release as early as 1858 (Gibb and Williams, 1994). They were well established by 1876 and between 1877 and 1884 an estimated 628,000 hectares of land was abandoned to them. While numbers have fluctuated over time, the European rabbit remains a major agricultural problem in New Zealand.

The reproductive biology and dispersal strategies of the rabbit make it an ideal invader of a new environment. The gestation period is 28 days, and immediately after giving birth the female is able to mate again, meaning litters of four to eight kittens can be weaned in monthly intervals. Sexual maturity may be reached at three months, resulting in more than one generation of progeny per breeding season (Myers and Poole, 1962). In addition, the Bilby, a native Australian rabbit-size marsupial with a similar burrowing habit, provided ideal warren systems that were readily taken over by the more aggressive rabbits (Read *et al.*, 2008) further accelerating the spread and displacing the Bilby and other native animals in most parts of mainland Australia.

Within 70 years rabbits spread across almost the southern two-thirds of the Australian continent, although excluded from very arid regions in the north-west and hot and humid areas north of the Tropic of Capricorn. Rabbits had turned from a prized game species to a major nuisance and agricultural pest (Rolls, 1969). Since their introduction into New Zealand and Australia in the 1850s, rabbits have exerted a huge impact on farm production, especially through competition with livestock for pastures and by damaging crops.

In both Australia and New Zealand various management approaches were attempted, including poisoning, shooting, trapping, harbour destruction, exclusion fencing, even the release of domestic cats (Rolls, 1969). None of these efforts proved satisfactory in the long term. By the end of the 1880s, in Australasia, a consensus emerged that biological control using a self-disseminating agent was going to be the only answer to landscape-scale rabbit control (Anon., 1890). Early biocontrol investigations even involved Louis Pasteur, who suggested using chicken cholera (*Pasteurella multocida*) to rid Australia of rabbits, but this was quickly abandoned as it was not species specific and did not spread between individuals (Fenner and Fantini, 1999; Roundtree, 1983). Meanwhile in New Zealand the introduction of rabbit predators, such as the stoat (ermine), as an alternative form of biological control contributed to the enormous devastation of New Zealand's native birds, many of which are flightless, while at the same time not reducing rabbit populations significantly. In the 1950s, mxyoma virus was introduced into Australia, and provided temporary relief with the deaths of millions of rabbits. However, genetic resistance was detected within 2 years (Fenner and Fantini, 1999) and the ongoing co-evolution of virus and host has led to development of partial resistance (Kerr and Best, 1998). In contrast to Australia, myxomatosis was never introduced successfully into New Zealand through lack of a suitable insect transmission vector. In both countries the rabbit remains a major economic and ecological pest and the emergence of RHDV generated significant interest.

The introduction of RHDV into Australia and New Zealand

A critical step towards the introduction of RHDV as a biocontrol agent was the testing of host range specificity. A total of 28 species of vertebrates were challenge tested under high security conditions in the Australian Animal Health Laboratories (Munro and Williams, 1994). Kiwi and bats from New Zealand have also been tested (Buddle *et al.*, 1997). The European rabbit proved to be the only host in which RHDV effectively replicated and caused disease. Even other lagomorphs that were experimentally exposed showed no clinical signs of disease; these included: the eastern cottontail (*Sylvilagus floridanus*), black-tailed jackrabbit (*Lepus californicus*) and volcano rabbit (*Romerolagus diazi*). The European brown hare (*Lepus europaeus*) and the Varying hare (*Lepus timidus*) could not be infected with RHDV in the challenge studies, but they proved susceptible to the closely related EBHSV that causes European brown hare syndrome. Once convinced that RHDV was specific to European rabbits, trials on Wardang Island off the coast of York Peninsula, South Australia, were carried out in an effort to confirm that the virus would spread and persist in relatively arid regions of Australia where rabbit impact was most severe. It was reasoned that there would be little point in releasing a virus that failed to spread in dry regions. Wardang Island had been previously used to assess myxoma virus in the 1930s and was chosen partly on this basis.

A quarantine area was established and within it smaller pens containing wild rabbits in natural warrens were maintained to follow disease spread. Experimentally infected rabbits and those in contact were monitored by direct observation from towers immediately outside the pens and the radio-collars were used to locate cadavers so that they could be tissue-sampled to confirm disease spread and removed from the pens as part of the quarantine protocol. Experimental evidence accumulated over several months showed that RHD spread was erratic, sometimes affecting most rabbits in a pen but occasionally not spreading beyond the infected rabbits. Part of the problem may have been due to the rapid removal of cadavers, but that changed suddenly in September 1995 when the virus began to spread into pens of rabbits that were being set up for future experiments. Although the experiments were closed down at that stage the virus escaped from the quarantine area and eventually reached the mainland.

The virus spread very rapidly across the continent (Kovaliski, 1998) and its impact was high, especially in arid areas (Mutze *et al.*, 1998). There was considerable evidence that insects, particularly flies, were involved in escape of the virus from the quarantine area and its rapid spread. The many millions of rabbits that died during the initial outbreak of RHD provided both a source of virus and abundant food for fly larvae.

The massive 'fly-wave' generated at that time may have assisted in moving the virus on a wide scale. As spring turned to summer and the virus spread over more of Australia, there were changes in the species of flies that potentially carried the disease. Although bush flies, *M. vetustissima*, were ubiquitous, summer-dominant blowfly species such as *Chrysomyia rufifacies* and *Lucilia cuprina* replaced species such as *C. dubia* and *C. stygia* to become more closely associated with the inland spread of the disease (McColl *et al.*, 2002).

Limited PCR studies at the time showed that fly maggots did not retain ingested RHDV upon entering pupation, so adult flies must become contaminated after alighting on rabbit carcasses and eating tissues. They produce faeces and other regurgita containing detectable RHDV for up to 9 days after feeding on RHDV-infected rabbit liver (Asgari *et al.*, 1998). As a result, if they fed on carcasses and deposited faeces on pastures subsequently eaten by rabbits, they could spread RHDV even without contacting live rabbits.

Despite the unintentional escape of the virus, it soon became apparent that RHDV was a highly effective biological control agent at least in the hotter inland areas of Australia where rabbits were generally reduced to less than 10% of their former abundance. There was a lesser reduction in cooler temperate areas but it was generally considered that Australia's rabbit population was reduced by about 60% overall.

Despite the New Zealand Government's close interest in the work on RHDV in Australia, it was decided not to introduce the virus on the basis of its unknown impact and likely limited benefit to the New Zealand economy. Nevertheless, the relative success of RHDV for rabbit control in Australia led to the illegal importation of RHDV into New Zealand in 1997 (Thompson and Clark, 1997). How this illegal importation occurred was never determined, however the process was clearly effective and led to the widespread dissemination of RHDV in New Zealand (Parkes *et al.*, 2002). As in Australia, RHDV rapidly caused significant drops in rabbit numbers with up to 80% of rabbits becoming infected with 70% being killed during the initial spread of RHDV (Parkes *et al.*, 2002). In the dry Central Otago area populations initially fell to possibly as low as 10% of former levels, though numbers have recovered

significantly, in some cases to pre-RHDV levels (Parkes and Norbury, 2004). Although rabbits still remain a major pest problem in New Zealand, the importation of the virus has been of major benefit to many farmers and continues to suppress rabbit populations.

The economic impact of RHDV can be broadly summarized by an investment of $12 million deriving benefits of $200 million–$400 million annually in Australia, primarily benefiting the livestock industries (McLeod, 2004; Vere *et al.*, 2004, Cooke unpublished, Gong, *et al.*, unpublished). Nevertheless, surveys in NSW, Victoria and South Australia found that residual rabbit problems remained: not only were there continuing losses in the livestock industries and agriculture due to forgone production and direct damage but there were also impacts on forestry, horticulture, conservation, transport and social amenities.

Following the introduction of RHD there was extensive regeneration of many species of trees and shrubs such as Mulga, *Acacia aneura*, in the Lake Eyre Basin (Collis, 2000). Similar regeneration was also widely recorded in semi-arid regions such as the Murray mallee in north-western Victoria. Buloke (*Allocasuarina leuhmanni*), regarded as a threatened species, regenerated for the first time since establishment of the Hattah-Kulkyne National Park when RHD reduced the rabbit population (Murdoch, 2005, unpublished). Murdoch (unpublished) has since shown that it takes less than one rabbit/hectare to remove all buloke seedlings, making rabbits a key obstacle to maintaining the buloke – pine woodlands of north-western Victoria.

Prior to the spread of RHD in South Australia's Flinders Ranges, red kangaroo numbers doubled on 3 km^2 experimental plots where rabbits were eliminated. However, kangaroos remained low on equivalent plots where rabbits persisted. After RHD, parallel increases of the same magnitude were observed on both experimental and control plots (Mutze, 2008). On the same Flinders Ranges study site, removal of rabbits allowed moderately palatable shrubs to increase; 105 new plants/hectare/year grew compared with a loss of 40 plants/hectare/year on rabbit grazed plots. After RHD, between 105 and 130 new shrubs/hectare appeared each year on all sites irrespective of their previous history of rabbit grazing. Mutze has pointed out that if the recruitment of moderately palatable shrubs applied to just 1% of the rangelands infested by rabbits then 100 million new shrubs must have grown in the rangelands each year in the immediate aftermath of RHD. Such increases are several orders of magnitude higher than attempts at revegetation through community schemes such as 'planting one billion trees' and impacts significantly upon objectives such as revegetating Australia's impoverished arid zone. The recovery of arid grasslands, a largely ignored issue in terms of arid zone management, has been associated with extensions of the known distributions of native rodents, the hopping mouse *Notomys alexis* and the plains rat *Pseudomys australis* (Mutze, 2008; Read, 2003)

The huge 'continent-wide' change triggered by RHDV in Australia has also brought with it a new environmental awareness. Experiments carried out at the time of its spread and the demonstration of tree and shrub regrowth in the virtual absence of rabbits brought with it a strong realization that rabbits were as much an environmental pest as an economic one.

Future biocontrol prospects for RHDV

Since 2002, rabbits have been noticeably increasing in some parts of Australia. In the Hattah-Kulkyne National Park, for example, night time spotlight counts of rabbits have increased from 0.5 rabbits/km to 10 rabbits/km despite expenditure on rabbit control increasing from about $30,000 to $200,000 annually (Sandell, 2006, and personal communication).

Only one variant of RHDV was introduced into Australia and New Zealand and some have asked whether more effective or alternative variants of RHDV might be available or whether increasing the genetic diversity of RHDV might enable recombination and allow the virus to rapidly counter any gains in host resistance. However, it is at least theoretically feasible that the same mechanisms could also promote viral attenuation, if such a strategy was beneficial for virus survival.

Nonetheless, this involves close understanding of a complex epidemiological system and the reasons underlying recent rabbit increases are not

well established. As the following sections show, things may be far more complex than the picture that emerged when Australian rabbits developed resistance to myxoma virus.

Endemic non-pathogenic lagoviruses in Australia

As part of the testing carried out prior to introducing RHDV as a biological control agent, the susceptibility of Australian wild rabbits to RHDV was assessed. Although the vast majority of animals died, some rabbits sourced from Bendigo, Victoria, survived. These animals were found to have pre-existing cross-reacting antibodies to RHDV prior to challenge (Nagesha *et al.*, 2000). Similar evidence of pre-existing antibodies was reported from other parts of Australia (Bruce *et al.*, 2004; Collins *et al.*, 1995; Cooke *et al.*, 2002; Richardson *et al.*, 2007; Robinson *et al.*, 2002), from New Zealand, before RHDV was illegally released in 1997 (O'Keefe *et al.*, 1999) and from Europe, before virulent strains of RHDV became apparent (Chasey *et al.*, 1997; Moss *et al.*, 2002; Rodak *et al.*, 1990).

In Australia, the occurrence of these cross-reacting antibodies is more frequent in the cooler and more humid parts of the country. When RHDV escaped from Wardang Island and reached mainland Australia in 1995, it was less effective in reducing rabbit numbers in these areas (Cooke *et al.*, 2002; Mutze *et al.*, 1998). Cooke *et al.* (2002) demonstrated that the RHDV mortality index was directly correlated with annual rainfall. These combined findings strongly suggested the presence of a related endemic virus in the more temperate zones, which was capable of providing some level of cross-protection to lethal RHDV challenge. Apart from Myxoma epidemics, no signs of other rabbit mass mortality were observed in Australia prior to the arrival of RHDV, so this putative virus was presumed to be non-pathogenic.

Attempts to detect this virus in the tissues of animals with antibody patterns that were scored as cross-reacting to RHDV were unsuccessful, indicating that the infection was either short lived and cleared from the system before a detectable antibody response was mounted, or tissues other than liver and spleen were the site of replication.

The application of a targeted screening strategy that was independent of the serology status of the rabbits eventually enabled the discovery of an endemic calicivirus. Because rabbits are most likely to acquire the infection as soon as they lose their maternal antibody protection (Cooke *et al.*, 2002; Richardson *et al.*, 2007), predominantly young animals from high rainfall areas were tested, and a variety of tissues were included in the analysis to allow the detection of a virus that causes an intestinal, respiratory or generalized infection. Although the genetic make-up of the putative virus was not known, the degree of serological cross-reaction strongly suggested a closely related lagovirus, which allowed screening using a universal PCR test for the detection of any member of the *Lagovirus* genus (Strive *et al.*, 2009).

The new endemic Australian calicivirus was termed 'Rabbit Calicivirus-Australia1' (RCV-A1). Unlike RHDV, and similar to a non-pathogenic rabbit calicivirus (RCV) discovered in Italy (Capucci *et al.*, 1995, 1996), there were no signs of disease in the affected rabbits and the highest concentration of viral RNA was found in intestinal tissues and contents, suggesting a faecal-oral transmission route. Viral RNA was barely detectable in the liver of only one rabbit (Strive *et al.*, 2009), suggesting that this benign virus differs from RHDV not only in terms of pathogenicity but also with regards to tissue tropism. The underlying causes for these differences in phenotype are yet to be determined.

Phylogenetic analyses show that RCV-A1 is distinct from other previously described members of the genus *Lagovirus* (Fig. 12.4). It is feasible that the ancestor of this virus came to Australia in the first rabbits and in the ensuing 150 years of geographical separation has evolved into RCV-A1 on the Australian continent, while the non-pathogenic rabbit caliciviruses (RCV) first identified in Italy have evolved in Europe (Capucci *et al.*, 1996; Forrester *et al.*, 2007; Moss *et al.*, 2002). The pathogenic variants RHDV and RHDVa most likely evolved more recently from the non-pathogenic European strains (Fenner and Fantini, 1999; Moss *et al.*, 2002).

It has been reported that homologous recombination occurs quite frequently in caliciviruses, and RHDV is no exception (Abrantes *et al.*, 2008;

Forrester, 2008). However, no recombination events between pathogenic and non-pathogenic lagoviruses have been described so far in Europe, where the situation is similar, although both viruses occur in the same populations. The junction between the polymerase gene and the capsid gene is also well conserved within the genus, and recombination between the two is at least theoretically feasible. A hypothetical recombinant with the biological properties of RCV-A1 and the antigenic make-up of RHDV could have potentially devastating effects for rabbit biocontrol in Australasia. In addition, since the arrival of pathogenic RHDV in 1996 there may be a positive selective advantage for the development of RCV-A1 variants that are antigenically more similar to RHDV, thus providing better cross-immunoprotection and weakening the potency of RHDV as a biocontrol agent.

Even without events as dramatic as recombination or directed evolution towards a more protective strain, interference with RHDV seems likely. Earlier studies based on serology suggest that RHDV mortality rates in adult rabbits previously exposed to the non-pathogenic calicivirus may be as low as 50–70% (McPhee *et al.*, 2002). While the exact extent of cross-immunoprotection conferred by RCV-A1 needs to be determined, even partial protection against the virulent virus promoting better survival of rabbits in the presence of RHDV, changes in RCV-A1 and/or increased distribution of this virus, as well as the possible development of genetic resistance to RHDV, may all be contributing separately or in combination to the recent increase in rabbit numbers being seen in Australia and New Zealand.

Are rabbits developing genetic resistance to infection with RHDV?

The possibility that wild rabbits are developing genetic resistance to RHD is being investigated (P. Elsworth, D. Berman and B. Cooke, unpublished) using a similar approach to that developed by Fenner and his team (Fenner and Ratcliffe, 1965) to assess developing genetic resistance to myxomatosis. Susceptible wild rabbits collected from different parts of south-eastern Australia have been experimentally challenged with standard doses of virus to see if they show a lowering

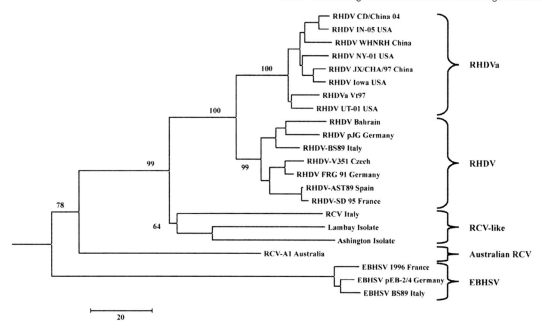

Figure 12.4 Phylogeny of benign calicivirus strain RCV-A1. The evolutionary history was inferred using the minimum evolution method and a fragment of the capsid protein sequence from nt 6163 to nt 6679 of the RCV-A1 genome. Bootstrap values of 1000 replicates are shown for the major branches only. The tree is drawn to scale, with branch lengths in the same units as those of the evolutionary distances used to infer the phylogenetic tree. The tree was searched using the close-neighbour-interchange (CNI) algorithm at a search level of 3. The neighbour-joining algorithm was used to generate the initial tree. Codon positions included were 1st + 2nd + 3rd + non-coding, and the complete deletion option was selected. GenBank sequences for the isolates were U54983, RHDV-V351 Czech; EU003579, RHDV Italy 90; M67473, RHDV FRG 91 Germany; Z29514, RHDV-SD 95 France; Z49271, RHDV-AST89 Spain; X87607, RHDV-BS89 Italy; EF363035, RHDV pJG Germany; DQ189077, RHDV 2006 Bahrain; EU003582, RHDV UT-01 USA; EU003581, RHDV NY-01 USA; DQ205345, RHDV JX/CHA/97 China; AF258618, RHDV Iowa USA; DQ280493, RHDV WHNRH China; EU003578, RHDV IN-05 USA; AY523410 RHDV CD/China 04; X96868, RCV Italy; AF454050, Ashington Isolate UK; EU871528, RCV-A1 Australia; NC_002615, EBHSV France; U09199, EBHSV pEB-2/4 Germany; U09199, EBHSV pEB-2/4 Germany; X9800, EBHSV BS89 Italy; AJ86699, rabbit vesivirus.

of the case mortality rate, longer survival times and lower virus titres in blood or the livers in comparison to unselected laboratory rabbits.

Wild rabbits from some localities in southeastern Australia appear to be much more resistant than domestic rabbits to oral infection with low doses (1:25 dilution) of Czech-strain 351 RHDV, the strain originally released in Australia. However, there was no evidence that rabbits are better able to recover from infection; case mortality rates remained high, over 90% of rabbits that become infected died, and there was no increase in survival times. Rabbits that did not become infected with a low experimental dose of virus were subsequently challenged with larger amounts of virus or by a different route and many died, confirming that they remained fully susceptible and had simply avoided initial infection.

Thus, avoidance of infection rather than dealing with the debilitating effects of RHD may offer rabbits some selective advantage. This suggests that virus-host co-evolution is likely to follow one of two pathways. On the one hand, rabbit resistance may eventually develop to such a point that the virus may no longer be able to circulate effectively and die out or become limited to only part of the rabbit's distribution in Australia. On the other hand, if RHDV is constantly co-evolving to find its way around host resistance in a biological 'arms race', then it might persist into the future as a useful biological control agent albeit with a reduced efficacy.

Additional work is needed to demonstrate whether or not the apparent resistance to RHDV infection has a genetic basis. This could be done through breeding experiments to show that resistance is heritable, as was done following the introduction of myxomatosis (Sobey, 1969). In the meantime, other approaches are being explored, including an investigation of the genetic background of rabbits to see if there are any genetic markers indicative of resistance.

Virus binding sites and resistance to infection

Early tests to detect RHDV (haemagglutination and haemagglutination-inhibition tests) were based on the observation that viral particles cause human O-type red blood cells to clump together. Subsequently, Le Pendu and colleagues gathered considerable evidence suggesting that the virus bound to the ABO histo-blood group antigens that, in rabbits, are expressed on the mucosal cells of the respiratory tract and intestine. These antigens have three attached sugars, forming a trisaccharide terminus – and RHDV exploits this to bind to cells on the rabbit's intestinal wall as a first step towards infection (Ruvoen-Clouet *et al.*, 2000).

It has been proposed that infection with a specific genogroup II genotype 4 (GII.4) norovirus strain was predictable if the genes determining an individual's blood group were known (Le Pendu *et al.*, 2006). The expression of blood group antigens on cells in the respiratory and gastric mucosa by people who are 'secretors' rather than 'non-secretors' is also important for infection. A single nucleotide mutation in the fucosyltransferase 2 gene (FUT2), G428A, which makes the secretor gene *Se* non-functional *se* leads to resistance to infection through lack of binding sites. People with this genotype make up about 20% of the Scandinavian population (Thorven *et al.*, 2005), however the norovirus can also change and there is a clear dynamic between virus and host (Lindesmith *et al.*, 2008).

Nuclear magnetic resonance studies with VLPs have confirmed that alpha fucosyl is the minimum requirement for RHDV to bind to the surface of rabbit cells (Rademacher *et al.*, 2008). This is a further development stemming from LePendu's original ideas (Le Pendu *et al.*, 2006) and since it is generic, it could be applied to other viruses and cell attachment sites and is a step towards identifying how drugs that block calicivirus virus attachment might be developed.

In rabbits there are six different alleles of the analogue of the human secretor gene (*Sec 1*) and five variants of FUT2 (Guillon *et al.*, 2009). Interestingly, these variants seem to actively enable binding and so there are no inactive or non-functional genes that would help the host resist infection as seen in some human noroviruses. Nonetheless, the *v5* allele of the *Sec 1* gene increased in a wild rabbit population in France following an outbreak of acute RHD suggesting that some natural selection may be occurring. Such studies are important steps towards an understanding of host-virus co-evolution and will potentially improve when more is known about virus genetics and interactions between the pathogenic viruses like RHDV, RHDVa and the non-pathogenic virus forms such as RCV and RCV-A1 discussed above.

DNA samples from Australian rabbits used in RHDV challenge experiments have also been analysed in France by J. Le Pendu and P. Guillon to determine whether *Fut2* and *Sec1* alleles may be useful indicators of susceptibility. While some alleles may influence susceptibility, there is no strong correlation implying a direct causal link. It is therefore anticipated that the link between virus cell-binding and resistance to infection will not be as clear as seen for some human noroviruses (Le Pendu *et al.*, 2006). This information makes it all the more likely that virus and host are involved in a constant arms-race where changes in host resistance are probably met with corresponding changes in the virus that allow new variants to spread despite host resistance. Such information is already being incorporated into mathematical models to predict long-term evolutionary outcomes (Fouchet, 2009).

In summary, a raft of factors are likely to influence the effectiveness of RHDV as a biocontrol agent, including genetic changes in the virus or its host, the affects of maternal antibody at an early age (Robinson *et al.*, 2002), the role of receptors and age when first exposed to virus (Ruvoen-Clouet *et al.*, 2000) and the presence of benign viruses in the environment (Strive *et al.*, 2009).

RHDV and international regulation of genetically modified organisms (GMOs)

As previously mentioned, recombinant vaccines to protect wild rabbits from both myxomatosis and RHD have already been developed. However, some of these 'live-vaccines' were deliberately designed to be capable of spreading from rabbit to rabbit in the field (Torres, 2000). Even if just a few wild rabbits were caught and inoculated there was potential for the live vaccine to spread and so protect a greater proportion of the rabbit population. Following a limited field trial on a small isolated Balearic island, it was reported that one recombinant virus met with these expectations.

Simultaneously, in Australia, it was proposed to use an attenuated, self-spreading myxoma virus to produce a recombinant virus that expressed zona pellucida proteins from rabbit's eggs to effectively sterilize wild rabbits. This viral-vectored immunocontraception would be a major achievement in an animal of such renowned fecundity. A suitable myxoma virus, called Uriarra/2-53/1, was selected as the vector based on low mortality in laboratory rabbits, and a number of recombinant viruses were made to test which of several *zona pellucida* proteins was most effective. One, called rabbit ZPC, stood out as being most effective and the recombinant virus also expressed IL-4 to enhance the final immune response. The recombinant antigen rZPC with IL-4 controlled by a p28 late promoter produced total sterility among all rabbits at least for the first expected litter (van Leeuwin and Kerr, 2007).

However, by the late 1990s, the two projects were causing growing concern (Angulo and Cooke, 2002) owing to the risk that recombinant viruses could arrive accidentally or be deliberately introduced into countries where they were not wanted. The arrival of the recombinant live-vaccine in Australia could potentially undo years of previous work while an immunocontraceptive virus could further set back any hope of recovery of RHD-ravaged wild rabbit populations in Spain and Portugal.

This was partly resolved when work in Australia to develop the immuno-sterilizing virus came to an end. Although the idea worked in principle, there were too many immediate problems to make it a useful field technique. The antibodies that attacked the egg membrane were not long-lasting; rabbits inoculated with the recombinant virus were only temporarily sterilized and some eventually went on to reproduce normally. Furthermore, from field experiments

in which rabbits were surgically sterilized to simulate the effects of the recombinant virus, it was also shown that because of the rabbit's fertility, 80% of adult females would need to be permanently sterilized to drive rabbit populations down (Twigg et al., 2002). On that basis, and the fact that a recombinant virus would be competing with virulent field strains of virus, the program was finally put aside (van Leeuwen and Kerr, 2007).

By contrast, the live myxoma-RHD recombinant vaccine in Spain still continues to be evaluated for potential field use, so efforts to explore ways of restricting the risk of international transfer remain extremely important. This has been taken up at a number of international workshops and conferences as an exceptionally clear example illustrating the problems raised by international transfer of genetically modified viruses. Because RHDV is seen as both detrimental and beneficial in different contexts its importance extends well beyond the ways in which viruses are normally viewed. In this instance it is clearly influencing public policy (Sagoff, 2008).

Conclusion

Our understanding of RHDV has been integral to many advances in our understanding of caliciviruses. RHDV has the unusual perspective of being both harmful and beneficial in different circumstances, whether that be in increasing our understanding of caliciviruses in areas such as genome replication, polyprotein processing, translation regulation and cellular receptors to the application of the virus in areas as diverse as biomolecule delivery, models for fulminant liver failure and the control of a vertebrate pest. This chapter presents an overview of the special place RHDV holds within the *Caliciviridae*. Perhaps it is true to conclude that one nation's pest is another nation's treasure and the contribution of RHDV to our understanding of caliciviruses at all levels looks set to continue into the future.

References

Abrantes, J., Esteves, P.J., and van der Loo, W. (2008). Evidence for recombination in the major capsid gene VP60 of the rabbit haemorrhagic disease virus (RHDV). Arch. Virol. *153*, 329–335.

Alonso, C., Oviedo, J.M., Martin-Alonso, J.M., Diaz, E., Boga, J.A., and Parra, F. (1998). Programmed cell death in the pathogenesis of rabbit hemorrhagic disease. Arch. Virol. *143*, 321–332.

Angulo, E., and Cooke, B. (2002). First synthesize new viruses then regulate their release? The case of the wild rabbit. Mol. Ecol. *11*, 2703–2709.

Anon. (1890). New South Wales, Royal Commission of Inquiry into Schemes for extermination of rabbits in Australasia. Progress Report, Minutes of Proceedings, Minutes of Evidence and Appendices. Government Printer Sydney.

Arguello Villares, J.L. (1991). Viral haemorrhagic disease of rabbits: vaccination and immune response. Rev. Sci. Tech. *10*, 459–480.

Asgari, S., Hardy, J.R., Sinclair, R.G., and Cooke, B.D. (1998). Field evidence for mechanical transmission of rabbit haemorrhagic disease virus (RHDV) by flies (Diptera:Calliphoridae) among wild rabbits in Australia. Virus Res. *54*, 123–132.

Barcena, J., Verdaguer, N., Roca, R., Morales, M., Angulo, I., Risco, C., Carrascosa, J.L., Torres, J.M., and Caston, J.R. (2004). The coat protein of rabbit hemorrhagic disease virus contains a molecular switch at the N-terminal region facing the inner surface of the capsid. Virology *322*, 118–134.

Bertagnoli, S., Gelfi, J., Le Gall, G., Boilletot, E., Vauterot, J.F., Rasschaert, D., Laurent, S., Petit, F., Boucraut-Baralon, C., and Milon, A. (1996a). Protection against myxomatosis and rabbit viral hemorrhagic disease with recombinant myxoma viruses expressing rabbit hemorrhagic disease virus capsid protein. J. Virol. *70*, 5061–5066.

Bertagnoli, S., Gelfi, J., Petit, F., Vauterot, J.F., Rasschaert, D., Laurent, S., Le Gall, G., Boilletot, E., Chantal, J., and Boucraut-Baralon, C. (1996b). Protection of rabbits against rabbit viral haemorrhagic disease with a vaccinia-RHDV recombinant virus. Vaccine *14*, 506–510.

Blancou, J., Kieny, M.P., Lathe, R., Lecocq, J.P., Pastoret, P.P., Soulebot, J.P., and Desmettre, P. (1986). Oral vaccination of the fox against rabies using a live recombinant vaccinia virus. Nature *322*, 373–375.

Boga, J.A., Casais, R., Marin, M.S., Martin-Alonso, J.M., Carmenes, R.S., Prieto, M., and Parra, F. (1994). Molecular cloning, sequencing and expression in *Escherichia coli* of the capsid protein gene from rabbit haemorrhagic disease virus (Spanish isolate AST/89). J. Gen. Virol. *75*, 2409–2413.

Boga, J.A., Marin, M.S., Casais, R., Prieto, M., and Parra, F. (1992). In vitro translation of a subgenomic mRNA from purified virions of the Spanish field isolate AST/89 of rabbit hemorrhagic disease virus (RHDV). Virus Res. *26*, 33–40.

Boga, J.A., Martin Alonso, J.M., Casais, R., and Parra, F. (1997). A single dose immunization with rabbit haemorrhagic disease virus major capsid protein produced in *Saccharomyces cerevisiae* induces protection. J. Gen. Virol. *78*, 2315–2318.

Bok, K., Prikhodko, V.G., Green, K.Y., and Sosnovtsev, S.V. (2009). Apoptosis in murine norovirus infected RAW264.7 cells is associated with survivin downregulation. J. Virol. *83*, 3647–3656.

Boniotti, B., Wirblich, C., Sibilia, M., Meyers, G., Thiel, H.J., and Rossi, C. (1994). Identification and characterization of a 3C-like protease from rabbit hemorrhagic disease virus, a calicivirus. J. Virol. *68*, 6487–6495.

Brochier, B., Kieny, M.P., Costy, F., Coppens, P., Bauduin, B., Lecocq, J.P., Languet, B., Chappuis, G., Desmettre, P., Afiademanyo, K., and *et al.* (1991). Large-scale eradication of rabies using recombinant vaccinia-rabies vaccine. Nature *354*, 520–522.

Bruce, J.S., Twigg, L.E., and Gray, G.S. (2004). The epidemiology of rabbit haemorrhagic disease, and its impact on rabbit populations, in south-western Australia. Wildl. Res. *31*, 31–49.

Buddle, B.M., de Lisle, G.W., McColl, K., Collins, B.J., Morrissy, C., and Westbury, H.A. (1997). Response of the North Island brown kiwi, *Apteryx australis mantelli* and the lesser short-tailed bat, *Mystacina tuberculata* to a measured dose of rabbit haemorrhagic disease virus. N. Z. Vet. J. *45*, 109–113.

Buller, R.M., Smith, G.L., Cremer, K., Notkins, A.L., and Moss, B. (1985). Decreased virulence of recombinant vaccinia virus expression vectors is associated with a thymidine kinase-negative phenotype. Nature *317*, 813–815.

Capucci, L., Fallacara, F., Grazioli, S., Lavazza, A., Pacciarini, M.L., and Brocchi, E. (1998). A further step in the evolution of rabbit hemorrhagic disease virus: the appearance of the first consistent antigenic variant. Virus Res. *58*, 115–126.

Capucci, L., Frigoli, G., Ronshold, L., Lavazza, A., Brocchi, E., and Rossi, C. (1995). Antigenicity of the rabbit hemorrhagic disease virus studied by its reactivity with monoclonal antibodies. Virus Res. *37*, 221–238.

Capucci, L., Fusi, P., Lavazza, A., Pacciarini, M.L., and Rossi, C. (1996). Detection and preliminary characterization of a new rabbit calicivirus related to rabbit hemorrhagic disease virus but nonpathogenic. J. Virol. *70*, 8614–8623.

Capucci, L., Scicluna, M.T., and Lavazza, A. (1991). Diagnosis of viral haemorrhagic disease of rabbits and the European brown hare syndrome. Rev. Sci. Tech. *10*, 347–370.

Carman, J.A., Garner, M.G., Catton, M.G., Thomas, S., Westbury, H.A., Cannon, R.M., Collins, B.J., and Tribe, I.G. (1998). Viral haemorrhagic disease of rabbits and human health. Epidemiol. Infect. *121*, 409–418.

Castanon, S., Marin, M.S., Martin-Alonso, J.M., Boga, J.A., Casais, R., Humara, J.M., Ordas, R.J., and Parra, F. (1999). Immunization with potato plants expressing VP60 protein protects against rabbit hemorrhagic disease virus. J. Virol. *73*, 4452–4455.

Chasey, D. (1997). Rabbit haemorrhagic disease: the new scourge of *Oryctolagus cuniculus*. Lab. Anim. *31*, 33–44.

Chasey, D., and Duff, P. (1990). European brown hare syndrome and associated virus-particles in the UK. Vet. Rec. *126*, 623–624.

Chasey, D., Lucas, M., Westcott, D., and Williams, M. (1992). European brown hare syndrome in the UK – a calicivirus related to but distinct from that of viral hemorrhagic-disease in rabbits. Arch. Virol. *124*, 363–370.

Chasey, D., Trout, R.C., and Edwards, S. (1997). Susceptibility of wild rabbits (*Oryctolagus cuniculus*) in the United Kingdom to rabbit haemorrhagic disease (RHD). Vet. Res. *28*, 271–276.

Chen, R., Neill, J.D., and Prasad, B.V. (2003). Crystallization and preliminary crystallographic analysis of San Miguel sea lion virus: an animal calicivirus. J. Struct. Biol. *141*, 143–148.

Collins, B.J., White, J.R., Lenghaus, C., Boyd, V., and Westbury, H.A. (1995). A competition ELISA for the detection of antibodies to rabbit haemorrhagic disease virus. Vet. Microbiol. *43*, 85–96.

Collis, B. (2000). Mulga rebirth begs fair-dinkum crack at the rabbit. ECOS *105*, 24–26.

Cooke, B.D. (2002). Rabbit haemorrhagic disease: field epidemiology and the management of wild rabbit populations. Rev. Sci. Tech. *21*, 347–358.

Cooke, B.D., McPhee, S., Robinson, A.J., and Capucci, L. (2002). Rabbit haemorrhagic disease: does a pre-existing RHDV-like virus reduce the effectiveness of RHD as a biological control in Australia? Wildl. Res. *29*, 673–682.

Da Silva, D.M., Pastrana, D.V., Schiller, J.T., and Kast, W.M. (2001). Effect of preexisting neutralizing antibodies on the anti-tumor immune response induced by chimeric human papillomavirus virus-like particle vaccines. Virology *290*, 350–360.

Delibes-Mateos, M., Redpath, S.M., Angulo, E., Ferrerasa, P., and Villafuerte, R. (2007). Rabbits as a keystone species in southern Europe. Biol. Conserv. *137*, 149–156

El Mehdaoui, S., Touze, A., Laurent, S., Sizaret, P.Y., Rasschaert, D., and Coursaget, P. (2000). Gene transfer using recombinant rabbit hemorrhagic disease virus capsids with genetically modified DNA encapsidation capacity by addition of packaging sequences from the L1 or L2 protein of human papillomavirus type 16. J. Virol. *74*, 10332–10340.

Farnos, O., Boue, O., Parra, F., Martin-Alonso, J.M., Valdes, O., Joglar, M., Navea, L., Naranjo, P., and Lleonart, R. (2005). High-level expression and immunogenic properties of the recombinant rabbit hemorrhagic disease virus VP60 capsid protein obtained in Pichia pastoris. J. Biotechnol. *117*, 215–224.

Farnos, O., Fernandez, E., Chiong, M., Parra, F., Joglar, M., Mendez, L., Rodriguez, E., Moya, G., Rodriguez, D., Lleonart, R., *et al.* (2008). Biochemical and structural characterization of RHDV capsid protein variants produced in *Pichia pastoris*: Advantages for immunization strategies and vaccine implementation. Antiviral Res. *81*, 25–36.

Farnos, O., Rodriguez, D., Valdes, O., Chiong, M., Parra, F., Toledo, J.R., Fernandez, E., Lleonart, R., and Suarez, M. (2007). Molecular and antigenic characterization of rabbit hemorrhagic disease virus isolated in Cuba indicates a distinct antigenic subtype. Arch. Virol. *152*, 1215–1221.

Farnos, O., Rodriguez, M., Chiong, M., Parra, F., Boue, O., Lorenzo, N., Colas, M., and Lleonart, R. (2006). The recombinant rabbit hemorrhagic disease virus VP60 protein obtained from *Pichia pastoris* induces a strong humoral and cell-mediated immune response

following intranasal immunization in mice. Vet. Microbiol. *114*, 187–195.

Fenner, F., and Fantini, B. (1999). Biological control of vertebrate pests: the history of myxomatosis, an experiment in evolution. Biological control of vertebrate pests: the history of myxomatosis, an experiment in evolution, xii + 339 pp CABI Publishing, Wallingford, Oxon.

Fenner, F., and Ratcliffe, F.N. (1965). Myxomatosis. Myxomatosis, xiv + 379 pp Cambridge University Press.

Fernandez-Fernandez, M.R., Mourino, M., Rivera, J., Rodriguez, F., Plana-Duran, J., and Garcia, J.A. (2001). Protection of rabbits against rabbit hemorrhagic disease virus by immunization with the VP60 protein expressed in plants with a potyvirus-based vector. Virology *280*, 283–291.

Ferreira, P.G., Costa-e-Silva, A., Monteiro, E., Oliveira, M.J., and Aguas, A.P. (2004). Transient decrease in blood heterophils and sustained liver damage caused by calicivirus infection of young rabbits that are naturally resistant to rabbit haemorrhagic disease. Res. Vet. Sci. *76*, 83–94.

Forrester, N., Moss, S.R., Turner, S.L., Schirrmeier, H., and Gould, E.A. (2008). Recombination in rabbit haemorrhagic disease virus: possible impact on evolution and epidemiology. Virology *376*, 390–396.

Forrester, N.L., Trout, R.C., and Gould, E.A. (2007). Benign circulation of rabbit haemorrhagic disease virus on Lambay Island, Eire. Virology *358*, 18–22.

Forrester, N.L., Trout, R.C., Turner, S.L., Kelly, D., Boag, B., Moss, S., and Gould, E.A. (2006). Unravelling the paradox of rabbit haemorrhagic disease virus emergence, using phylogenetic analysis; possible implications for rabbit conservation strategies. Biol. Conserv. *131*, 296–306.

Fouchet, D., Le Pendu J., Guitton, J.S., Guiserix, M., Marchandeau, S., and Pontier, D. (2009). Evolution of microparasites in spatially and genetically structured host populations: the example of RHDV infecting rabbits. J. Theor. Biol. *21*, 212–227.

Fuchs, A., and Weissenbock, H. (1992). Comparative histopathological study of rabbit hemorrhagic-disease (RHD) and European brown hare syndrome (EBHS). J. Comp. Pathol. *107*, 103–113.

Gavierwiden, D., and Morner, T. (1993). Descriptive epizootiological study of European brown hare syndrome In Sweden. J. Wildl. Dis. *29*, 15–20.

Gibb, J.A., and Williams, J.M. (1994). The rabbit in New Zealand, In The European rabbit: the history and biology of a successful colonizer, H.V. Thompson, and C.M. King, eds. (Oxford: Oxford University Press), pp. 245.

Gil, F., Titarenko, E., Terrada, E., Arcalis, E., and Escribano, J.M. (2006). Successful oral prime-immunization with VP60 from rabbit haemorrhagic disease virus produced in transgenic plants using different fusion strategies. Plant Biotechnol. J. *4*, 135–143.

Gregg, D.A., House, C., Meyer, R., and Berninger, M. (1991). Viral haemorrhagic disease of rabbits in Mexico: epidemiology and viral characterization. Rev. Sci. Tech. *10*, 435–451.

Gromadzka, B., Szewczyk, B., Konopa, G., Fitzner, A., and Kesy, A. (2006). Recombinant VP60 in the form of virion-like particles as a potential vaccine against rabbit hemorrhagic disease virus. Acta Biochim. Pol. *53*, 371–376.

Guillon, P., Ruvoen-Clouet, N., Le Moullac-Vaidye, B., Marchandeau, S., and Le Pendu, J. (2009). Association between expression of the H histo-blood group antigen, alpha 1,2 fucosyltransferases polymorphism of wild rabbits, and sensitivity to rabbit hemorrhagic disease virus. Glycobiology *19*, 21–28.

Jung, J.Y., Lee, B.J., Tai, J.H., Park, J.H., and Lee, Y.S. (2000). Apoptosis in rabbit haemorrhagic disease. J. Comp. Pathol. *123*, 135–140.

Kerr, P.J., and Best, S.M. (1998). Myxoma virus in rabbits. Rev. Sci. Tech. *17*, 256–268.

Kieny, M.P., Lathe, R., Drillien, R., Spehner, D., Skory, S., Schmitt, D., Wiktor, T., Koprowski, H., and Lecocq, J.P. (1984). Expression of rabies virus glycoprotein from a recombinant vaccinia virus. Nature *312*, 163–166.

Koopmans, M.K., Green, K.Y., Ando, T., Clarke, I.N., Estes, M.K., Matson, D.O., Nakata, S., Neill, J.D., Smith, A.W., Studdert, M.J., and Thiel, H.J. (2005). Caliciviridae, In Virus Taxonomy: Eighth Report of the International Committee on Taxonomy of Viruses, C.M. Fauquet, M.A. Mayo, J. Maniloff, U. Desselberger, and L.A. Ball, eds. (San Diego: Elsevier Inc.), pp. 843–851.

Kovaliski, J. (1998). Monitoring the spread of rabbit hemorrhagic disease virus as a new biological agent for control of wild European rabbits in Australia. J. Wildl. Dis. *34*, 421–428.

Laurent, S., Kut, E., Remy-Delaunay, S., and Rasschaert, D. (2002). Folding of the rabbit hemorrhagic disease virus capsid protein and delineation of N-terminal domains dispensable for assembly. Arch. Virol. *147*, 1559–1571.

Laurent, S., Vautherot, J.F., Le Gall, G., Madelaine, M.F., and Rasschaert, D. (1997). Structural, antigenic and immunogenic relationships between European brown hare syndrome virus and rabbit haemorrhagic disease virus. J. Gen. Virol. *78*, 2803–2811.

Laurent, S., Vautherot, J.F., Madelaine, M.F., Le Gall, G., and Rasschaert, D. (1994). Recombinant rabbit hemorrhagic disease virus capsid protein expressed in baculovirus self-assembles into virus-like particles and induces protection. J. Virol. *68*, 6794–6798.

Le Gall, G., Huguet, S., Vende, P., Vautherot, J.F., and Rasschaert, D. (1996). European brown hare syndrome virus: molecular cloning and sequencing of the genome. J. Gen. Virol. *77*, 1693–1697.

Le Gall-Recule, G., Zwingelstein, F., Laurent, S., de Boisseson, C., Portejoie, Y., and Rasschaert, D. (2003). Phylogenetic analysis of rabbit haemorrhagic disease virus in France between 1993 and 2000, and the characterisation RHDV antigenic variants. Arch. Virol. *148*, 65–81.

Le Pendu, J., Ruvoen-Clouet, N., Kindberg, E., and Svensson, L. (2006). Mendelian resistance to human norovirus infections. Semin. Immunol. *18*, 375–386.

Lindesmith, L.C., Donaldson, E.F., Lobue, A.D., Cannon, J.L., Zheng, D.P., Vinje, J., and Baric, R.S. (2008). Mechanisms of GII.4 norovirus persistence in human populations. PLoS Med. *5*, e31.

Liu, G., Ni, Z., Yun, T., Yu, B., Chen, L., Zhao, W., Hua, J., and Chen, J. (2008). A DNA-launched reverse genetics system for rabbit hemorrhagic disease virus reveals that the VP2 protein is not essential for virus infectivity. J. Gen. Virol. *89*, 3080–3085.

Liu, S.J., Xue, H.P., Pu, B.Q., and Qian, N.H. (1984). A new viral disease in rabbits. J. Vet. Diagn. Invest. *16*, 253–255.

Liu, G., Zhang, Y., Ni, Z., Yun, T., Sheng, Z., Liang, H., Hua, J., Li, S., Du, Q., and Chen, J. (2006). Recovery of infectious rabbit hemorrhagic disease virus from rabbits after direct inoculation with in vitro-transcribed RNA. J. Virol. *80*, 6597–6602.

Marcato, P.S., Benazzi, C., Vecchi, G., Galeotti, M., Della Salda, L., Sarli, G., and Lucidi, P. (1991). Clinical and pathological features of viral haemorrhagic disease of rabbits and the European brown hare syndrome. Rev. Sci. Tech. *10*, 371–392.

Martin Alonso, J.M., Casais, R., Boga, J.A., and Parra, F. (1996). Processing of rabbit hemorrhagic disease virus polyprotein. J. Virol. *70*, 1261–1265.

Martin-Alonso, J.M., Castanon, S., Alonso, P., Parra, F., and Ordas, R. (2003). Oral immunization using tuber extracts from transgenic potato plants expressing rabbit hemorrhagic disease virus capsid protein. Transgenic Res. *12*, 127–130.

McColl, K.A., Merchant, J.C., Hardy, J., Cooke, B.D., Robinson, A., and Westbury, H.A. (2002). Evidence for insect transmission of rabbit haemorrhagic disease virus. Epidemiol. Infect. *129*, 655–663.

McCormick, C.J., Salim, O., Lambden, P.R., and Clarke, I.N. (2008). Translation termination reinitiation between open reading frame 1 (ORF1) and ORF2 enables capsid expression in a bovine norovirus without the need for production of viral subgenomic RNA. J. Virol. *82*, 8917–8921.

McIntosh, M.T., Behan, S.C., Mohamed, F.M., Lu, Z., Moran, K.E., Burrage, T.G., Neilan, J.G., Ward, G.B., Botti, G., Capucci, L., and Metwally, S.A. (2007). A pandemic strain of calicivirus threatens rabbit industries in the Americas. Virol. J. *4*, 96.

McLeod, R. (2004). Counting the cost: impact of invasive animals in Australia 2004 (Canberra: Cooperative Research Centre for Pest Animal Control).

McPhee, S.R., Berman, D., Gonzales, A., Butler, K.L., Humphrey, J., Muller, J., Waddington, J.N., Daniels, P., Koch, S., and Marks, C.A. (2002). Efficacy of a competitive enzyme-linked immunosorbent assay (cELISA) for estimating prevalence of immunity to rabbit haemorrhagic disease virus (RHDV) in populations of Australian wild rabbits (*Oryctolagus cuniculus*). Wildl. Res. *29*, 635–647.

Meyers, G. (2003). Translation of the minor capsid protein of a calicivirus is initiated by a novel termination-dependent reinitiation mechanism. J. Biol. Chem. *278*, 34051–34060.

Meyers, G. (2007). Characterization of the sequence element directing translation reinitiation in RNA of the calicivirus rabbit hemorrhagic disease virus. J. Virol. *81*, 9623–9632.

Meyers, G., Wirblich, C., and Thiel, H.J. (1991a). Genomic and subgenomic RNAs of rabbit hemorrhagic disease virus are both protein-linked and packaged into particles. Virology *184*, 677–686.

Meyers, G., Wirblich, C., and Thiel, H.J. (1991b). Rabbit hemorrhagic disease virus – molecular cloning and nucleotide sequencing of a calicivirus genome. Virology *184*, 664–676.

Meyers, G., Wirblich, C., Thiel, H.J., and Thumfart, J.O. (2000). Rabbit hemorrhagic disease virus: genome organization and polyprotein processing of a calicivirus studied after transient expression of cDNA constructs. Virology *276*, 349–363.

Mocsari, E., Meder, M., Glavits, R., Ratz, F., and Sztojkov, V. (1991). Rabbit viral hemorrhagic-disease. 2. Study on the susceptibility according to the age. Magyar Allatorvosok Lapja *46*, 351–355.

Morales, M., Barcena, J., Ramirez, M.A., Boga, J.A., Parra, F., and Torres, J.M. (2004). Synthesis in vitro of rabbit hemorrhagic disease virus subgenomic RNA by internal initiation on (−)sense genomic RNA: mapping of a subgenomic promoter. J. Biol. Chem. *279*, 17013–17018.

Morisse, J.P., Le Gall, G., and Boilletot, E. (1991). Hepatitis of viral origin in Leporidae: introduction and aetiological hypotheses. Rev. Sci. Tech. *10*, 269–310.

Moss, S.R., Turner, S.L., Trout, R.C., White, P.J., Hudson, P.J., Desai, A., Armesto, M., Forrester, N.L., and Gould, E.A. (2002). Molecular epidemiology of rabbit haemorrhagic disease virus. J. Gen. Virol. *83*, 2461–2467.

Moussa, A., Chasey, D., Lavazza, A., Capucci, L., Smid, B., Meyers, G., Rossi, C., Thiel, H.J., Vlasak, R., Ronsholt, L., and et al. (1992). Haemorrhagic disease of lagomorphs: evidence for a calicivirus. Vet. Microbiol. *33*, 375–381.

Munro, R.K., and Williams, R.T. (1994). Rabbit haemorrhagic disease: issues in assessment for biocological control. (Canberra: Bureau of Resource Science).

Mutze, G., Bird, P., Cooke, B.D., and Henzell, R. (2008). Geographic and seasonal variation in the impact of rabbit haemorrhagic disease on European rabbits, Oryctolagus cuniculus, and rabbit damage in Australia. In Lagomorph Biology: Evolution, Ecology and Conservation, P.C. Alves, N. Ferrand and K. Hackländer, eds. (Berlin-Heidelberg: Springer), pp. 279–293.

Mutze, G., Cooke, B.D., and Alexander, P. (1998). The initial impact of rabbit hemorrhagic disease on European rabbit populations in South Australia. J. Wildl. Dis. *34*, 221–227.

Myers, K. (1970). The rabbit in Australia, In Dynamics of Populations, P.J. den Boer, and G.R. Gradwell, eds. (Oosterbek, the Netherlands: Centre for Agricultural publishing and documentation), pp. 478–506.

Myers, K., and Poole, W.E. (1962). Oestrous cycles in the rabbit Oryctolagus cuniculus (L.). Nature *195*, 358–359.

Nagesha, H.S., McColl, K.A., Collins, B.J., Morrissy, C.J., Wang, L.F., and Westbury, H.A. (2000). The presence of cross-reactive antibodies to rabbit haemorrhagic disease virus in Australian wild rabbits prior to the escape of virus from quarantine. Arch. Virol. *145*, 749–757.

Nagesha, H.S., Wang, L.F., and Hyatt, A.D. (1999). Virus-like particles of calicivirus as epitope carriers. Arch. Virol. *144*, 2429–2439.

Nagesha, H.S., Wang, L.F., Hyatt, A.D., Morrissy, C.J., Lenghaus, C., and Westbury, H.A. (1995). Self-assembly, antigenicity, and immunogenicity of the rabbit haemorrhagic disease virus (Czechoslovakian strain V-351) capsid protein expressed in baculovirus. Arch. Virol. *140*, 1095–1108.

Nieba, L., and Bachmann, M.F. (2000). A new generation of vaccines. Modern Aspects Immunobiol. *1*, 36–39.

Noad, R., and Roy, P. (2003). Virus-like particles as immunogens. Trends Microbiol. *11*, 438–444.

O'Keefe, J.S., Tempero, J.E., Motha, M.X.J., Hansen, M.F., and Atkinsona, P.H. (1999). Serology of rabbit haemorrhagic disease virus in wild rabbits before and after release of the virus in New Zealand. Vet. Microbiol. *66*, 29–40.

Ohlinger, V.F., Haas, B., Ahl, R., and Weiland, F. (1989). Rabbit haemorrhagic disease – a contagous disease caused by a calicivirus. Tierarztliche Umschau *44*, 284–294.

Ohlinger, V.F., Haas, B., Meyers, G., Weiland, F., and Thiel, H.J. (1990). Identification and characterization of the virus causing rabbit hemorrhagic disease. J. Virol. *64*, 3331–3336.

Ohlinger, V.F., Haas, B., and Thiel, H.J. (1993). Rabbit hemorrhagic disease (RHD): characterization of the causative calicivirus. Vet. Res. *24*, 103–116.

Opgenorth, A., Graham, K., Nation, N., Strayer, D., and McFadden, G. (1992). Deletion analysis of two tandemly arranged virulence genes in myxoma virus, M11L and myxoma growth factor. J. Virol. *66*, 4720–4731.

Parkes, J. and Norbury, G. (2004). Consequences of successful biocontrol. What do we get for fewer rabbits? Prim. Indust. Manag. *7*, 16–18.

Parkes, J.P., Norbury, G.L., Heyward, R.P., and Sullivan, G. (2002) Epidemiology of rabbit haemorrhagic disease (RHD) in the South Island, New Zealand, 1997-2001. Wildl. Res. *29*, 543–555.

Parra, F., Boga, J.A., Marin, M.S., and Casais, R. (1993). The amino terminal sequence of VP60 from rabbit hemorrhagic disease virus supports its putative subgenomic origin. Virus Res. *27*, 219–228.

Peacey, M., Wilson, S., Baird, M.A., and Ward, V.K. (2007). Versatile RHDV virus-like particles: Incorporation of antigens by genetic modification and chemical conjugation. Biotechnol. Bioeng. *98*, 968–977.

Peacey, M., Wilson, S., Perret, R., Ronchese, F., Ward, V.K., Young, V., Young, S.L., and Baird, M.A. (2008). Virus-like particles from rabbit hemorrhagic disease virus can induce an anti-tumor response. Vaccine *26*, 5334–5337.

Perez-Filgueira, D.M., Resino-Talavan, P., Cubillos, C., Angulo, I., Barderas, M.G., Barcena, J., and Escribano, J.M. (2007). Development of a low-cost, insect larvae-derived recombinant subunit vaccine against RHDV. Virology *364*, 422–430.

Plana-Duran, J., Bastons, M., Rodriguez, M.J., Climent, I., Cortes, E., Vela, C., and Casal, I. (1996). Oral immunization of rabbits with VP60 particles confers protection against rabbit hemorrhagic disease. Arch. Virol. *141*, 1423–1436.

Poyry, T.A., Kaminski, A., Connell, E.J., Fraser, C.S., and Jackson, R.J. (2007). The mechanism of an exceptional case of reinitiation after translation of a long ORF reveals why such events do not generally occur in mammalian mRNA translation. Genes Dev. *21*, 3149–3162.

Prasad, B.V., Hardy, M.E., Dokland, T., Bella, J., Rossmann, M.G., and Estes, M.K. (1999). X-ray crystallographic structure of the Norwalk virus capsid. Science *286*, 287–290.

Prasad, B.V., Matson, D.O., and Smith, A.W. (1994). Three-dimensional structure of calicivirus. J. Mol. Biol. *240*, 256–264.

Rademacher, C., Krishna, N.R., Palcic, M., Parra, F., and Peters, T. (2008). NMR experiments reveal the molecular basis of receptor recognition by a calicivirus. J. Am. Chem. Soc. *130*, 3669–3675.

Ramiro-Ibanez, F., Martin-Alonso, J.M., Garcia Palencia, P., Parra, F., and Alonso, C. (1999). Macrophage tropism of rabbit hemorrhagic disease virus is associated with vascular pathology. Virus Res. *60*, 21–28.

Read, L.J. (2003). Red Sand, Green Heart: Ecological adventures in the outback, 1st edn (Lothian Books).

Read, J.L., Carter, J., Moseby, K.M., and Greenville, A. (2008). Ecological roles of rabbit, bettong and bilby warrens in arid Australia. J. Arid Environments *72*, 2124–2130.

Richardson, B.J., Phillips, S., Hayes, R.A., Sindhe, S., and Cooke, B.D. (2007). Aspects of the biology of the European rabbit (*Oryctolagus cuniculus*) and rabbit haemorrhagic disease virus (RHDV) in coastal eastern Australia. Wildl. Res. *34*, 398–407.

Robinson, A.J., Kirkland, P.D., Forrester, R.I., Capucci, L., Cooke, B.D., and Philbey, A.W. (2002). Serological evidence for the presence of a calicivirus in Australian wild rabbits, *Oryctolagus cuniculus*, before the introduction of rabbit haemorrhagic disease virus (RHDV): its potential influence on the specificity of a competitive ELISA for RHDV. Wildl. Res. *29*, 655–662.

Rodak, L., Smid, B., Valicek, L., Vesely, T., Stepanek, J., Hampl, J., and Jurak, E. (1990). Enzyme-Linked-Immunosorbent-Assay of antibodies to rabbit hemorrhagic disease virus and determination of its major structural proteins. J. Gen. Virol. *71*, 1075–1080.

Rolls, E.C. (1969). They all ran wild (Angus and Robertson, Australia).

Ros Bascunana, C., Nowotny, N., and Belak, S. (1997). Detection and differentiation of rabbit hemorrhagic disease and European brown hare syndrome viruses by amplification of VP60 genomic sequences from fresh and fixed tissue specimens. J. Clin. Microbiol. *35*, 2492–2495.

Roundtree, P.M. (1983). Pasteur in Australia. The Australian Microbiologist, historical review, 1–5.

Roy, P., and Noad, R. (2008). Virus-like particles as a vaccine delivery system – Myths and facts. Hum. Vacc. *4*, 5–12.

Ruvoen-Clouet, N., Ganiere, J.P., Andre-Fontaine, G., Blanchard, D., and Le Pendu, J. (2000). Binding of rabbit hemorrhagic disease virus to antigens of the ABH histo-blood group family. J. Virol. *74*, 11950–11954.

Sagoff, M. (2008). Third-generation biotechnology – A first look. Issues Sci. Technol. *25*, 70–74.

Sanchez-Campos, S., Alvarez, M., Culebras, J.M., Gonzalez-Gallego, J., and Tunon, M.J. (2004). Pathogenic molecular mechanisms in an animal model of fulminant hepatic failure: rabbit hemorrhagic viral disease. J. Lab. Clin. Med. *144*, 215–222.

Sandell, P. (2006). Promoting woodland recovery in the Victorian Mallee Parks. Proc. Royal Soc. Victoria, 118, 313–321.

San-Miguel, B., Alvarez, M., Culebras, J.M., Gonzalez-Gallego, J., and Tunon, M.J. (2006). N-acetyl-cysteine protects liver from apoptotic death in an animal model of fulminant hepatic failure. Apoptosis *11*, 1945–1957.

Sobey, W.R. (1969). Selection for resistance to myxomatosis in domestic rabbits (*Oryctolagus cuniculus*). J. Hyg. *67*, 743–754.

Strive, T., Wright, J.D., and Robinson, A.J. (2009). Identification and partial characterisation of a new lagovirus in Australian wild rabbits. Virology *384*, 97–105.

Thompson, J., Clark, G. (1997). Rabbit calicivirus disease now established in New Zealand. Surveillance *24*, 5–6.

Thorven, M., Grahn, A., Hedlund, K.O., Johansson, H., Wahlfrid, C., Larson, G., and Svensson, L. (2005). A homozygous nonsense mutation (428G →A) in the human secretor (FUT2) gene provides resistance to symptomatic norovirus (GGII) infections. J. Virol. *79*, 15351–15355.

Thouvenin, E., Laurent, S., Madelaine, M.F., Rasschaert, D., Vautherot, J.F., and Hewat, E.A. (1997). Bivalent binding of a neutralising antibody to a calicivirus involves the torsional flexibility of the antibody hinge. J. Mol. Biol. *270*, 238–246.

Thumfart, J.O., and Meyers, G. (2002). Rabbit hemorrhagic disease virus: identification of a cleavage site in the viral polyprotein that is not processed by the known calicivirus protease. Virology *304*, 352–363.

Torres, J., Ramírez, M.A., Morales, M., Bárcena, J., Vázquez, B., Espuña, E., Pagès-Manté A., and Sánchez-Vizcaíno J.M. (2000). Safety evaluation of a recombinant myxoma-RHDV virus inducing horizontal transmissible protection against myxomatosis and rabbit haemorrhagic disease. Vaccine *19*, 174–182.

Tunon, M.J., Sanchez-Campos, S., Garcia-Ferreras, J., Alvarez, M., Jorquera, F., and Gonzalez-Gallego, J. (2003). Rabbit hemorrhagic viral disease: characterization of a new animal model of fulminant liver failure. J. Lab. Clin. Med. *141*, 272–278.

Twigg, L.E., Lowe, T.J., Martin, G.R., Wheeler, A.G., Gray, G.S., Griffin, S.L., O'Reilly, C.M., Robinson, D.J., and Hubach, P.H. (2000). Effects of surgically imposed sterility on free-ranging rabbit populations. J. Appl. Ecol. *37*, 16–39.

van de Bildt, M.W., van Bolhuis, G.H., van Zijderveld, F., van Riel, D., Drees, J.M., Osterhaus, A.D., and Kuiken, T. (2006). Confirmation and phylogenetic analysis of rabbit hemorrhagic disease virus in free-living rabbits from the Netherlands. J. Wildl. Dis. *42*, 808–812.

van Leeuwen, B.H., and Kerr, P.J. (2007). Prospects for fertility control in the European rabbit (*Oryctolagus cuniculus*) using myxoma virus-vectored immunocontraception. Wildl. Res. *34*, 511–522.

Vere, D.T., Jones, R.E., and Saunders, G. (2004). The economic benefits of rabbit control in Australian temperate pastures by the introduction of rabbit haemorrhagic disease. Agric. Econ. *30*, 143–155.

Villafuerte, R., Calvete, C., Gortazar, C., and Moreno, S. (1994). First epizootic of rabbit hemorrhagic disease in free living populations of *Oryctolagus cuniculus* at Donana National Park, Spain. J. Wildl. Dis. *30*, 176–179.

Wiktor, T.J., Macfarlan, R.I., Reagan, K.J., Dietzschold, B., Curtis, P.J., Wunner, W.H., Kieny, M.P., Lathe, R., Lecocq, J.P., Mackett, M., and *et al.* (1984). Protection from rabies by a vaccinia virus recombinant containing the rabies virus glycoprotein gene. Proc. Natl. Acad. Sci. U.S.A. *81*, 7194–7198.

Wirblich, C., Meyers, G., Ohlinger, V.F., Capucci, L., Eskens, U., Haas, B., and Thiel, H.J. (1994). European brown hare syndrome virus: relationship to rabbit hemorrhagic disease virus and other caliciviruses. J. Virol. *68*, 5164–5173.

Wirblich, C., Sibilia, M., Boniotti, M.B., Rossi, C., Thiel, H.J., and Meyers, G. (1995). 3C-like protease of rabbit hemorrhagic disease virus: identification of cleavage sites in the ORF1 polyprotein and analysis of cleavage specificity. J. Virol. *69*, 7159–7168.

Wirblich, C., Thiel, H.J., and Meyers, G. (1996). Genetic map of the calicivirus rabbit hemorrhagic disease virus as deduced from in vitro translation studies. J. Virol. *70*, 7974–7983.

Xu, Z.J., and Chen, W.X. (1989). Viral haemorrhagic disease in rabbits: a review. Vet. Res. Commun. *13*, 205–212.

Young, S.L., Wilson, M., Wilson, S., Beagley, K.W., Ward, V., and Baird, M.A. (2006). Transcutaneous vaccination with virus-like particles. Vaccine *24*, 5406–5412.